CARBON AND ITS DOMESTICATION

Carbon and Its Domestication

by

A.M. MANNION

University of Reading,
Reading, U.K.

A C.I.P. Catalogue record for this book is available from the Library of Congress.

ISBN 10 1-4020-3957-3 (PB)
ISBN 13 978-1-4020-3957-7 (PB)
ISBN 10 1-4020-3956-5 (HB)
ISBN 13 978-1-4020-3956-0 (HB)
ISBN 10 1-4020-3958-1 (e-book)
ISBN 13 978-1-4020-3958-4 (e-book)

Published by Springer,
P.O. Box 17, 3300 AA Dordrecht, The Netherlands.

www.springer.com

Printed on acid-free paper

Printed in the Netherlands.

Hamlet, Prince of Denmark
William Shakespeare

KING CLAUDIUS
Now, Hamlet, where's Polonius?
HAMLET
At supper.
KING CLAUDIUS
At supper! Where?
HAMLET
Not where he eats, but where he is eaten: a certain
convocation of politic worms are e'en at him. Your
worm is your only emperor for diet: we fat all
creatures else to fat us, and we fat ourselves for
maggots: your fat king and your lean beggar is but
variable service, two dishes, but to one table:
that's the end.
KING CLAUDIUS
Alas, alas!
HAMLET
A man may fish with the worm that hath eat of a
king, and eat of the fish that hath fed of that worm.
KING CLAUDIUS
What dost you mean by this?
HAMLET
Nothing but to show you how a king may go a
progress through the guts of a beggar.

Act IV Scene III

Contents

List of Figures

List of Tables

Preface

I began my academic life as a specialist in Quaternary environments together with a broad interest in physical geography and people-environment relationships. I have ended this career and begun a new one with a deep interest in physical geography and a major focus on people-environment relationships while still retaining a passion for all things Quaternary. The polymathic disposition of geography as a subject is what led me to university in the first place and now I have come full cycle and embraced once again this strength, having stepped in the real and literal mires of specialism on the journey. Some of my broad-spectrum interests have been thrust upon me due to teaching responsibilities (in the days when academics taught), not least of which are the carbon and other biogeochemical cycles and the full gamut of biogeography and biotic resources. All roads have led to carbon, that fundamental unit of life, atmosphere and socio-economic activity. I could resist no longer. Moreover, I no longer have any need to resist, so almost four years into retirement this book is my catharsis. As with my previous books, I have learnt a great deal and the path of discovery remains compelling and alluring, perhaps these are the second and third most important reasons for writing at all, with the first being the generation of an idea and its culmination in the receipt of the first copy of the printed work of that idea; this represents the fruits of a considered and considerable labour.

For this book there is no prescribed audience. I have written what I wanted to write given the reasonable word limits granted by Springer. My chapter headings are conventional. The chemistry, biology, geology, geography and history of carbon and its domestication, plus the politicization of carbon, are each vast but interrelated topics. My synopses are but thumb-nail sketches though I have attempted to provide sufficient depth to capture the excitement of this carbon-based world. I have no doubt that there will be criticisms; some sections could be expanded, other topics are themselves worthy of being turned into books and indeed some already have been so treated, and other topics could have been included on more than a cursory basis, e.g. the historical and present role of disease in society, petropolitics and war, the role of genetic engineering in relation to society and environment, and carbon management at local, national and international scales. So be it. However, there is a long list of references on all topics as well as internet addresses for those who wish to further their interests. I am also aware that much of what I have written will change relatively rapidly due to many exciting developments in science as well as in the social and environmental sciences. Keeping up with the flow of information is itself a major task. However, there was a time when it took two years from receipt of the manuscript for a book to be published; now it is c. six months; for many academic journals it may take eighteen months for the publication of a paper. Consequently, books can now provide up-to-date syntheses of subject matter, and as in my previous books I have made a considerable effort to share my broad-based perusal of the literature.

In my new career in semi-retirement, in reality this is at most only 10 percent retirement, I am fully occupied with book chapters, editing and encyclopaedia contributions. I learn more geography, a euphemism for all manner of subjects, everyday and never cease to be engaged, entertained and enlightened: long may it last. I hope that 'Carbon and its Domestication', the underpinning of almost all environmental, ecological and economic function educates, enthuses and excites the reader just as the subject matter has inspired me. That 'it all boils down to carbon in the end', an oft-used phrase of mine, will serve me well as an epitaph!

Antoinette M Mannion
Reading
May 2005.

Acknowledgements

All books have inputs from people other than the author. In this case I should like to express my gratitude to Dr Sophie Bowlby and Dr Peter Pearson for advice and information while Judith Fox has helped to produce diagrams. In addition, Figures 4.4, 4.5 and 4.6 have been reproduced with permission from BIODIDACT, a bank of digital resources for teaching Biology at the University of Ottawa, and Figure 4.3 has been reproduced with permission from the United States Geological Survey. I am especially cognisant of my debt to Dr Mike Turnbull who has drawn the chemistry diagrams of Chapters 2 and 3 and redrawn many other diagrams. He has also undertaken the least creative but time-consuming task of compiling the index and helped to prepare the manuscript in camera-ready form. In respect of the latter task, I also wish to acknowledge the assistance of Betty van Herk from Springer and their copy editor, Edwin Beschler.

Chapter 1

1 INTRODUCTION

Carbon may be a commonly occurring element but it has an uncommon place in human and environmental history. This is because it is a central component in the vital energy transfers that facilitate the survival of all plants and animals. For organisms other than humans, the exchange of carbon with the atmosphere and between organisms involves the manufacture of food through photosynthesis, the transfer of food energy in food chains, and respiration. These functions and relationships can be described as passive insofar as the objective is maintenance and survival of the species. In contrast, for most of their history humans have sought to improve their lot by actively manipulating and harnessing carbon transfers to provide not only food energy but also fuel energy. This process can be considered as domestication and it has been achieved through scientific and technological innovations. These have underpinned the development processes in society and are a major cause of environmental change. Their significance is sometimes implicit in the documentation of both human history and environmental change but is rarely explicit despite the continuing, sometimes vehement and often polarized, debate on people-environment relationships and the now accepted view that the sciences and the humanities are interdependent. While spatial and temporal dimensions are common to both the social and environmental sciences, historians have concentrated on people, with an emphasis on social, economic and political factors that have shaped society, while environmentalists have focused on the physical, chemical and biological elements of the environment. Even geographers, whose raison d'être is the study and explanation of people and place in time and space, have not succeeded in identifying a truly unifying theme within their polymathic subject with which to link human and physical geography. Fragmentation and specialization of both the humanities and the sciences has its benefits insofar as in-depth research provides ground-breaking insights. Nevertheless the broader cross-disciplinary synthesis also has a vital role in providing a unifying narrative that describes and explains the links between science, society and environment. This is the primary objective of this book. It highlights the role of carbon as a unifying theme linking the social and natural sciences through technology and within a temporal and spatial context.

Carbon makes both society and the world tick, and its domestication has been the objective of most technological developments since the primate ancestors of modern humans evolved from the apes some seven million years ago. From the first threshold of carbon domestication, which involved the controlled use of fire more than one million years ago, to the twenty-first century developments in the deciphering of genetic codes and genetic modification, humans have manipulated carbon not only for their survival but also for wealth generation and social advance. Often, this has been to the disadvantage of other organisms. Of particular significance is the harnessing of fire, the domestication of various plants and animals as part of the technology of agriculture, the annexation of new lands by Europeans between the fifteenth and nineteenth centuries and the Industrial Revolution of the eighteenth century that grew from and encouraged the exploitation of fossil fuels, as well as the current gene-based discoveries. All represent carbon-based thresholds in social, political, scientific and environmental developments.

1.1 What is carbon?

Carbon is one of some 92 naturally-occurring elements; as an atom it is a fundamental entity that cannot be broken down to produce other chemical substances. Like all atoms, carbon has a physical structure which comprises sub-atomic particles known as neutrons, protons and electrons. An atom of carbon consists of a nucleus containing six neutrons (neutral particles with no electrical charge), six protons (positively charged particles) and six electrons (negatively charged particles), as shown in Figure 1.1. The electrons are arranged in so-called shells surrounding the nucleus; in the case of carbon there are two shells, an inner containing two electrons and an outer containing four electrons. Carbon has an atomic number of 6, which denotes the number of units of positive charge, i.e. the protons, and an atomic mass number of 12, i.e. the total number of protons and neutrons.

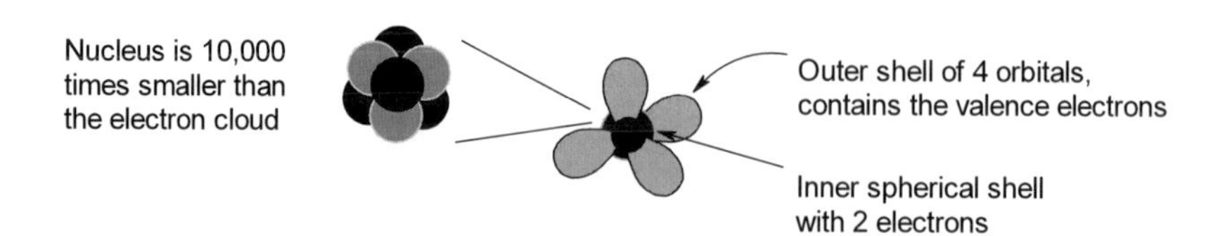

Figure 1.1. Constituents of a carbon atom

The atomic number of a given element reflects the properties of that element and all elements are arranged in order of increasing atomic number in the Periodic Table shown in Table 1.1. The horizontal rows are referred to as periods and so called because specific properties recur on a regular (periodic) basis as atomic numbers increase and consequently those elements with similar properties fall into vertical groups. This similarity occurs because of the configuration of electrons in the outer shells around the atom; in the case of carbon, and all those elements in the same column, there are four electrons in this shell, also known as the valence shell. As Table 1.1 shows, there is a left to right gradation from metals, to metalloids (non metals with some properties of metals) to non metals. Carbon is thus a non metal with a low atomic number. The presence of four electrons in a shell that can accommodate eight means that carbon can share electrons with as many as four different atoms, including other carbon atoms. Consequently, carbon can combine with many different atoms to produce many different compounds. These are properties that help to define why this element is so important (see Section 1.2).

Carbon exists in five isotopic forms. That is to say there are atoms of carbon with five different nuclear characteristics, specifically with different numbers of neutrons. These variations are denoted as carbon-11 (C-11), carbon-12 (C-12), carbon-13 (C-13), carbon-14 (C-14) and carbon-15 (C-15) in which the numbers of neutrons vary from 5 to 9 respectively. Of these isotopes, C-12 and C-13 are stable while the others are unstable and radioactive. In order to achieve stability, radioactive decay occurs involving the emission of particles from the nucleus. This occurs at a constant rate for each isotope; for C-11 the half-life, i.e. the time required for half the original number of atoms to decay is 20.3 minutes, for C-15 it is 2.5 seconds and for C-14 it is 5730 years. The long half life of the latter isotope allows it to be used as a means of estimating the age of carbon-rich substances that have been preserved through burial in environments that prohibit decomposition. The approximate age is determined by comparing the amount of radioactive carbon in the preserved sample with that in a modern sample. While the physical properties of the isotopes vary in this way, their chemical properties are the same, as reflected by their atomic number of six.

Carbon has been known since ancient times and its name derives from *carbo*, the Latin word for coal, reflecting its association with fire. As an element, it occurs in five different forms or allotropes. These include amorphous carbon which is formed when substances containing carbon are burnt in an atmosphere with little oxygen to produce soot. This has various names such as lamp-black, gas black, channel black and carbon black and it is used in the production of inks and paints.

Table 1.1. The Periodic Table of the elements

Metals (unshaded) | Non-metals (shaded)

1	2	3	4	5	6	7	8	9	10	11	12	13	14	15	16	17	18
1 H																	2 He
3 Li	4 Be											5 B	6 C	7 N	8 O	9 F	10 Ne
11 Na	12 Mg											13 Al	14 Si	15 P	16 S	17 Cl	18 Ar
19 K	20 Ca	21 Sc	22 Ti	23 V	24 Cr	25 Mn	26 Fe	27 Co	28 Ni	29 Cu	30 Zn	31 Ga	32 Ge	33 As	34 Se	35 Br	36 Kr
37 Rb	38 Sr	39 Y	40 Zr	41 Nb	42 Mo	43 Tc	44 Ru	45 Rh	46 Pd	47 Ag	48 Cd	49 In	50 Sn	51 Sb	52 Te	53 I	54 Xe
55 Cs	56 Ba	57 La	72 Hf	73 Ta	74 W	75 Re	76 Os	77 Ir	78 Pt	79 Au	80 Hg	81 Tl	82 Pb	83 Bi	84 Po	85 At	86 Rn
87 Fr	88 Ra	89 Ac															

Series														
Lanthanide series	58 Ce	59 Pr	60 Nd	61 Pm	62 Sm	63 Eu	64 Gd	65 Tb	66 Dy	67 Ho	68 Er	69 Tm	70 Yb	71 Lu
Actinide Series	90 Th	91 Pa	92 U	93 Np	94 Pu	95 Am	96 Cm	97 Bk	98 Cf	99 Es	100 Fm	101 Md	102 No	103 Lr

Graphite is another allotrope. Soft and greasy with a low melting point, it comprises carbon atoms combined to produce sheets (Figure 1.2). Graphite is used for lubrication, in steel production and paints and is the 'lead' in pencils. The characteristics of graphite contrast with those of diamond, a third allotrope of carbon, which is the hardest known naturally-occurring substance. The difference between the two is illustrated in Figure 1.2 which shows that the atoms of carbon are linked via single rather than double bonds. Diamond has a rigid, crystalline rather than malleable structure and a high melting point. It is highly prized as a gem stone but it also has industrial applications for precision tools.

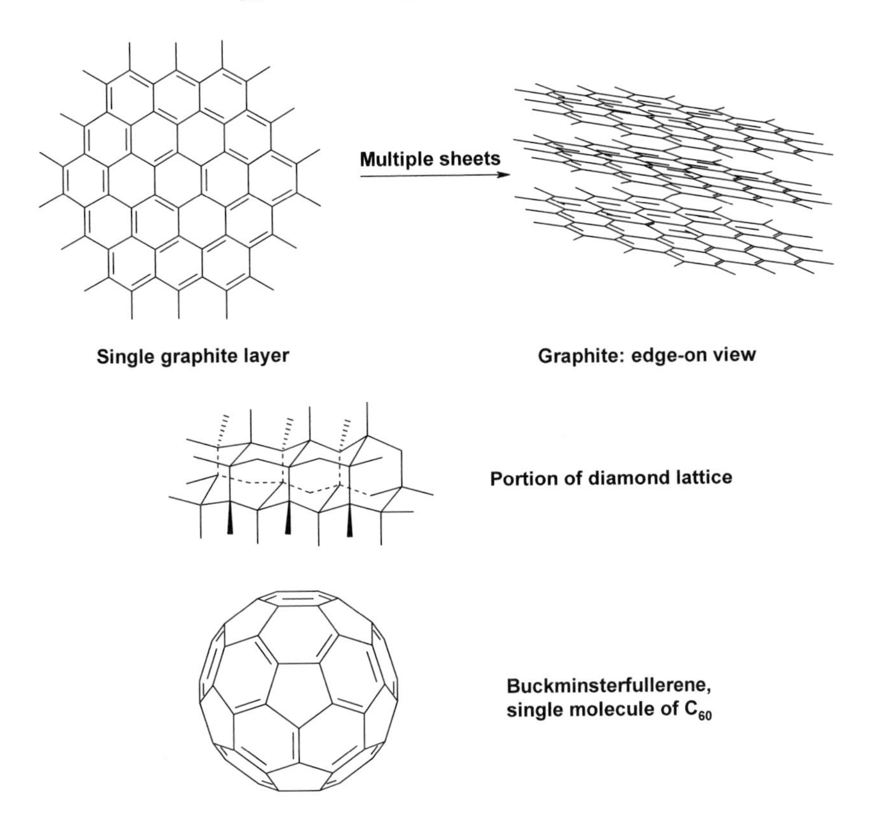

Figure 1.2. Some allotropes of carbon

So-called buckminsterfullerenes ('buckyballs'), only identified in the 1990s, are the manifestation of carbon atoms joined in yet a different way. As shown in Figure 1.2 an individual bucky ball comprises 60 or 70 carbon atoms linked through a series of pentagons and hexagons in a structure shaped like a soccer ball. These substances can act as superconductors but their range of applications has yet to be determined.

Finally, in the late 1960s, a fifth allotrope of carbon was created in the laboratory. This substance is transparent and is produced at low pressure from graphite. Called 'white carbon', it is crystalline and birefringent, i.e. it can split a beam of polarised light in two.

1.2 Why is carbon so important?

The importance of carbon relates to the chemistry, physics and biology of this versatile element. Chemically, carbon is highly reactive and so will combine with many other elements to produce a wide variety of compounds. So important and varied are these compounds that they are the subject matter of the field of organic chemistry. They can be described as organic because life forms are or have been involved in their formation. They include substances with a relatively simple structure such as calcium carbonate, the major component of calcareous rocks like chalk and limestone. More complex carbon compounds are involved in essential processes such as photosynthesis, the oxygen-carrying capacity of blood cells, the structural compounds of plants and animals e.g. fats, proteins, cellulose, vitamins and the genetic components ribonucleic and deoxyribonucleic acids. These are all natural compounds and examples are discussed in Chapter 2. There is another group of complex organic compounds which have been synthesized in laboratories and which have no natural counterparts. Examples include DDT, chlorofluorocarbons, artificial fibres such as nylon, and plastics as well as a wide range of pharmaceuticals.

Apart from its chemical ubiquity, carbon can adopt many structural forms, as discussed in Section 1.1 above. These allotropes of carbon are possible because of the varied ways carbon can combine with itself. Such substances cannot, however, be described as organic as they are the products of geological rather than biological processes. Nevertheless they remain important in terms of human and environmental history because they have industrial and social uses and so are components of wealth generation.

Carbon is also inextricably linked with processes in the living world. This is due to its presence in chlorophyll, a green pigment in plant leaves. This has the unique capacity to trap solar energy which in combination with water from the soil and carbon dioxide from the atmosphere produces carbohydrates. These are carbon-rich organic molecules from which all other chemicals needed by plants are produced. They also provide a source of energy which is necessary for metabolic processes, growth and reproduction. This energy is released in the presence of oxygen during respiration when carbon dioxide is returned to the atmosphere. Thus the world's vegetation communities can be described as stores of carbon compounds with potential energy. They can also be considered as purveyors of energy insofar as they provide a carbon-based energy source for organisms that have no means of generating energy for their metabolic and reproductive requirements, notably the animal kingdom of which humans are a component.

The chemistry and biology of carbon thus conspire to provide a vast source of potential energy in present vegetation communities and in the remains of past vegetation communities which have created peat, coal, oil and natural gas through geological processes. This attribute has been exploited in relation to both food and fuel energy. The manipulation of organisms, first through scavenging, then hunting and gathering and eventually through agriculture has provided food energy for modern humans (*Homo sapiens*) and their hominid ancestors. The domestication, or harnessing, of specific plants and animals, a process begun some 10,000 years ago, has underpinned the development and spread of agricultural systems throughout the world and transformed wildscapes (wilderness) into landscapes. Agriculture not only provided a means of survival but also, through the production of surpluses, a means of barter which encouraged trade and wealth and the resulting exchange of ideas. The production of surplus food allowed the diversification of human labour and contributed to the exploitation of other resources such as metals. Food security also provided, as it continues so to do, a significant element of power and agriculture is a major cause of environmental change. The exploitation of fuel energy resources, notably extant sources such as wood and biomass fuels, and the geological reservoirs of coal, oil and natural gas, has also played a major role in cultural and environmental change. Wood and other types of vegetation have a long history of use, extending back at least 2 million years (see Sections 1.4 and 1.5). The large scale exploitation of fossil fuels has provided energy for industrial development and associated demographic and socio-economic changes since the 1700s. These carbon-based resources also confer power on their host nations and a means of wealth generation. They are, however, the cause of environmental change through the extractive process and especially through the emissions of carbon dioxide and other heat-trapping gases which contribute to climatic change. This reflects another reason why carbon is so important and that is its combination with oxygen to produce carbon dioxide. This gas is present in the atmosphere where it has a warming effect by reducing the radiation of heat (received initially from the sun) from the Earth's surface into space. This is the so-called greenhouse effect. As discussed in Section 1.3, carbon dioxide has played and continues to play a significant role in regulating global climate.

1.3 Carbon and climate

Carbon is chemically reactive and can combine with many other elements to produce simple and complex compounds. Some of these are gases. For example carbon combines with oxygen to produce carbon monoxide and carbon dioxide and with hydrogen to produce methane. The chemical structures of these compounds are given in Figure 1.3.

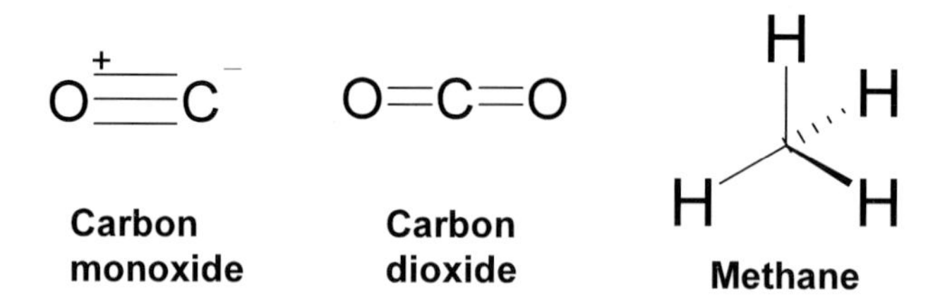

Figure 1.3. Structures of some simple carbon compounds

Many carbon-based gases have also been created artificially, including the chlorofluorocarbons or CFCs. All are heat-trapping gases and are thus involved in the greenhouse effect. In terms of geological time, only carbon dioxide and methane are important while CFCs have become significant only since the 1930s, when they were developed for refrigeration and aerosols. CFCs have a role in recent climatic change because of their heat-trapping abilities and they also have a role in damaging the ozone layer in the upper atmosphere because of the halogens, notably chlorine, fluorine and bromine, they contain.

In terms of the geological history of the Earth, there have been continuous chemical, physical and biological changes throughout its 5,000 million year existence. All the elements which are present on the Earth's surface when it was formed are still present today. They have, however, combined, separated and recombined to produce compounds that have circulated spatially and temporally. The major structural features of the Earth today are illustrated in Figure 1.4a which shows that the molten core has an outer layer of solid rock. Above this, the atmosphere comprises carbon dioxide, oxygen and nitrogen with some trace gases such as argon, and at the interface between the lithosphere and atmosphere is the biosphere comprising living organisms in soil or water as shown in Figure 1.4b. The water creates its own environment known as the hydrosphere which occupies a similar position as the biosphere. All of these interact as energy and matter are redistributed. Consequently, the chemistry of these structural components has altered over geological time.

Many elements have developed circular patterns of movement known as biogeochemical cycles, a term that highlights their links with life and the geology of the lithosphere. For several elements, including carbon, the atmosphere has been or is an important reservoir or pool. The resulting composition of the atmosphere has been a major factor in regulating global climate throughout Earth history, especially in relation to carbon dioxide.

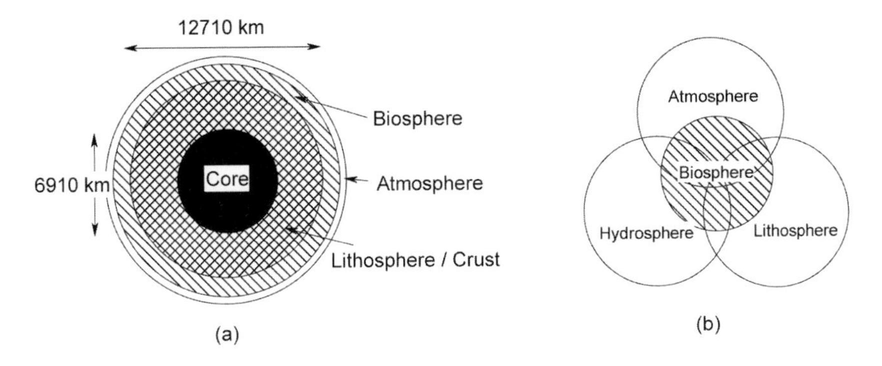

Figure 1.4. (a) Major structural features of the Earth
(b) The relationships between the biosphere and other components

A comparison of the atmospheric compositions of the Earth and its neighbours, Mars and Venus, which were most likely formed at the same time, is given in Table 1.2. This shows that the main differences relate to the concentrations of carbon dioxide in the atmospheres of Mars and Venus when compared to the atmospheric composition of the Earth, and the relative abundance of oxygen and nitrogen in the atmosphere of the latter. In view of a likely common origin of all three planets, these differences are surprising. However, Earth has had life forms almost since its inception.

Table 1.2. A comparison of the atmospheric compositions of Earth, Venus and Mars (based on data in Wayne, 2003).

%s	EARTH	MARS	VENUS
Carbon dioxide	0.035	95.30	96.50
Nitrogen	78.10	2.70	3.50
Oxygen	20.90	0.13	0.00
Argon	0.93	1.60	0.00007

Why Mars and Venus are, as far as is known, devoid of life forms is unclear but on Earth life forms have diversified and multiplied over time, despite some periods of extinction, as the evolution and the radiation of organisms have occurred. As life has evolved it has exerted tremendous influence over biogeochemical cycles, especially those of carbon, oxygen and nitrogen, and these, in turn, have influenced climate. The role of carbon in the atmosphere, as carbon dioxide (see Section 1.2), is particularly important in the context of temperature control. What happens to carbon in life forms on the Earth's surface is thus central to climatic characteristics, not least because temperature relates to energy transfer in the context of heat energy in the atmosphere and the production of food energy via photosynthesis. These relationships emphasise the organic rather than inorganic nature of carbon compounds. They also reflect the interdependence of the structural components of the Earth, as shown in Figure 1.4, and the dynamic nature of the Earth system.

This coupling or interdependence of life and geomorphic and geological processes, i.e. the organic and the inorganic components of the Earth system, is the essence of the Gaia hypothesis which was proposed in the early 1970s by James Lovelock (1972 and 1995). It was initially controversial insofar as it questioned the traditional view of life and the physics and chemistry of the Earth as separate with the latter determining the course of evolution etc. However, it is now widely accepted (see, for example, Volk, 2003 and Schneider *et al.*, 2004) and indeed the essence of this relationship is well illustrated by the carbon biogeochemical cycle (see Figure 2.16) and its changes over time. In its youth, the atmospheric composition of the Earth was much closer to that of Mars or Venus but over time carbon dioxide concentrations have decreased and oxygen concentrations have increased (see Table 1.2). The overall result has been to maintain temperatures at the Earth's surface so that they fluctuate within a limited range which favours the maintenance of life. The removal of carbon dioxide from the atmosphere reduces its warming effect while the release of free oxygen from the crustal rocks and through photosynthesis into the atmosphere, transforming it into an oxidizing rather than a reducing milieu, facilitated the evolution of organisms such as the mammals, including humans.

The carbon was removed from the atmosphere through inorganic and organic processes and became incarcerated in the lithosphere. Limestone and chalk rocks, for example, comprise calcium and magnesium carbonates and bicarbonates. These are produced in oceans by inorganic carbonate precipitation and by organisms, such as foraminifera and ammonites, which absorb the components from sea water to produce shells or outer skeletons. On death these organisms are buried in sediments on the ocean floor, the soft tissue decomposes but the shell persists and compresses as shells and sediments accumulate. The resulting limestone /chalk thus sequesters a vast amount of carbon. If uplift due to Earth movements occurs the sediments become part of the continents. Carbon has also been incarcerated in sedimentary basins when the remains from extensive swamp forests, which accumulated through photosynthesis, were buried in subsiding areas. The increasing pressure caused by the deposition of sediments on top of these organic remains resulted in the formation of coal. Much of this formation occurred during a geological period known as the Carboniferous which lasted from 360 million years ago to almost 286 million years ago (see Figure 4.1). Oil and natural gas formed in marine environments where the abundant remains of various micro-organisms became buried in sedimentary sequences belonging to various geological periods. Coal, oil and natural gas comprise the fossil fuels which are so highly valued by society. Variations in climate which are related to natural changes in the atmospheric concentrations of carbon dioxide have also occurred on much shorter timescales, notably during the last three million years. Moreover, the Earth appears to be experiencing warming currently due to the accumulation of carbon dioxide and other heat-trapping gases in the atmosphere since the Industrial Revolution in the mid-1700s. These issues are referred to in Section 1.4.

1.4 Carbon and the environment in the last 3 million years

There have been various times in Earth history when the extent of ice has been considerably greater than it is now. There were five such periods in which great ice ages occurred, as listed in Table 1.3. All were characterized by extensive polar ice sheets and cooler global temperatures than at present.

Table 1.3. The major ice ages in Earth history and when they occurred

ICE AGE(S)	Approximate age x 10^6 Years BP
Cenozoic (Tertiary/Quaternary)	3-present
Carboniferous-Permian	350-250
Silurian	460-430
End Precambrian	800-600
Mid-Precambrian	2400-2600

BP: Before Present

The Earth today remains in a cold period which began c. 3 million years ago during which time numerous ice advances and retreats have occurred. These are referred to as glacial stages and the intervening warm stages are known as interglacials. The current stage is an interglacial and it is likely that another glacial stage will occur in the next c. 10,000 years. The evidence for the events of the last 3 million years are relatively fresh in comparison with the evidence for the older ice ages and so a great deal of knowledge has accrued about how and why the Earth began to cool and what the patterns of warming and cooling were like in terms of geological, climatic and biotic characteristics.

The causes of the long-term past and recent ice ages are founded on two factors. First, there are the astronomical factors which relate to the way the Earth revolves around the Sun and, second, there are factors involving the carbon cycle. The important astronomical factors are now considered to be those originally proposed by Milutin Milankovich, a Serbian mathematician, in the 1930s. These are illustrated in Figure 1.5.

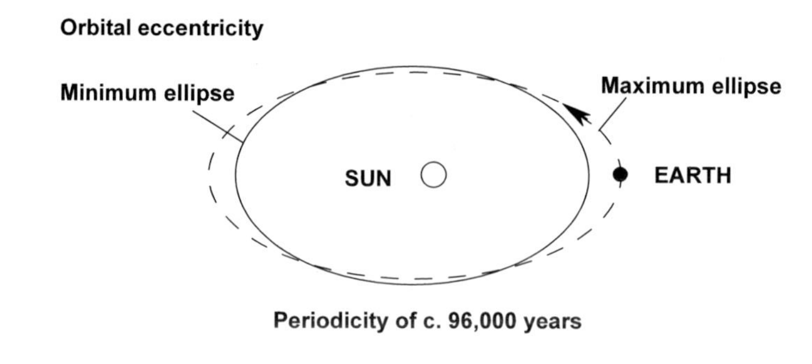

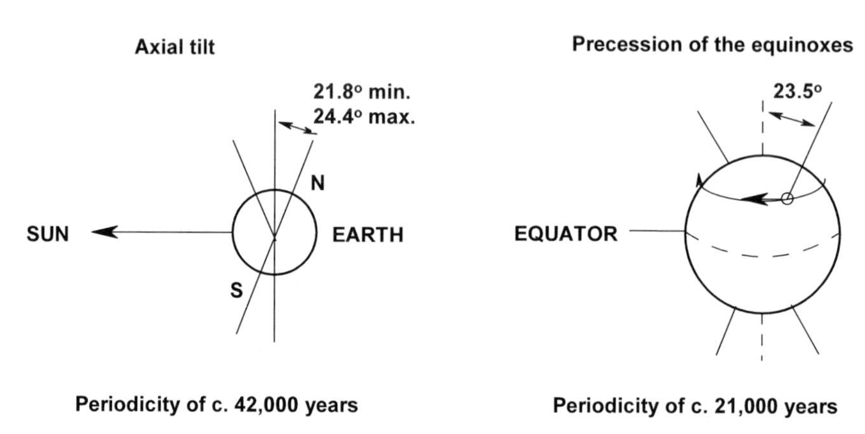

Figure 1.5. Astronomical factors in orbit of the Earth

Milankovich noted that the Earth's path around the sun is elliptical rather than circular, so when the Earth is at the furthest point of the ellipse it receives less heat from the Sun than it does at the nearest point of the ellipse. The periodicity of occurrence is c. 96,000 years. In the 1970s, emerging research on ocean sediments identified just such a periodicity in fossil assemblages and sediment stratigraphy, and so interest in Milankovich's ideas were renewed. Ocean sediment cores from all over the world confirm the existence of this periodicity and this orbital parameter is now judged to be a major trigger of cooling and the onset of a major ice advance. There are other orbital parameters which exhibit periodicities and which are also considered to influence global climate, as shown in Figure 1.5. However, the orbital factor is considered to be of insufficient magnitude alone to cause the global cooling necessary for an ice age. The other major factor appears to be the carbon cycle, compelling evidence for which derives from the polar ice caps. The extraction of deep ice cores from these inhospitable regions began in the 1980s and allowed the analysis of the composition of air entrapped in bubbles in the ice. This provided an important window on the past and allowed direct analysis of ancient atmospheres. The periodicity revealed in the ocean cores was also present in the ice core record of carbon dioxide. Specifically, in ice deposited during ice advances or cold periods the amount of carbon dioxide was found to be c. 25 percent less than that in ice deposited during (warm) interglacials. This pattern is repeated for several cold-warm oscillations with supporting evidence for cold stages of approximately 100,000 years and warm stages of c. 20,000 years.

So, where did all the carbon go? What mechanisms caused the drawdown of carbon dioxide and the removal of a substantial volume of carbon from the actively circulating part of its biogeochemical cycle? These questions have not yet been satisfactorily explained but it is likely that life forms increased their uptake of atmospheric carbon dioxide substantially. Possibly, algae in the oceans were responsible. Alternatively or additionally, increased productivity by tropical and semitropical ecosystems may have been stimulated when they increased in extent as land emerged from the continental shelf sea-levels due to ice accumulation and sea-level fall. The drop in temperature caused by the depletion of atmospheric carbon dioxide combined with the astronomical position of the Earth at the furthest point of its elliptical path around the sun conspired to generate an ice age. Why ice ages of the past (Table 1.3) came to an end is another question that must relate to the carbon cycle, at least to some extent; equally enigmatic are the issues of when and why the current ice age will end.

As the last ice advance drew to a close c. 12,000 years ago, the global warming that initiated ice retreat also brought about major changes in the composition and distribution of the world's ecosystems. In turn, this means that the above ground stores of carbon were reconfigured. Vast tracts of land vacated by retreating ice, especially in the northern hemisphere, were recolonised by shrubs and herbs. The world's tundra lands of northern Eurasia and North America emerged. They are characterised by ponds and lakes as well as extensive peatlands which comprise the remains of vegetation in various stages of decay. Today, such areas are huge stores of carbon. As ice retreated trees moved into grassland and shrub-dominated regions except in the continental interiors, the species composition of tropical forests altered as did desert margins and as sea levels rose marine ecosystems replaced terrestrial communities. By c. 6,000 years ago, the present configuration of the world's major ecosystems had come into being but these wildscapes were soon subject to another agent of change: agricultural societies.

The last three million years are unique in geological history for another reason: the emergence of humans and their transformation into a major agent of environmental change. It is likely that the environmental changes of this period contributed to the evolution of humans and their radiation from their region of emergence in Africa. Once hominids had mastered the control of fire, probably c. 2 million years ago, they began to control ecosystems rather than being integral components on a par with other animals. Fire relies on carbon so its harnessing was a first step in domesticating carbon (see Section 1.5). Another threshold was passed some 10,000 years ago, as the present interglacial opened, when agriculture emerged from hunter-gathering activities. This marks this period, the Holocene, as unique amongst the interglacials of the last 3 million years. So too is the impact of human activity on the Earth's carbon stores, notably forest destruction and the loss of other natural ecosystems, and the exploitation of geologically sequestered carbon, especially the fossil fuels. The latter returned stored carbon to the actively circulating carbon cycle at a time when the living stores of carbon are also being depleted with consequences for global climatic change. It has been argued that this provides interglacial conditions with no precedents; there is now more carbon dioxide, and probably methane, in the atmosphere than in any earlier interglacial. This anthropogenic effect is causing global warming. Will this prohibit another ice advance, postpone it or create an entirely different set of unpredictable conditions? There are no answers to these and other questions posed in this section, reflecting an inadequate understanding of the complexities of the carbon cycle.

1.5 Carbon and technology through prehistory and history

Humans differ from the rest of the animal kingdom because of their ability to manipulate their environment including other organisms. In this way organic and inorganic components of the environment are transformed into useful substances or resources. It is most likely that early hominids were integral components of ecosystems, surviving in much the same way as other animals by consuming foods within their ecological niche. Early hominids were vegetarian and fed on fruits, nuts and leaves but whether they simply grazed, as other animals do, or collected is not known. As new species evolved there was a shift to an omnivorous diet, with the meat element being acquired through scavenging and eventually through the active strategy of hunting while plant foods were actively collected. The first technology to be developed, by the hominid *Homo habilis,* about 2.5 million years ago was the making of stone tools. In itself, this did not involve carbon manipulation but it provided the means for early humans to more effectively manipulate plants and animals than hitherto. This together with the shift to a more planned than random acquisition of food not only marks the demarcation of selected materials and plant and animal species as resources but also a significant manipulation of the carbon cycle. From c. 2 million years ago fire played an important role in this manipulation as it provides a mechanism for herding animals and for influencing the distribution of plant types. Fire requires carbon, and so humans pit carbon against carbon to obtain resources, i.e. the process employs carbon present in wood and vegetation to harness the carbon of living resources. This use of tools and the domestication of fire must be considered as the first major threshold in people – environment relationships (see Section 1.4). From this point onward this relationship becomes tripartite, comprising people – carbon – environment.

By at least 12,000 years ago the first domestication of a wild species had occurred. This was the dog, domesticated from the grey wolf (*Canis lupus*) probably in the Near East. Hunter – gatherer communities were widespread globally at this time. Rapid environmental change was occurring as the last major ice advance of the Cenozoic ice age was drawing to a close. The distributions of land and sea, forest and grassland, animal populations and water resources were adjusting to global warming. The period 13,000 to 10,000 years ago witnessed another major threshold in people – environment relationships. This was the advent of agriculture which emerged, probably gradually rather than suddenly, from its hunter-gatherer precursors in the Near East and possibly in the Yangtze Valley of the Far East. The reasons for this development are unknown but probably relate to environmental change and/or cultural change. The earliest plants and animals to be domesticated in the Near East were wheat, barley, lentil and broad bean, sheep, cattle and goats. In the Yangtze basin, new evidence suggests that rice may have been domesticated as long as 15,000 years ago. All of the plants and animals harnessed then are amongst the most important agricultural products today. Additional centres of innovation arose later in other parts of the world, notably the northern Andes and Mesoamerica where crops such as potato, tobacco and maize were domesticated. From these centres of innovation agriculture spread worldwide with momentous repercussions for nature and culture. The modification and destruction of the world's natural ecosystems began, affecting their capacity to store carbon while food security led to division of labour in society facilitating the development of other resources such as metals. The carbon-capturing juggernaut that is agriculture was firmly ensconced on the road to today's myriad agricultural systems which range from the subsistent to the industrialized.

Another significant threshold in the people – carbon – environment relationship began with the great era of exploration in the fifteenth and sixteenth centuries. The primary objective of European colonists was the acquisition of riches. They came, not always in the form of sought-after precious metals, but as biomass or carbon-based resources. For example, wood was prized by British colonists and traders because of a dearth of tall trees in the deforested land of England, furs were coveted by the British and the French, while they and the Spanish quickly realised the potential of tobacco and of cotton. An exchange of crops and domesticated animals ensued. Maize, potatoes and gourds were introduced to Europe while wheat, barley, various legumes, sheep, cattle, pigs and horses spread rapidly throughout the Americas and later to Australia, New Zealand and parts of Asia and Africa. As new agricultural systems were developed in the newfound lands natural ecosystems were modified and replaced and once again the carbon cycle was altered. Moreover, Europe found a means of dispersing its population to assist its development.

By c. 1750 another momentous development was taking place, again in Europe and especially in Britain. This was the Industrial Revolution. The key factor in terms of carbon was the mechanisation of industries such as wool and cotton weaving and the production of metal goods which was founded on the use of steam power from coal. Thus the use of the first fossil fuel intensified, as did the industries which became concentrated in areas with coal reserves or those to which coal could be easily transported. Once again carbon was employed to manipulate carbon; in this case coal power was used to manufacture the protein that is wool and the carbohydrate that is cotton into saleable cloth for home and foreign markets. Coal power became synonymous with industrial development in the eighteenth and nineteenth centuries. As discussed in Section 1.6 industrialisation, which spread throughout Europe and North America, prompted major changes in socio-economic conditions. It brought new forms of pollution caused by coal use, industrial plant, and waste products. It also led to a decline in British and, to some extent, in European agriculture which, in turn, prompted the intensification of crop and animal production in the colonies; increasing volumes of carbon as food products thus crossed the oceans as Britain and Europe intensified their importation of their colonies' soil and water resources.

Beginning in the 1950s a related type of revolution occurred in the people – carbon – environment relationship. This involved the start of the large scale exploitation of the alternative fossil fuels of natural gas and oil for industrial and domestic use, and the start of the decline in coal use. One form of geologically-sequestered carbon gradually replaced another as industrialisation spread worldwide. Significant additional industries based on carbon had already developed by this time, notably the industrial production of artificial nitrate fertilizers and the embryonic chemical industries, such as that developed from war-time research, involved with crop and animal protection (agrochemical), and pharmaceuticals. The production of artificial fertilizers is significant because the conversion of inert atmospheric nitrogen into nitrate fertilizer requires large quantities of energy. The use of such fertilizers, representing a major input of fossil-fuel energy into agricultural systems, enhances carbon capture in the harvest. The widespread use of such fertilizers marked the industrialisation of agriculture. The production of artificial compounds such as plastics and artificial fibres such as nylon and Terylene, which are based on oil-derived chemicals (petrochemicals), added to the growing agrochemical and pharmaceutical businesses. Oil had thus become not just a source of energy but also a source of raw materials for specific, usually high-value, commodities.

Finally, another carbon – people – environment threshold is currently being crossed. This focuses on carbon at the microscale as compared with the macroscale developments described above. Its manipulation is the subject of biotechnology, and especially the post-1980s science of genetic modification or genetic engineering. To date, most innovations using genetic modification have been in agriculture; already transgenic cotton, maize, rape (canola) and wheat, engineered to display a variety of traits to enhance productivity, are being widely cultivated, especially in North America. There are concerns about repercussions for human health and possible adverse environmental effects but there are also many advantages, not least being increased food production. Biotechnology is also being deployed via the use of organisms for the remediation of environmental contamination with chemicals and oil. Moreover, and possibly the most momentous manipulation of carbon, is the issue of animal and especially human engineering through genetic modification.

As the twenty-first century begins the manipulation of carbon is occurring at the macroscale, through agriculture, the alteration of natural ecosystems, and fossil fuels, and at the microscale through plant, animal and human genetic manipulation. Although this technology facilitates the manipulation of carbon at the microlevel of genetic characteristics, it has serious implications for agriculture, the environment and human development.

1.6 Carbon and development

The acquisition of carbon throughout prehistory and history has been accompanied by other, often momentous, changes in resource manipulation and social organization. The domestication of fire, for example, conferred advantages on those hominid communities who could manipulate it. At the very least the use of fire enhanced food procurement. For most of human history food acquisition was through hunting and gathering at a subsistence level. Any surplus produced would have been available for trade for other resources such as stone or stone tools. A food surplus, thus, provides a degree of flexibility and of power. The shift to agriculture at the beginning of the present interglacial involved not only the harnessing of selected plant and animal species but also the increasing control over the environment; both involve the carbon cycle as carbon stores in natural vegetation are manipulated, e.g. through grazing, or removed as land is cleared for agriculture. Although particular plants and animals were important to hunter gatherers, the actual domestication of selected species marked a significant development in the human control of their environment. It also marked changing attitudes to food procurement. For whatever reason, it was considered to be beneficial to become reliant on a proximal food source rather than a dispersed wild food source. Possibly, changing resource distributions and/or climatic change forced the issue, or it may have become advantageous to have a planned surplus of food for trade. Certainly, food security confers power and the wherewithal to support population increase and the development of non-food resources by a component of the population that has become freed from food production. Division of labour, a reliable source of adequate food and planning give rise to a more heterogeneous society, and possibly a more innovative society. However, arable farming in particular is hard work; not only does it require planning it also requires labour for field preparation, weeding and harvesting. The benefits and/or the necessity must have been considerable for this type of commitment to have been adopted over a large area. In Europe, it took c. 5,000 years for the practice to spread from the Near East to the periphery e.g. Britain and southern Scandinavia. Nor can it have been a necessity everywhere. Overall, the benefits must have outweighed the disadvantages.

In terms of other developments, the initiation of agriculture in the Near East was followed rapidly by metallurgy. First, copper ores were exploited, giving rise to a relatively brief Copper Age. Then came the Bronze Age, so named because of the production of alloys of copper and tin or copper and arsenic. Bronze was more durable than copper but it required smelting at high temperatures. This and subsequent iron exploitation required an energy source: wood. While agriculture relied on solar, human and probably animal labour for energy, metallurgy presented a new use for the stored energy of photosynthesis. The resulting tools (and weapons) assisted with the acquisition of food energy in agricultural systems. There is evidence based on the genetic characteristics of modern humans that people actually migrated from the centre of agricultural innovation in the Near East into Europe, but the technology also spread through the communication of ideas between groups. Migration may have been the result of population pressure but there is no way this hypothesis can be tested, while the spread of ideas indicates contact and trade between groups. Subsequent innovations in agriculture, e.g. rotations, the use of lime as fertilizer, animal manure, improved crop and animal species through breeding etc., all enhanced the capture of carbon through agriculture and facilitated the exploitation of other non-carbon based resources.

The age of exploration of the sixteenth and seventeenth centuries was made possible through wood. European adventurers, church and royalty encouraged exploration for many reasons but it was only possible because of the skill of shipbuilders to create robust sailing vessels from wood and because of the ability of seafarers to navigate via the stars. The repercussions of European annexation of the Americas, parts of Africa and the Far East, and later of Australasia, were momentous socially and environmentally for both the conquerors and the conquered. European colonies contributed significantly to the wealth and development of Europe through the wide range of resources imported while the indigenous agriculture and settlement patterns were altered or even dismantled in favour of those of European-style. New lifestyle practices spread in Europe, such as the use of tobacco which was introduced from the Americas; its nicotine rapidly became as important a drug as alcohol. New beverages such as coffee and tea were introduced, with attendant social implications. The price paid was heavy: entire peoples were wiped out, impoverished or enslaved, and their religions and beliefs belittled. This was the start of globalization.

Soon after the age of exploration came the Industrial Revolution. The shift to coal as the major source of energy enabled industrialisation to proceed. Industries provided an alternative source of employment for an increasing population who abandoned the countryside and flocked to towns and cities many of which were emerging centres based on the supply of resources such as coal, iron and agricultural produce. The first truly urban societies emerged in Britain, Europe and North America. In the former agriculture became subordinate as a source of wealth generation as industry came to dominate the economy. This was achieved through the exploitation not only of home-based resources such as coal, iron and farm produce for food processing but also of resources from the colonies, notably raw wool, cereals, sugar, various metals, beverages and cotton. Demand for such commodities also prompted changes in the colonies such as the expansion of agriculture. The advent of refrigerated ships in the late 1800s, for example, instigated the expansion of cattle and sheep ranching in the Americas because meat could then be exported to an expanding market in Europe. This resulted from rising wealth and an increasing population. On a negative front, the burgeoning urban populations of mostly impoverished citizens were housed in appalling housing conditions. Overcrowding was endemic, disease epidemics were rife, and the quality of life was notably worse than in the countryside. The emergence, due to wealth generation from trade and industry, of a wealthy non-royal elite, set in motion urban and rural construction, museum collections, scientific research and municipal infrastructures. This was an era of unparalleled socio-economic and scientific change. It paved the way for all manner of educational, health and housing change.

The shift to oil use in the 1950s, and the development of related petrochemical industries, also influenced the changes outlined above and initiated others, including the invention and large-scale production of new commodities such as plastics, artificial fibres, agrochemicals and pharmaceuticals provided additional employment. Perhaps the most important development was the political aspect of oil reserves and production, the so-called petropolitics. Countries, notably those of the Middle East which had little international influence hitherto, gained credence, and a new industry involving oil exploration and oilfield development was spawned. Mechanisation continued apace, especially in the developed world. Not only did it expand in agriculture, which was becoming less and less labour intensive, but also on an individual basis with the mass production of cars. Oil consumption soared and the amount of carbon dioxide and methane emitted to the atmosphere increased dramatically. Post World War II populations also increased and massive building programmes began to provide homes and to rebuild bomb-damaged areas. In some developing nations industrialisation was beginning and everywhere in the developing world populations were growing rapidly. Agriculture was expanding to accommodate this growth and to provide crops for export. Mining and logging, to generate resources for export, also increased. Such activities were undertaken at the expense of natural ecosystems; deforestation and the loss of savannas and grasslands accelerated. Thus the carbon stores of the biosphere were being further depleted just at a time when terrestrial sinks were required to counteract the increasing carbon emissions from fossil fuels. Moreover, the harnessing of carbon rarely occurs in isolation. Agriculture, deforestation, industrialisation etc. involve manipulation, directly or indirectly, of other biogeochemical cycles and this causes various types of pollution and environmental problems. Examples include the nitrogen and sulphur cycles to which problems of eutrophication and acid rain relate while the use of CFCs has altered the chemistry of the stratospheric ozone layer. Similarly, the acquisition of carbon affects physical processes in the biosphere/lithosphere which can result in soil erosion and desertification. Indeed, most environmental problems result from the domestication of carbon.

By the 1980s these problems had been recognized not only by environmentalists but also by politicians. The world was waking up to environmental change, especially climatic change and the role therein of fossil-fuel carbon. International conferences and agreements have sought to instigate measures to protect the environment and to manage the carbon cycle which have been acknowledged as global enterprises. The Montreal Protocol in 1987, aimed at curtailing CFC use, was the first international agreement. Currently, the latest protocol, the Kyoto agreement to limit carbon dioxide emissions, remains a matter of some dispute, especially as the USA, a major emitter of heat-trapping gases, has rejected it. Efforts to preserve the world's remaining ecosystems are also underway though their destruction continues. Many conservation measures are inadequate or ineffectively enforced while necessity and poverty continue to drive land-cover change. That the world is not a set of unrelated nations is now accepted environmentally and politically but the practicality of ensuring fairness, security in its many senses and sustainability for all is a long way from becoming reality.

1.7 Carbon and the future

Carbon management must be a central focus of environmental policies at scales from the local to the global. First, there is a need to curtail emissions of carbon to the atmosphere; reducing consumption, absorbing emissions and finding alternative fuels will all play a role in carbon management. New developments in technology can assist and it is important the industrializing nations benefit from such advances so that they can improve standards of living without compromising their environment or the global commons. The industrialized countries can help to provide solutions. Alternative fuels, ideally, should not involve carbon emissions but their overall impact must be low when compared with fossil fuels. Improvements in agricultural productivity, especially in nations where food supplies are limited and unpredictable, are important. This is not only to improve standards of living in the poorest parts of the world but also to protect remaining natural ecosystems and the stores of carbon therein. Agricultural biotechnology has a role to play in this context as improved crops and animal health could contribute to food security, improved health and wealth generation. This also requires technology transfer between the developed and developing nations. Equally, biotechnology has a role in the agriculture of developed nations where increased yields could facilitate the withdrawal of land from cultivation for alternative uses including its return to nature. However, agricultural biotechnology, as with all technologies, is likely to have its drawbacks; adequate laboratory and field experiments must be employed to ascertain the risks to both the environment and humans. There is also scope for the use of organisms through biotechnology in remediation programmes to redress various forms of pollution.

The control of land cover and land use is another way in which the carbon cycle can be managed. First, it is imperative that the remaining forests of the world are preserved, not only to store carbon but also to maintain biodiversity, including micro-organisms as well as plants and animals. Afforestation programmes are also important, mainly as carbon stores and as providers of raw materials for future use; the sound management of such resources is essential to ensure sustainable forests and forestry. The preservation of other natural ecosystems is desirable; again natural ecosystems are more biodiverse and store more carbon than their disturbed counterparts or if they are turned into agricultural systems. Thus it is essential that, as the world's population continues to grow, the productivity of existing agricultural systems is enhanced rather than creating new agricultural systems. These measures will also contribute to the conservation of soil and water.

Other aspects of people – carbon – environment relationships are less direct but nevertheless significant. They focus on education, including the enhancement of environmental awareness through media attention and the exchange of information and ideas via the internet. Moreover, the use of computers should facilitate improved environmental management. Politically, corruption with respect to resource appropriation must be addressed and income from carbon-based resources (and other resources) should be used for the benefit of all.

Overall, human intervention in the carbon cycle must be minimized. The relationships between society, plants, animals and the milieu in which they live need to be re-examined apropos carbon conservation which is, directly or indirectly, a major tenet of sustainable development.

Chapter 2

2 THE CHEMISTRY OF CARBON

The fact that carbon has a valency of four, i.e. in the outermost shell surrounding the nucleus of the atom there are four electrons (see Section 1.1 and Figure 1.1), is the key to why carbon is so versatile. This outermost shell can accommodate eight electrons so each carbon atom has the capacity to share electrons with as many as four different atoms, including other carbon atoms. Thus carbon has the capacity to combine with many different elements to produce a vast array of compounds. This is one reason why carbon is so important in nature and to society.

Although most carbon compounds are associated with life and are thus referred to as organic, some carbon compounds are described as mineral because they are mainly, but not always exclusively, the product of geological rather than biological processes. Truly mineral forms of carbon include its allotropes, comprising pure carbon in various structural forms, e.g. graphite, diamond and soot (see Section 1.1 and Figure 1.2). Such substances are, however, important because they have decorative or industrial uses and their extraction from geological environments is a wealth generating activity. Many other compounds containing carbon have also been formed geologically. Such minerals include calcite (calcium carbonate), magnesite (magnesium carbonate) and dolomite (calcium magnesium carbonate). They can also be produced biologically as shells of marine organisms, which, as they become compressed through geological processes, produce limestone and chalk. Throughout Earth history the formation and weathering of these materials has played an important role in global climate because of the link between carbon dioxide in the atmosphere and shell-producing marine organisms. These are discussed in Chapter 4 because of their importance in the geology of carbon burial and change in atmospheric composition. The fossil fuel reservoirs are also carbon-rich minerals which are linked with life. Their formation sequestered carbon from the active biosphere, a process that also had an impact on global climate because of the withdrawal of carbon dioxide from the atmosphere. They are also discussed in Chapter 4. However, another relatively minor group of carbonates, including malachite (copper carbonate) and rhodochrosite (manganese carbonate), are considered here because they form through geological rather than biological processes.

In contrast there are millions of organic compounds. These can be classified in many different ways including a basic subdivision into two groups: naturally occurring and synthetic compounds. The former are found in living organisms, or in the environment having been produced by living organisms whilst synthetic compounds are so-called because they are artificial, i.e. they have been created in the laboratory and do not occur in nature. They have usually been created for a specific purpose, such as pharmaceuticals, crop protection chemicals and plastics. Usually they are complex molecules and comprise many carbon atoms combined with more than one other atom. The huge range of organic compounds means that they can also be classified as simple or complex. Simple compounds contain one or two carbon atoms combined with between one and four atoms of other elements; examples include carbon dioxide, which is a combination of carbon and oxygen, and methane, which is a combination of carbon and hydrogen (Figure 1.3). The many carbon atoms of complex compounds may be arranged in long chains or rings, in combination with several additional atoms; examples include the hydrocarbons, cellulose, vanilla, nicotine and the alcohols. Organic compounds, whether naturally occurring or synthetic, can be classified according to the chemical composition and structural arrangement of the defining atoms. This is a universally accepted classification scheme which was originally devised by the German chemist Beilstein in the early 1900s. It has the advantage that compounds yet to be discovered in nature or synthesised in the laboratory can eventually be included because it identifies three major structural types which can be combined with 14 different functional groups. This scheme is used here for the sake of complementarity with chemistry practice and examples of familiar compounds within the various categories are given by way of illustration. Many organic compounds have practical applications. They include pharmaceuticals, agricultural chemicals (e.g. pesticides and animal health products) and many products which are, today, taken for granted; examples include soaps, detergents, dyes, plastics and various fibres. Indeed, such commodities feature strongly in the daily lives of most people worldwide.

It is also important to note that carbon, as various compounds, is continually being circulated in the environment and like many other elements it has a biogeochemical cycle. As this name suggests there is a transfer of carbon between life and the environment, i.e. between the biosphere, the lithosphere and the atmosphere (Figure 1.4). The major pools or reservoirs of carbon are the atmosphere, the oceans, the biosphere in plants, animals and soils, and in the Earth's crust as carbonate rocks and fossil fuels. Transfers of carbon between these pools are the fluxes, the rates of which vary from minutes to millions of years. Humans have altered these reservoirs and fluxes as they have domesticated and harvested carbon and the consequences are now being felt as increased carbon in the atmosphere is a major factor causing global climatic change.

2.1 The mineral world

The four naturally-occurring forms of carbon, the allotropes of graphite, diamond, soot and buckminsterfullerenes, have been described in Section 1.1. Forms which are elementally identical but structurally different, as in the case of the soft graphite and the hard diamond, occur because of the way in which the atoms are linked as shown in Figure 1.2. Each allotrope has industrial and wealth-generating value. All are mineral insofar as they have no association with life.

The structure of each allotrope reflects the mode of geological formation. Graphite is formed when sedimentary rocks containing carbon-rich material are subject to heat and pressure so that metamorphosis occurs. The resulting metamorphic rock, usually a schist or marble relatively close to the Earth's surface, will contain quartz, and other rock-forming substances, as well as graphite. Veins of graphite may also occur in igneous rocks in which it has replaced other minerals. Most graphite is extracted by open-cast, near surface mining though in Sri Lanka deep mining is necessary to extract it from veins. In contrast, diamond is formed well below the Earth's crust, usually in excess of 150 km in the mantle (see Figure 1.4), where enormous heat and pressure occur. Carbon in the molten rock cools very slowly allowing large crystals to develop. Diamonds are formed in rocks known as kimberlites which occur in pipes rarely more than 50 m in diameter and which have reached the surface, along with rock fragments, due to explosive events as volcanic gases escape. Few kimberlites occur worldwide and they are thus highly prized; most are in Africa with others in Russia, Canada and Australia. Only a small proportion of the output comprises diamonds of gemstone quality with the remainder being used industrially as abrasives for cutting, polishing etc. alloys and ceramics. Australia is the largest producer of diamonds, with the Argyle Mine of the Kimberley region of northeast Western Australia being the largest source. Recent developments some 300 km north of Yellowknife in Canada's Northwest Territories may, however, allow the Diavik and Ekati mines of the lac de Gras area to usurp this position. The first kimberlites were discovered here in the early 1990s since when more than 200 have been discovered; each being only 200 to 300 m in diameter. According to Holmes (2005) these two competing mines "have produced more wealth than a century of gold mining, the previous mainstay of the north [of Canada]". The Diavik mine may prove to be the richest diamond lode in the world; it produces more than five times the world average and c. 33 percent are of gem quality; there is the potential for the extraction of 7×10^6 carats annually with a value of c. US$408 $\times 10^6$ (Holmes, 2005). Diamonds also occur in secondary deposits, known as placer deposits resulting from the erosion of kimberlites by rivers. Such deposits occur off the coast of

Namibia from which diamonds are recovered by the dredging and filtration of sediment in specially adapted ships.

The occurrence geologically of amorphous carbon relates to the occurrence of naturally-occurring fire in the geological past and the combustion of organic material in an oxygen-poor environment. Such conditions are termed reducing conditions resulting in carbon production rather than carbon dioxide production which would occur in oxygen-rich conditions. Amorphous carbon is widespread and as stated in Section 1.1, it has a variety of names and commercial uses. Buckminsterfullerenes are enigmatic insofar as they do occur in nature but to date they have been identified only in the outer atmospheres of stars. This was how they were inadvertently discovered in the mid 1980s! There would appear to be no earth-based geological history!

All these allotropes of carbon can be manufactured, though as discussed in Section 1.1 a fifth allotrope, white carbon, is entirely synthetic. Laboratory/manufacturing conditions, notably, heat and pressure, can be varied to produce these substances. This can be less expensive than attempting to recover them from the Earth's crust. For example, the annual production of artificial diamonds, mostly industrial rather than gemstone quality, is more than double that from mining. The manufacture of carbon forms unavailable in nature, such as diamond films, has also opened up new industrial applications. Other lines of enquiry are focussing on the potential applications of Buckminsterfullerenes. For example, so-called 'buckytubes' have been produced comprising hexagonal tubes which may be a single wall or a series of nested tubes. The importance of this concerns their capacity to conduct electricity and they can be manipulated to act as either semiconductors or as metals. They may have a significant role to play in the emerging science of nanotechnology, i.e. engineering concerned with entities of less than 100 nanometers, including individual molecules. One example of an application of Buckminsterfullerenes is in nanocomputers.

Limestone, dolomite, coal, oil and natural gas are also described as minerals. This term is used in the context of their value as resources rather than in relation to their chemical and geological formation which is intimately linked with life forms and thus makes them organic. However, the inorganic formation of limestones, dolomites and other calcareous, sedimentary rocks is also important geologically, as discussed in Sections 4.1.2 and 4.2. These are non-renewable resources, i.e. they are not being replaced by natural processes, or are not being replaced at the rate at which they are being used. Moreover, they are mostly the products of photosynthesis in the geological past and all have resulted in the removal of vast quantities of carbon dioxide from the atmosphere. This reflects the close links between life and climate throughout Earth history, as discussed in detail in Chapter 4. Such links also explain why recent escalated consumption of fossil fuels is causing climatic change (see Chapter 7). Many other carbonates may be formed by mainly geological rather than biological processes. Examples include the carbonates of reactive metals such as copper, manganese, strontium, lead, barium, magnesium and iron etc. Descriptions of these minerals are given in Table 2.1. These carbonate minerals are formed under two possible geological scenarios: either carbonate-rich waters come into contact with ore veins in igneous rocks, or water with a high metal contact originating in the crust where the metal ores have been deposited comes into contact with carbonate rocks such as limestone. Many of these minerals, notably malachite and rhodochrosite, are used for jewellery and decoration.

Table 2.1. Carbonate minerals (based on Pellant, 1990)

MINERAL	CHEMISTRY	COLOUR	SOURCE
Smithsonite	Zinc carbonate	Grey/white	Around zinc ores
Rhodochrosite	Manganese carbonate	Pink	Argentina
Witherite	Barium carbonate	White/yellow	USA, Russia
Siderite	Iron carbonate	Brown/grey	Widespread
Malachite	Copper carbonate	Green	Chile, DRCongo, Zimbabwe, Russia
Azurite	Copper carbonate	Blue	As above
Cerrusite	Lead carbonate	Grey/white	Widespread
Artinite	Magnesium carbonate	White/grey	New Jersey and New York
Aurichallite	Zinc copper carbonate	Pale blue-green	Scotland, France, Namibia
Strontianite	Strontium carbonate	White/pale green/yellow	Scotland, England, Germany

2.2 Organic compounds

More than 6.5 million organic compounds are known to exist and since there are so many organisms that have not yet been described it is inevitable that there are many more organic compounds yet to be discovered. Many of these will be of economic and social value and the potential of these and related but not naturally-occurring compounds has led research chemists to conceptualize the potential as chemical space (see Dobson, 2004; Clardy and Walsh, 2004). All such possibilities testify to the versatility of carbon. Similarly, there are endless possibilities for the creation of new substances with no counterparts in nature. Classifying and describing such a vast array of substances is daunting. Elementary classifications include a subdivision into natural and synthetic compounds, or into simple and complex compounds. A further classification is one based on functional groups, i.e. the presence of a distinct grouping of atoms which confers a specific capability or function. For example, all compounds with an oxygen and hydrogen atom joined together as the hydroxyl ion are alcohols and all have similar properties. The presence of a halide such as chlorine or fluorine (the organohalides) represents another functional group as in chlorofluorocarbons which are used in refrigeration and which are known to damage the stratospheric ozone layer. The 12 major functional groups are listed in Table 2.2 along with example drawings of a typical structure from each class.

Alternatively, a basic structural classification may be employed: compounds comprising carbon atoms linked in chains are the acyclic compounds while those with carbon atoms (and possibly other atoms) arranged in rings are known as cyclic compounds of which aromatic compounds are an interesting sub-class. As their name suggests, the latter are characterised by distinctive odours. Cyclic compounds can be further subdivided according to whether the rings contain only carbon, or include other atoms; these are the isocyclic (also carbocyclic) and heterocyclic categories. The structural differences between these three groups are illustrated in Figure 2.1. In general, acyclic compounds are sources of fuel while cyclic compounds are widely used as pharmaceuticals and agrochemicals.

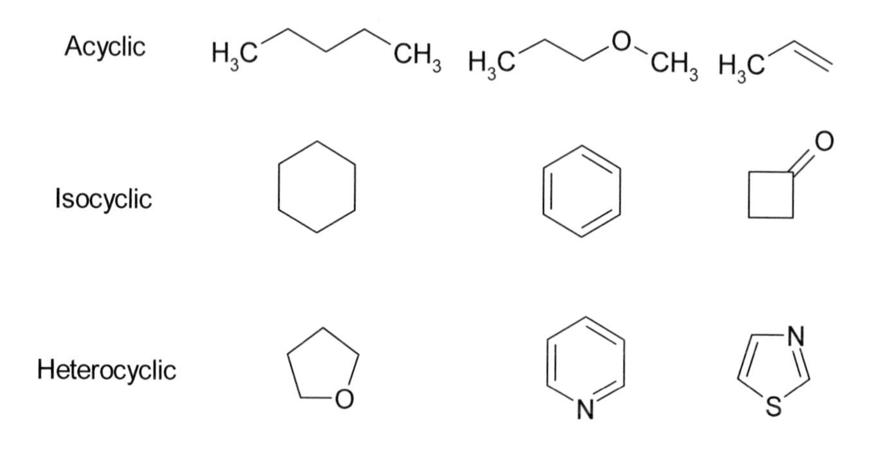

Figure 2.1. Classification of compounds, based on their carbon framework

Table 2.2. A classification of organic chemicals based on functional groups

CLASS OF COMPOUND	FUNCTIONAL GROUP	EXAMPLE STRUCTURE
Alkenes	Double bond between carbon atoms	H_3C CH_3
Alkynes	Triple bond between carbon atoms	HC_3 CH_3
Alcohols	Oxygen and hydrogen joined as hydroxyl group	O H
Ethers	Presence of oxygen atom between two carbon atoms	H_3C O CH_3
Organohalides	Presence of halogens: fluorine, chlorine, bromine or iodine	Cl F Cl
Aldehydes	Double bond between carbon and oxygen and a hydrogen atom attached to same carbon	O H_3C H
Ketones	Double bond between carbon and oxygen	O H_3C CH_3
Carboxylic acids	Double bond between carbon and oxygen plus hydroxyl attached to same carbon	O H_3C O H
Amines	Presence of nitrogen combined with hydrogen	H_3C N H H
Nitro compounds	Presence of nitrogen attached to two oxygens	H_3C N^+ O^- O
Nitriles	Triple bond between carbon and nitrogen	HC_3 N
Organometallic compounds	Metal atom bonded to one or more carbons	H_3C Zn CH_3

In the late 1800s, the German chemist Friedrich Konrad Bielstein, by then Professor of Organic Chemistry at the Imperial Technical Institute of St Petersburg, Russia, devised a classification system which combined both structural and functional characteristics of organic compounds. This was first published as the *Beilstein Handbook of Organic Chemistry* in 1881–1883; it described and classified some 1500 compounds and even then it ran to 2200 pages in two volumes. By 1984 the 5th edition comprised 400,000 pages in 480 volumes and in 1988 it appeared in electronic form (see details in Heller, 1998). It remains the most widely used classification scheme today and it is given in Table 2.3.

A brief survey of the characteristics of organic compounds is given below following the three structural divisions of Beilstein.

Table 2.3. Beilstein's classification scheme for organic chemicals

FUNCTIONAL GROUPS	STRUCTURAL DIVISIONS
1. No functional group	Acyclic
2. Hydroxy compounds	Isocyclic
3. Oxo compounds	Heterocyclic
4. Carboxylic acids	
5. Sulphinic acids	
6. Sulphonic acids	
7. Seleninic, Selenonic, Tellurinic acids	
8. Amines	
9. Hydroxylamines, Dihydroxyamines	
10. Hydrazines	
11. Azo compounds	
12. Diazonium compounds	
13. 3 or more N atoms present	
14. Carbon directly bonded toSi,Ge,Sn	
15. Carbon directly bonded to P,As,Sb,Bi	
16. Carbon directly bonded to elements of 3rd-1st A-groups of periodic table	
17. Carbon directly bonded to elements of 1st-8th B-groups of periodic table	

2.2.1 Acyclic compounds

In acyclic organic compounds, the carbon atoms are joined together in chains. This arrangement contrasts with the other two structural groups, the isocyclic and heterocyclic compounds in which the carbon atoms are joined together in rings (see Sections 2.2.2 and 2.2.3). As Table 2.3 shows, acyclic compounds may contain a wide variety of atoms other than carbon and may have any type of functional group listed in Table 2.2. They may be natural or synthetic and many are in widespread use. Examples from the major groups of acyclic compounds, and their common uses, are given in Table 2.4, which shows that methane, ethane, butane and propane are hydrocarbons (see Olah and Molnar, 1995). They are naturally occurring alkanes present in natural gas, the product of ancient ecosystems whose carbon has become sequestered in geological deposits (see Chapter 4). Natural gas comprises between 80 and 95 percent methane while the remainder is a mixture of ethane, butane, propane and other hydrocarbons in varying amounts. The latter are also obtained from petroleum, another carbon-rich geological deposit (Chapter 4). The chemical structures of these compounds are given in Figure 2.2.

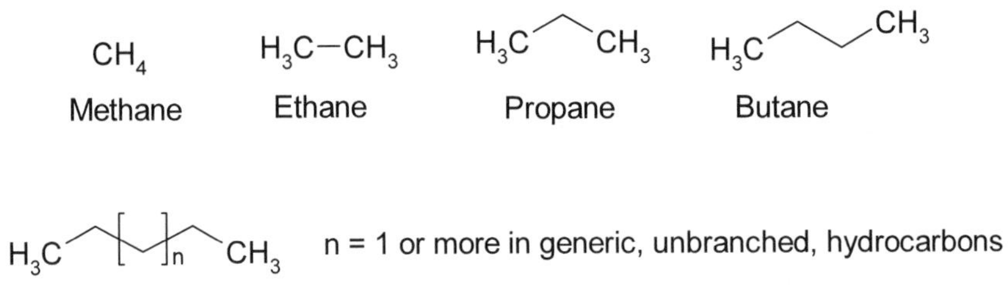

CH_4 Methane

H_3C-CH_3 Ethane

H_3C CH_3 Propane

H_3C CH_3 Butane

H_3C [ ]n CH_3 n = 1 or more in generic, unbranched, hydrocarbons

Figure 2.2. Chemical structure of some simple acyclic hydrocarbons

The C_1 to C_4 hydrocarbons are colourless, odourless gases and all are used as sources of fuel, especially methane where natural gas is supplied for industrial and domestic purposes. Methane is also produced from marshlands and peatlands, and from landfill sites, as a result of vegetation decay. Ethane and propane are employed as refrigerants while butane is used in the manufacture of synthetic rubber. Moreover, many hydrocarbons from crude oil are the basis of the petrochemical industry (see Section 5.4) which provides the starting chemicals for many widely-used synthetic compounds such as plastics, artificial fibres, soaps, resins and detergents.

Table 2.4. Some common acyclic carbon compounds and their uses

CHEMICAL NAME	**COMMON NAME** (if applicable)	**USE**
HYDROCARBONS:		
Methane		Fuels
Ethane		
Propane		Propellants for aerosols
Butane		
HALIDES:		
Carbon tetrachloride		Solvent
Chlorofluorocarbons	CFCs	Propellants for aerosols
ESTERS	Soap	Cleaning product
	Terylene (polyester)	Fabric
ALCOHOLS:		
Methanol	Wood alcohol	
Ethanol	Alcohol	
Ethylene glycol		Antifreeze
ETHERS:		
Dimethyl ether		
Diethyl ether		Solvent
CARBONYL COMPOUNDS:		
Formaldehyde	Formalin	Preservative
Acetone	Nail-varnish remover	Solvent
CARBON–NITROGEN COMPOUNDS:		
Methylamine		
Acetonitrile		Solvent
Polyamides	E.g. Nylon	Fabric
ORGANIC ACIDS		
Acetic acid	Vinegar	Culinary uses
N-(phosphonomethyl) glycine	Glyphosate	Herbicide

Acetic acid, its name derived from the Latin word *acetum* meaning vinegar, is another acyclic compound in widespread use. Its structure is shown in Figure 2.3. It has numerous uses other than as a culinary condiment, which is a solution in water containing between 4 and 8 percent acetic acid. In nature, acetic acid is formed by the action of bacteria on wine or other fermented alcohol but because of its many industrial applications it is manufactured from carbon monoxide and methanol in the presence of a catalyst. Its many uses include the manufacture of the pharmaceutical known as aspirin, photographic films, rayon, various glues and certain types of plastics.

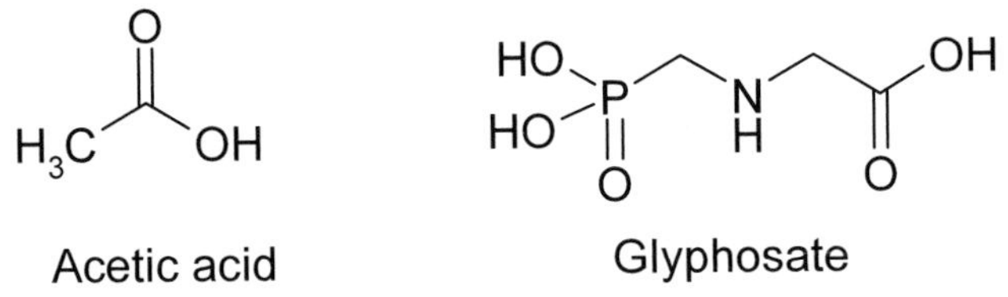

Acetic acid Glyphosate

Figure 2.3. Chemical structure of acetic acid and glyphosate

Glyphosate is another organic acid but it is chemically more complex than acetic acid, as shown in Figure 2.3. It is a broad spectrum herbicide in widespread use. It kills most types of plant and is thus non selective. Glyphosate is produced industrially and was introduced as Roundup® in 1974 by the chemical company Monsanto. There are several formulations, including one for use in aquatic environments. It has a systemic mode of action involving the inhibition of aromatic amino acid formation in the plant, which prevents the biosynthesis of essential proteins. According to Tu *et al.*(2001) glyphosate is highly soluble in water, though it binds with soil particles which restricts movement in the environment, and it is not toxic to animals. It is usually sprayed to eradicate weeds prior to crop emergence and kills them within a few days of application. Its widespread use in crop protection has prompted Monsanto to produce genetically-engineered crop plants, such as maize, with resistance to glyphosate. Thereby glyphosate can be used in post-emergent stages of crop growth and the resistant crop seeds and the herbicide can be marketed together.

Soaps are acyclic compounds in everyday use and are soluble both in water and in fat. Chemically, soaps are the alkali salts of fatty acids. The latter include stearic, oleic and palmitic acids which are derived from oils and fats. When mixed with a strong alkali such as potassium hydroxide (potash) or sodium hydroxide (caustic soda) a soap is formed, as shown in Figure 2.4.

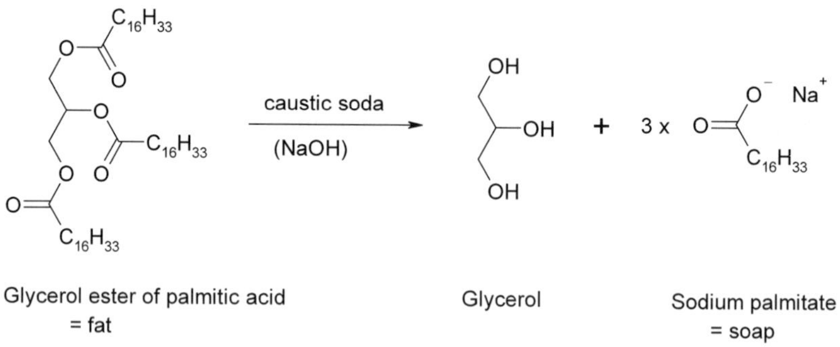

caustic soda (NaOH)

3 x

Glycerol ester of palmitic acid = fat

Glycerol

Sodium palmitate = soap

Figure 2.4. The chemistry of soap formation from natural fats

The fatty ester or fatty acid raw materials may derive from plants or animals. According to the Soaps and Detergents Association (2004) soap can be produced in two ways. Saponification involves heating fats/oils with a liquid alkali to produce soap, water and glycerine. Alternatively, the fatty acids may be neutralised with an alkali. This requires the chemical separation of the fatty acids using high-pressure steam. The glycerine produced is removed and the remaining fatty acids are distilled to purify them and then neutralised with an alkali. The resulting soap may be hard or soft depending on which alkali is used: sodium hydroxide yields hard soaps and potassium hydroxide yields soft soaps used in liquid soaps and shaving creams. Soap is a cleansing agent because part of the molecule, the long chain hydrocarbon, is fat-soluble but the other part of the molecule, the carboxylic acid, prefers to dissolve in water. In use, an emulsion is formed in which the hydrocarbon component dissolves the greasy grime of soiled clothes, making them soluble in water. Hence during the agitation of washing the grease and grime are removed by the soap and left in the water.

2.2.2 Isocyclic (carbocyclic) compounds

Isocyclic compounds contain ring structures in which all the ring components are carbon atoms, as shown in Figure 2.1. There are millions of isocyclic compounds, including many naturally occurring compounds with a variety of uses. Many have tastes and/or smells and are used as sweeteners or flavourings or are present in specific crop types selected for that very property. Others have medicinal properties, the best known example of which is aspirin.

Table 2.5. Isocyclic compounds of plant origin (based on Senese, 2004, with additions)

	PROPERTY	**SOURCE**	**USES**	**ORIGIN**
Vanillin	Sweetish but subtle taste and aroma	The tropical orchid *Vanilla planifolia*	Food flavouring, production of pharmaceutical for treating Parkinson's disease	Mexico
Eugenol	Stronger aroma and taste than vanilla	Bay leaves, allspice and oil of cloves	Food flavouring, antiseptic	Mediter-ranean Caribbean China
Zingerone	Strong hot spicy aromatic flavour	Root and stems of the plant *Zingiber officinalis*	Food flavouring, traditional medicinal uses	South China
Capsaicin	Hot flavour of variable degree, odourless	Plants of the *Capsicum* family, including the pepper and chilli	Food flavouring	Central America

Some of the most well known food flavours belong to a group known as the vanilloids, the components of which are listed in Table 2.5. All are phenols, meaning that they containing a benzene ring of six carbon atoms to which hydroxyl groups are attached (see category 2 in Table 2.3). Their chemical structures are illustrated in Figure 2.5. Vanillin is the smallest of these molecules. It is soluble in water and highly prized as a food flavouring. Its natural source is *Vanilla planifolia,* a tropical orchid native to Mexico which was introduced into Europe by the Spanish conquistador Hernan Cortes in the early sixteenth century. It was later introduced into many tropical islands and for many, e.g. Tahiti, Madagascar and the West Indies, it remains an important cash crop. Vanillin is also found in lignin and is the reason why wines aged in oak barrels sometimes have a faint vanilla taste. Today, vanillin is also produced synthetically. It can be manufactured from eugenol, another food flavouring (see above) which is characterised by a short hydrocarbon chain (Figure 2.6). This means that it is only slightly soluble in water (see comments in Section 2.2.1 in relation to soap) but soluble in fats and oils, and volatile. In cooking this chemical composition imparts a stronger flavour than vanilla.

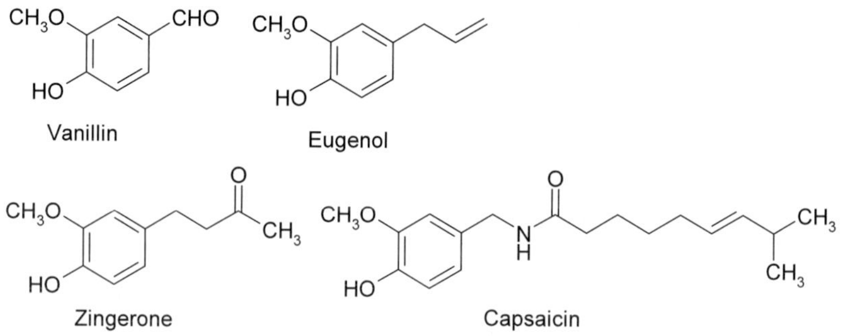

Vanillin Eugenol

Zingerone Capsaicin

Figure 2.5. The structures of selected vanilloids

Eugenol also has an analgesic affect, a reason that oil of cloves is used as a traditional 'cure' for toothache, and is a mild antiseptic. Zingerone, a component of ginger, especially the root, and mustard oil, has a carbonyl-containing chain (see Figure 2.5). The presence of the carbonyl group causes zingerone molecules to be attracted to each other, which reduces volatility but generates intense flavour. Hence ginger is widely used as a food flavouring especially in oriental and Indian recipes, which reflects its domestication and abundance in the eastern tropics. Capsaicin has a longer hydrocarbon chain than the other vanilloids; this plus the presence of an amide group means that volatility is low and so there is little smell from the plants that produce it. Apart from the *Capsicum* group, it occurs in small amounts in oregano, cinnamon and coriander, all of which have culinary uses. This isocyclic compound is responsible for the varying intensities of 'hotness' that characterise the various species of peppers and chillies. Senese (2004) reports that the long hydrocarbon chain renders capsaicin insoluble in water and is the cause of its strong flavour because it facilitates strong binding with lipoprotein receptors in human taste buds. Apart from its use in food, capsaicin is used in the manufacture of self-defence sprays against muggers etc. and as a treatment for arthritis and similar conditions.

Another commonly used flavouring is that of menthol which is a terpene. Terpenes are a large group of organic molecules, most of which are isocyclic though a few are heterocyclic. They comprise five carbon atoms and eight hydrogen atoms or multiples thereof, and they are widely found in plants. Many terpenes are used as flavourings because of their taste and as constituents of perfumes because of their distinctive odours (see Bauer *et al.*,

2001). Some examples are given in Table 2.6 and their chemical structures are given in Figure 2.6.

Table 2.6. Some commonly occurring terpenes.

No. of C atoms	NAME	SOURCE
10	Menthol	*Mentha arvensis* (field mint)
10	Camphor	Laurel tree
10	Oils of lemon, orange	Lemons, oranges
15	Oil of cloves	Cloves
15	Oil of cedarwood	*Juniperus virginiana* (Red cedar)
20	Vitamin A	Liver, dairy products
40	Beta carotene	Green, orange and yellow vegetables

As Figure 2.6 shows, menthol has a carbon ring with a hydroxyl group and three methyl (CH_3) groups. It is present in the common mint from which the majority of menthol is still obtained. Its 'cooling' effect is widely used in foods and chewing gum as well as decongestants, pharmaceuticals, cosmetics, toothpastes and mouthwashes. The fruit-derived oils listed in Table 2.6 are also employed in the food industry. Oil of cedar is one of many terpenes used in the pharmaceutical industry: it is distilled from the leaves of the red cedar and is a constituent of lineaments for treating rheumatic conditions. Oil of cedar wood, containing the terpene cedar camphor, is a related oil obtained from the wood of red cedar. It is used in the perfume industry, especially for perfuming soap. Beta carotene is an essential dietary requirement and is one of a group of carotenoids which are present in the red, yellow and orange pigments found in many fruits and vegetables. It is an essential component of nutrition because vitamin A is manufactured from it in the human body and this is essential for growth and development as well as being important for the body's immune system.

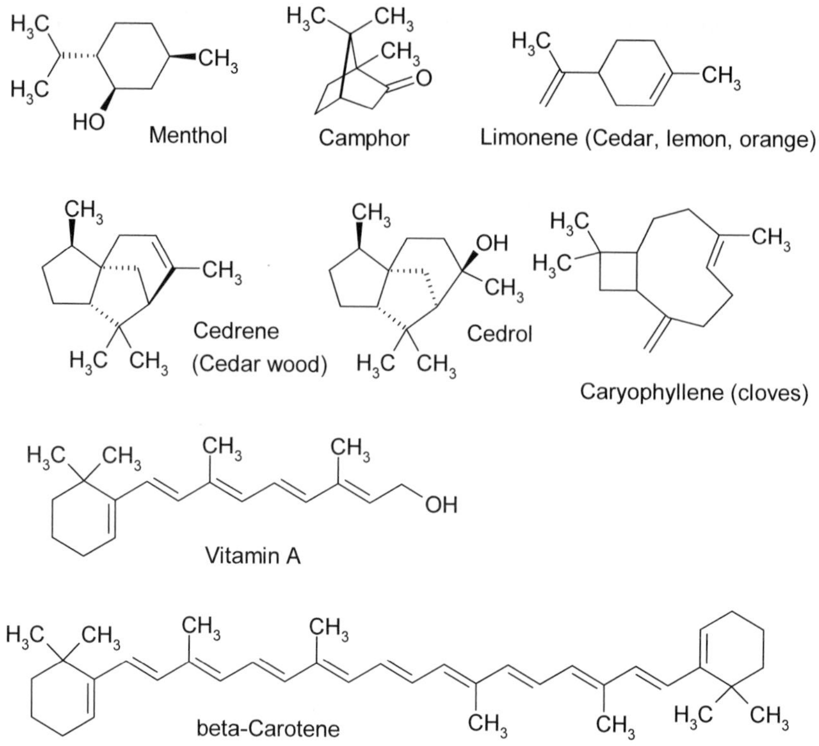

Figure 2.6. The structures of selected terpenes

Figure 2.7 gives the structure of aspartame (NutraSweet) which is a synthetic artificial sweetener in widespread use. According to Walters (2001) it was discovered accidentally in the mid 1960s in the search for an anti-ulcer compound. Chemically, aspartame is a methyl ester of an amino acid combination which is synthesised from phenylalanine and aspartic acid. The methyl ester is the $COOCH_3$ group shown in Figure 2.7. Aspartame is c. 180 times sweeter than sucrose (see Section 2.2.3), the sweet molecule in sugar, and is processed in the human body as a protein. It is widely used as a substitute for the more calorie-rich sugar in soft and carbonated drinks, desserts etc. though it is not suitable for cooking because it breaks down and loses its sweetness at high temperatures; currently, aspartame is used in more than 6,000 products and is consumed by c. 100 x 10^6 people worldwide (Aspartame Information Center, 2004).

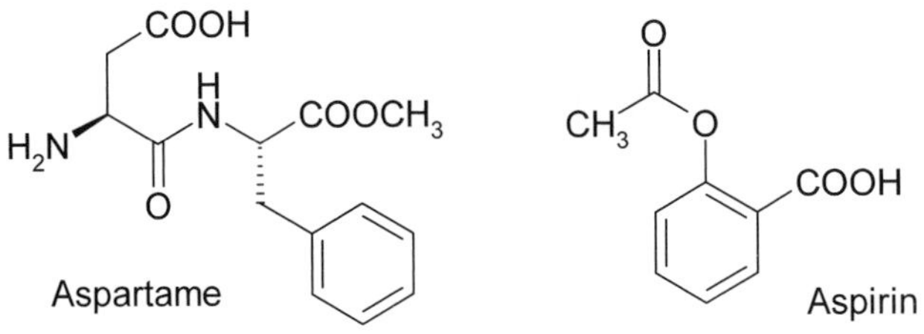

Figure 2.7. The structures of aspartame and aspirin

Another well-known isocyclic compound is aspirin. This is acetylsalicylic acid, a carboxylic acid, as illustrated in Figure 2.7. According to the Aspirin Foundation (2004), aspirin is the most widely used pharmaceutical in the world with annual production/consumption amounting to 35,000 tonnes or more than 100×10^9 standard tablets. It is synthesised from salicylic acid and acetic anhydride. Aspirin has a long history of use (Bellis, 2004) with written records ascribed to Hippocrates, a Greek physician often referred to as 'the father of medicine' who lived c. 400 years BCE. He noted pain-relief treatments involving a powder made from the bark and leaves of willow (*Salix* spp.) which was used to treat headaches and fevers. In 1829 the active compound, salicin, was identified and commercialisation commenced. The development of aspirin occurred by the late 1880s when a buffering compound was combined with the salicin (salicylic acid) and a patent was granted in 1889. Aspirin has many applications; it is an analgesic, anti-inflammatory, antipyretic and anti blood platelet clotting.

2.2.3 Heterocyclic compounds

Heterocyclic compounds have ring structure(s) that, unlike isocyclic compounds, contain atoms of elements other than carbon (see Figure 2.1). There are millions of heterocyclic compounds in existence, which can be divided into six groups, as defined by Beilstein (see Section 2.2 above) and as listed in Table 2.7. The groups are characterised by the number of non-carbon (= "hetero") atoms they contain. Commonly these are atoms of oxygen or nitrogen, which form stable cyclic combinations with carbon, especially in 5- and 6-membered rings.

Table 2.7. Beilstein's classification of heterocyclic organic compounds, with examples

GROUPING	EXAMPLES	SOURCES / USES
1. One oxygen atom	Glucose Fructose Galactose	Sugar in the blood A sugar in honey and fruit A sugar in milk
2. Two or more oxygen atoms	2,3,7,8-tetrachlorodibenzo-p-dioxin Artemisinin	Formed by burning chlorine-containing compounds with hydrocarbons, hazardous substances Anti-malarial pharmaceutical
3. One nitrogen atom	Nicotine Cocaine Paraquat Indigo	In tobacco (*Nicotiana tabacum*) In leaves of *Erythroxylon coca* Synthetic herbicide Blue dye
4. Two nitrogen atoms	Histidine Sildenafil (Viagra®) Caffeine	An amino acid found in protein manufactured in the human body Pharmaceutical A stimulant in coffee
5. Three or more nitrogen atoms	Atrazine	A synthetic herbicide
6. One or more nitrogen atoms with one or more oxygen atoms	Acesulfame Clavulanic acid	A sweetener Beta-lactamase inhibitor

Glucose, galactose and fructose are monosaccharides, or simple sugars, which are members of a group of foods known as carbohydrates. They are described as monosaccharides because they contain only one ring of carbon atoms in which there is also one oxygen atom, as shown in Figure 2.8. Other carbohydrates include the disaccharide sugars sucrose, lactose and maltose, so called because they each comprise two different monosaccharides linked together. Polysaccharides are more complex structures formed when monosaccharide units are combined into polymers (many molecules linked together, see Section 2.2.4) arranged in various ways to produce starches, glycogen and cellulose.

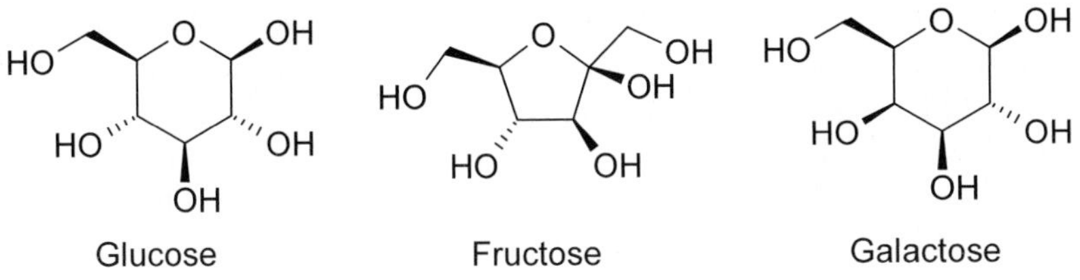

Glucose Fructose Galactose

Figure 2.8. The structures of selected monosaccharides

Glucose is manufactured by green plants from carbon dioxide and water in the process of photosynthesis (see Section 3.3) which requires light energy. It is a vital process to all forms of life and it underpins energy transfers within food chains and webs. Fructose has the same chemical formula as glucose but it has a different structure (Figure 2.8) and so is called an isomer. Isomerism is common in organic chemicals and occurs in general when different structures share the same elemental composition. Fructose's different structure confers additional sweetness. Both sugars are important in animal and human diets and both are components of disaccharides and polysaccharides which reflects their significance as building blocks in life. All three types of saccharides are present in honey, which contains c. 38 percent fructose and c. 31 percent glucose, as well as proteins, amino acids, vitamins, minerals, antioxidants and organic acids making it a beneficial source of food energy. The chemistry of carbohydrates is examined in detail by Davis and Fairbanks (2002).

The dioxin TCDD is, in contrast, one of the most harmful chemicals known. It contains two oxygen atoms in one of its carbon rings, as shown in Figure 2.9, and is one of a group of dioxins which are all hazardous. TCDD is a synthetic compound that is produced when some chemicals containing chlorine are combusted with hydrocarbons. The health problems caused by dioxin were exposed in the 1970s in the context of the Vietnam War and the use of the herbicide known as Agent Orange for defoliation, in which dioxin was a contaminant. It has become linked with birth defects, reproductive problems, skin problems and some forms of cancer.

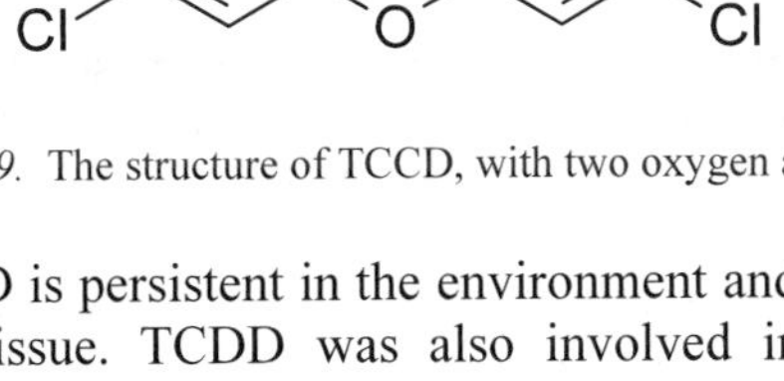

TCCD

Figure 2.9. The structure of TCCD, with two oxygen atoms included in a ring

TCDD is persistent in the environment and it accumulates in animal and human tissue. TCDD was also involved in a major accident at a chemical factory in Seveso, Italy, in 1976. A cloud of toxic chemicals, including dioxins, was released into the atmosphere and eventually contaminated an area of 15 square kilometres with a population of 37,000 people. TCCD enters the environment through the incineration of waste containing chlorine and is one of many reasons why waste incineration is strongly opposed. It is also associated with paper mills and other industries that use chlorine compounds for bleaching, and with the production of Polyvinyl Chloride, the PVC plastics. For additional information on TCDD and other dioxins consult Crummett (2002).

Another compound containing three oxygen atoms in its ring structure is artemisinin, as shown in Figure 2.10. This is a naturally occurring compound found in a plant called sweet wormwood (*Artemisia annua).* This is native to western Asia, notably China, though it now occurs worldwide. Wormwood has a long history of use in Chinese herbal medicine for the treatment of fevers and one of the earliest written records of its use as a herbal medicine brewed from leaves is in the 'Handbook of Prescriptions for Emergency Treatments' compiled by Ge Hong who lived between 281 and 340 AD (O'Neill and Posner, 2004). It was not until the early 1970s that its active ingredient was isolated. This is artemisinin, a sesquiterpene, which has the Chinese name of *qinghaosu.*

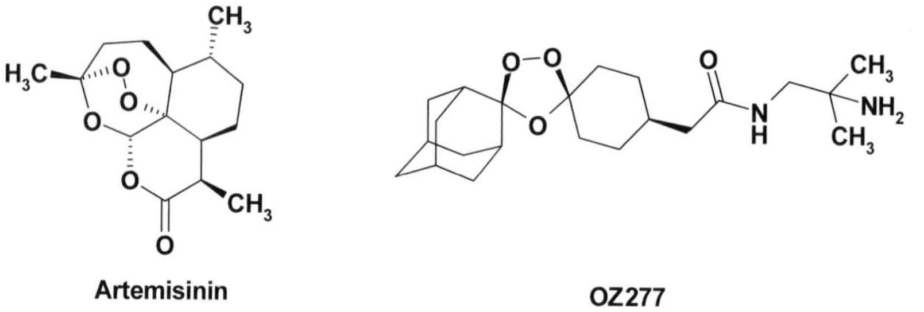

Figure 2.10. The structure of natural artemisinin and synthetic OZ277

The significance of artemisinin is its use as a model for the development of new anti-malarial pharmaceuticals. New treatments are urgently needed because many existing pharmaceuticals are no longer effective because of the development of resistance in mosquito populations. Nevertheless, as O'Neil and Posner point out, some 2×10^9 people inhabit areas where malaria is endemic; the incidence of the disease and the death rate from it are increasing. Two derivatives of artemisinin have been synthesized: arteether and artesunate; they work by killing the blood stages of the parasite which causes malaria (*Plasmodium falciparum).* However, there are drawbacks with these compounds and research has been focussed on developing synthetic alternatives based on the same mode of action. Vennerstrom *et al.* (2004) have now produced such a prototype clinical candidate, OZ277, and it is anticipated that an inexpensive and efficacious product will soon be widely available.

The addition of one nitrogen atom to a carbon ring yields quite a different range of substances. The examples given in Table 2.7 have very different properties though nicotine and cocaine are both alkaloids. These latter are a large group of naturally occurring plant products with both harmful and beneficial properties; the group also includes morphine and codeine which contain carbon rings only, as well as caffeine which has two nitrogen atoms in the carbon ring (see below). Nicotine is generally a harmful substance, cocaine can be harmful but its derivatives can be used beneficially in medicine for local anaesthesia, while the herbicide paraquat is widely used to increase crop yields but is toxic if swallowed. Figure 2.11 gives the chemical structure of nicotine and cocaine. Nicotine comprises two carbon rings containing nitrogen plus a methyl (CH_3) group. It is found in the tobacco plant and has a distinct pungent taste and smell which reflects its role as an insecticide in the natural world.

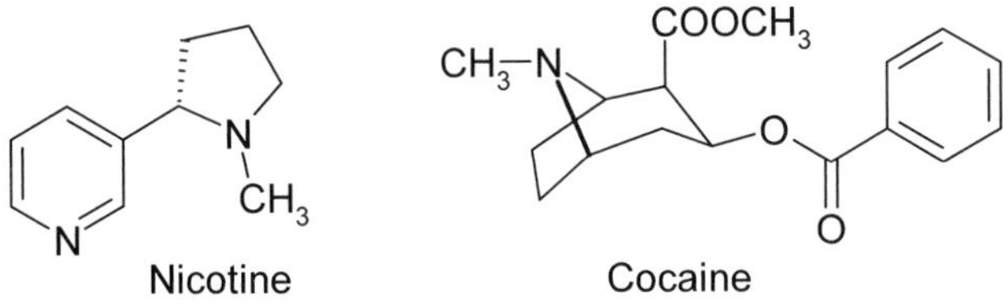

Figure 2.11. Examples of structures with a nitrogen atom included in a ring

The tobacco plant and nicotine itself are named after Jean Nicot de Villemain, ambassador of the King of France in Portugal, who introduced tobacco to the French court in 1560. As Gately (2001) has highlighted, tobacco has a chequered history beginning with its domestication in the northern Andes (modern Peru) and its subsequent spread throughout central and North America whence it was introduced to Europe in the sixteenth century. Today, it is a major economic crop of tropical and sub tropical regions though it is controversial because of the proven health risks that its consumption through smoking carries.

Cocaine is a naturally-occurring alkaloid in several plants known as coca (*Erythroxylon spp.)* which are native in the wet forests of the eastern Andes. According to the United Nations Office on Drugs and Crime (UNODC, 2004) there are fifteen species in this family which contain cocaine though only two species have a relatively high cocaine content and which are cultivated for the purpose of producing the illicit drug. These are *E. coca* and *E. novogranatense.* The drug works by releasing abundant dopamine, a neurotransmitter, from brain neurons. Continual use can lead to addiction with attendant behavioural and social problems. However, the leaves of the coca plant are also widely chewed socially throughout the South American Andes, a practice which provides only a tiny amount of the alkaloid. In this context, and as a tea, it is a stimulant and used to suppress hunger, to increase physical endurance and, in the Andes, to help cope with high altitudes.

Paraquat is a synthetic herbicide with a similar action to glyphosate which was discussed in Section 2.2.1. Its structure is illustrated in Figure 2.12 which shows that it comprises two carbon rings containing nitrogen plus a methyl group attached to the nitrogen. According to the Pesticide News (1996, quoted by the Pesticide Action Network, 2004), paraquat was first synthesised in 1882 but its action as a herbicide was only discovered in 1955 and it was produced commercially in 1962. Since then it has become one of the world's leading herbicides (see Tomlin, 2004 for details). It is a non-selective herbicide and so combats broad-leaved weeds and grasses. It has no effect on mature trees and so is used to clear ground weeds in orchards, plantations etc. worldwide and clearing weeds in open fields prior to crop planting. It is toxic to animals and humans but as it associates strongly with soil particles and soil organic matter, environmental and water contamination is limited.

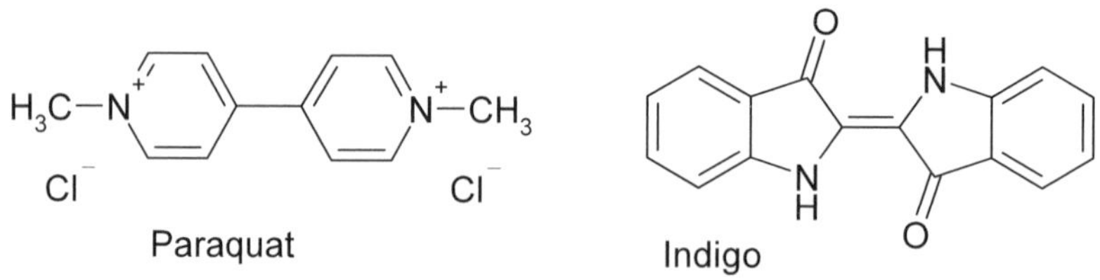

Figure 2.12. The structures of paraquat and indigo

Indigo, the structure of which is shown in Figure 2.12, is another compound with nitrogen and carbon in a ring. It occurs naturally in several plants, including the temperate woad (*Isatis tinctoria)* and dyer's knotweed (*Polygonum tinctorum*), as well as species of the tropical *Indigofera* family. It has been used as a dye since prehistoric times and until the late 1800s indigo was obtained by processing these plants. Traditionally, dried leaves were placed in a vat of water maintained at c. 50°C for several months; anaerobic conditions were established by maintaining a dense leaf mass in order to encourage the growth of the anaerobic bacterium *Clostidium isatidis* which is responsible for the fermentation process. Nutrients were added to encourage bacterial proliferation as well as wood ash to maintain alkalinity. Hydrogen sulphide was also produced and so the production of indigo was a smelly process! The fermentation product, indoxyl, is water soluble but colourless; it oxidises to blue on exposure to air. The Prussian chemist, Adolf Von Baeyer discovered the chemical structure of indigo in the 1870s and was responsible for its synthesis in the laboratory. In 1897 industrial synthesis began and is now the major source of this widely-employed blue dye. For an account of indigo and its history consult Balfour-Paul (1998).

Of the many compounds with two nitrogen atoms in the carbon ring histidine, sildenafil (Viagra®) and caffeine are particularly familiar. Their structures are given in Figure 2.13. Histidine is an essential amino acid and like all amino acids it contains a carboxyl group (COOH) and an amino group (NH_2). It is present in human blood and is necessary for growth, tissue repair and the manufacture of red and white blood cells as well as the manufacture of histamine which can cause allergic reactions. Medicinally, histidine is used in the treatment of arthritis, ulcers and allergies.

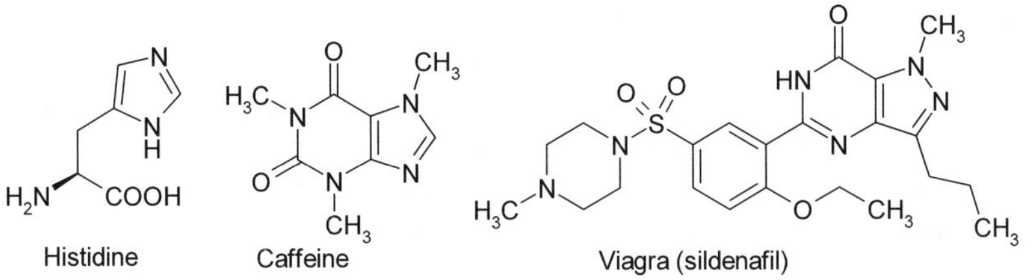

Histidine Caffeine Viagra (sildenafil)

Figure 2.13. Examples of structures with two nitrogen atoms in a ring

Sildenafil has four carbon-containing rings, three of which also contain two nitrogen atoms. It became available in 1998 as a prescription pharmaceutical designed to counteract impotency in men by improving blood flow to the genitalia (see Pfizer, 2003 for details). The third example of a compound containing a carbon ring with two nitrogen atoms is caffeine which is probably the most widely used 'drug' today (see Reid, 2005). Its history, geography and social relevance have been documented by Weinberg and Bealer (2002). As Figure 2.12 shows, it has two fused rings and three methyl groups. Caffeine occurs naturally in tea, cocoa and especially in coffee. The coffee plant (*Coffea arabica*) is native to north Africa and is thought to have been domesticated at least two thousand years ago in what is now Ethiopia whence it was introduced to Yemen where it was discovered by Europeans in the late 1500s. The first coffee house in England was established in 1652 and by the late 1700s coffee houses had became widely established in Europe. The first coffee plants were introduced to the Caribbean islands in 1723 and to Brazil in 1727, giving rise to an important source of income which remains important today. Caffeine tablets are also now available and many soft and so-called energy drinks contain caffeine; the source of the compound is coffee which is then sold as decaffeinated. Caffeine acts as a stimulant on the central nervous system and has a rapid effect soon after ingestion. It can cause insomnia and headaches and is considered to be addictive. Along with other sources of caffeine, such as cocoa and its manufactured form in chocolate and some soft drinks, coffee is very popular and is produced as a cash crop in many tropical countries, notably Costa Rica, Brazil, Indonesia and Kenya.

Atrazine is an example of a compound with three nitrogen atoms in the carbon ring, as shown in Figure 2.14. It has only one carbon-nitrogen ring with attached groups that include chlorine. Atrazine is one of a group of herbicides known as triazines which were discovered in the 1950s (see Tomlin, 2004 for details). They are the most widely used herbicides today, notably for the control of broadleaved and grass plant species in agriculture and recreation. The chemical action inhibits photosynthesis which leads ultimately to death of the target plants. Some crop plants are not susceptible to atrazine and so it is used to control weeds in such crops, notably maize, sorghum, sugarcane, pineapple, and some conifer plantations. It has no effects on birds or bees, is slightly toxic to fish and moderately toxic to animals and humans. There may be other disadvantages as atrazine has been linked with several human health problems, including birth and reproductive defects and some types of cancer. One disadvantage is the persistence of atrazine in the environment as it is mobile, albeit to a limited extent, in soil and moderately soluble in water. It is also implicated in the increased occurrence of deformation in amphibians.

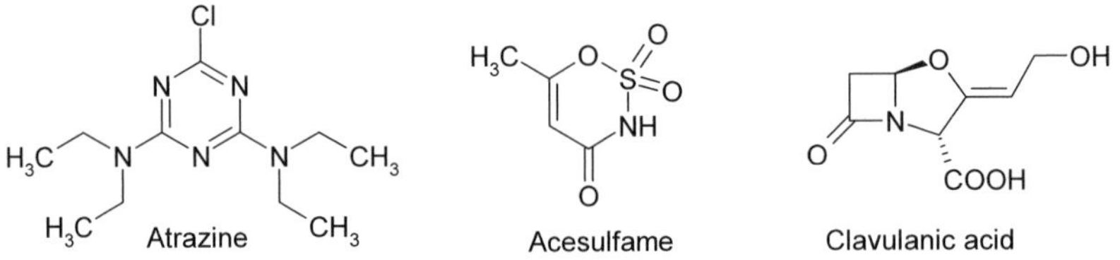

Atrazine Acesulfame Clavulanic acid

Figure 2.14. Structures with two or more non-carbon atoms in a ring

The compounds listed in Table 2.7 which are composed of carbon rings containing both oxygen and nitrogen are less well known than the compounds in the other Beilstein groups. Nevertheless acesulfame and clavulanic acid, whose structures are given in Figure 2.14, are in general use. According to the Calorie Control Council (2004) acesulfame is a sweetener which is 200 times as sweet as sucrose and is used in a wide variety of foods. Discovered in 1967, it has an advantage over aspartame (see Section 2.2.2) insofar as it does not break down at high temperatures and so can be used in cooking. Clavulanic acid is a natural product obtained from the fermentation of the fungus *Streptomyces clavuligerus.* It is added to antibiotics such as amoxicillin, an antibiotic similar to penicillin, in order to boost their effect against bacteria which have developed resistance (GlaxoSmithKline, 2004).

2.2.4 Polymers

A polymer is a large molecule comprised of repeated units of smaller molecules, the monomers, joined together in chains or sheets. Many polymers occur in nature, e.g. the starches, glycogen and cellulose referred to in Section 2.2.3, while many artificial polymers have also been synthesised, such as nylon, Terylene and the plastics. A list of polymers which are important in everyday life is given in Table 2.8. In many respects polymers epitomise the domestication of carbon since several of the naturally occurring polymers derive from crops which have become important agriculturally while the synthetic polymers are largely generated from petrochemicals derived from fossil fuels, especially oil. For an account of polymers and their production and uses see Morawetz (1995).

Table 2.8. Examples of natural and synthetic polymers

NATURAL POLYMERS	TYPE OF CHEMICAL
Wool	Protein
Hair	Protein
Fur	Protein
Silk	Protein
Cotton	Carbohydrate
Starch	Carbohydrate
Cellulose	Carbohydrate
Latex	Acyclic terpene (Alkene)
Chitin (Insect hard parts)	Carbohydrate
Sisal	Carbohydrate
Jute	Carbohydrate
Hemp	Carbohydrate
Flax	Carbohydrate
SYNTHETIC POLYMERS	**TYPE OF CHEMICAL**
Nylon	Amide
Terylene	Ester
Rayon	Carbohydrate
Various plastics	Various
Rubber (neoprene)	Chloroalkene
Rubber (Thiokol)	Chloroalkane

The natural polymers listed above are organic compounds; wool and fur are animal products whose chemical structure is very similar but fur reflects greater density of protein expression, silk is the product of an insect (the silk worm *Bombyx mori*), while cotton and the various fibres are mainly cellulose which is a carbohydrate of plant origin. Silk consists of a thread-like protein known as fibroin which occurs as heavy and light chains held together by disulphide bonds. The heavy chain imparts most of the characteristics of silk and comprises an amino acid sequence of glycine, serine and alanine; serine is replaced by tyrosine in some units and adds to the strength of the fibre. Wool and hair, as well as nail, horn and claw, are forms of the protein keratin. This comprises polypeptide chains of three classes of protein of varying thickness. The peptide chains are coiled and the number present determines the nature of the material. For example, hair has three strands and sheep wool has seven.

Cellulose is a polysaccharide and a major constituent of plants since it is the structural element of cell walls. Unlike the fibres discussed above which comprise various combinations of proteins it consists of repeated units of the monomer glucose, as do glycogen and starch (see Section 2.2.3). In the latter, the glucose monomers (see Figure 2.9) are bonded in such a way that the polymer forms spirals whereas in cellulose and chitin the units are straight. The cellulose of plants is an important store of carbon in the biosphere (see Section 2.3) and as wood and various plant fibres it is a valuable commodity. Wood and flax comprise c. 40 to 50 percent and 80 percent cellulose respectively while cotton is almost entirely cellulose. It has many uses including the manufacture of paper, explosives, plastics, viscose, rayon and lacquers. Latex is another natural polymer with commercial value. It is the precursor of rubber, the fascinating history of which is documented by Slack (2002), and is produced as particles suspended in the milky white sap of the rubber tree (*Hevea brasiliensis*), a native of Amazonia. Other plant species also produce latex but this is not important commercially. Chemically, latex has a carbon chain structure as shown in Figure 2.15, and is similar to an acyclic terpene (isocyclic terpenes are discussed in Section 2.2.2). The rubber is produced from the latex by diluting it with water and adding acid which causes rubber particles to coagulate. The bundles are then pressed between rollers to produce sheets prior to air-or smoke-drying.

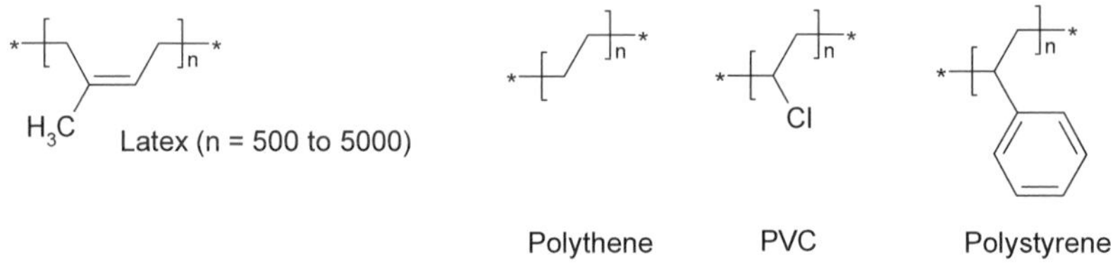

H_3C Latex (n = 500 to 5000)

Polythene PVC Polystyrene

Figure 2.15. The structure of some polymers

Vulcanisation is then carried out using sulphur, which creates cross-links between the carbon chains to turn the sheets into a strong but flexible solid. Latex is produced in tropical rubber plantations though synthetic rubber is also widely manufactured. Synthetic rubber has several forms depending on the starting chemical. The two main forms are neoprene and Thiokol. Both are made from chemicals present in petroleum using techniques that promote polymerisation. The starting chemical for neoprene is chloroprene and that for Thiokol is ethylene chloride. In contrast to natural rubber, neither of the artificial rubbers is flammable so they can be used for a wide range of goods, especially automobile tires, footwear, hoses, pipes, insulating materials etc.

Table 2.9. Categories of plastics (based on British Plastics Federation, 2004)

	Properties	**Applications**
THERMOPLASTICS		
Acrylics (perspex)	Tough	Glass replacement
Acrylo-nitrile (nylon)	Rigid, tough	Telephone handsets, luggage, handles
Polyethylene (polythene)	Flexible, translucent, weatherproof	Household and kitchenware
Polypropylene	Semi-rigid, tough	Packaging
Poly Vinyl Acetate (PVA)	Versatile	Adhesives, paints, paper and textile coatings
Poly Vinyl Chloride (PVC)	Clear or coloured, rigid or flexible	Window frames, storage bags, cling film, LP records
Polystyrene	Brittle, hard, rigid	Toys, food containers
Polycarbonate	Strong, stiff, tough	Blends with other polymers
Teflon	High melting point, chemical resistance	Non-stick surfaces in cookware
THERMOSETS		
Bakelite	Hard, heat resistance	Toasters, clocks, car components
Epoxies	Clear, rigid, tough, chemical resistance	Adhesives, electrical components
Melamine	Hard, opaque, tough	Decorative laminates, bottle caps, toilet seats
Polyesters		Clothing
Polyurethane	Elastic, abrasion and chemical resistant	Solid tyres, car bumpers

Plastics are another group of artificial polymers in widespread use. As in the case of artificial rubbers and artificial fibres (see below) the starting chemicals derive from petrochemicals. Table 2.9 shows that there are two main groups of plastics: thermoplastics, which soften on heating and harden on cooling, and thermosets which are hardened by heat. The former can be repeatedly melted and then moulded so they can be recycled which is not the case for thermosetting plastics. Within these groups there are many different plastics depending on the basic monomer present.

The versatility of plastics is reflected in the many uses listed in Table 2.9. The chemical structures of several plastics in common use are given in Figure 2.15 which shows that polystyrene and polyvinyl chloride comprise long chain carbon atoms with aromatic rings or chlorine respectively attached. The first true plastic was Bakelite which was invented in the late 1800s and which was in general use by the 1920s. Further plastics were invented between the two world wars, including nylon, a polyamide which has many uses including fabric. Plastics developed in the post war period include polyurethane, Teflon, and polycarbonate. The search is on for plastic polymers with the capacity to conduct electricity. However, plastics have a major disadvantage in that they are not environmentally friendly. They do not degrade easily and if incinerated they may produce toxic fumes. Reuse and recycling are sometimes possible but plastic-based rubbish is a serious problem.

Artificial fibres are as much in common use as the plastics and have equally interesting histories, as documented in Jenkins (2003). One of the first artificial fibres was rayon, the starting point for which is cellulose from wood pulp. It was first discovered in 1855 and was commercially produced by 1924 in the USA. Rayon is produced in two stages. First, cellulose is mixed with sodium hydroxide and carbon disulphide to produce viscose which, as its name suggests, is a thick, sticky liquid. As in the case of Terylene, the viscose is then forced through fine holes in a cylinder (a spinneret) as a jet and into dilute sulphuric acid to produce rayon. This is a fine, almost silky thread with multiple uses which include textiles, tyres and carpets.

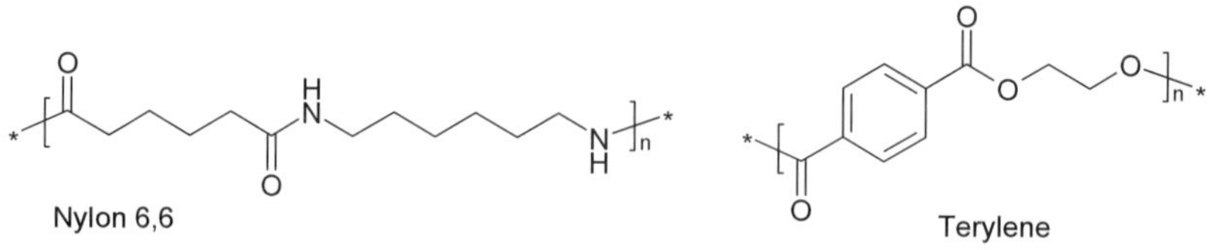

Nylon 6,6

Terylene

Figure 2.16. The structure of some artificial fibres

Several other artificial fibres are related to plastics; indeed, nylon is classified as a plastic (see Table 2.9). It was the first truly artificial fibre and was discovered in the early 1930s, patented in 1935 and was commercially available by 1939. It was initially used to produce bristles for toothbrushes and later nylon stockings. The chemical structure of nylon is given in Figure 2.16 which shows that it comprises a long chain of carbon atoms with the functional group known as an amide (see Table 2.2). It is thus a polyamide and it is used not only for clothing but for a wide range of products, e.g. machine parts, fishing nets, carpets, that require flexibility and strength. The starting materials for its industrial manufacture are adipic acid and hexamethylene diamine which derive from petroleum.

The polyesters are another group of synthetic organic chemicals which are used as fibres for fabric manufacture. They were discovered as an extension of the work on nylon production and Terylene itself was first produced in 1941 (see Jenkins, 2003). Esters ($-CO_2CH_2-$) are present as the functional group (see Table 2.2): when these units are replicated in polymers the chemicals are known as polyesters. Terylene is one such polyester which is made from tetraphthalic acid (a dicarboxylic acid) and ethylene glycol which are derived from petroleum. The resulting compound mass is jetted through mesh-like holes to produce Terylene fibres. Chain polymers like Terylene have great strength but remain flexible and combine well with natural fibres such as cotton or wool to produce fabrics for clothing. Terylene is also used to make conveyor belts and boat sails. Other artificial fibres include lycra and elastane.

2.3 The biogeochemistry of carbon

Amongst the most common chemical elements that make up the living and non-living components of the Earth are carbon, hydrogen, oxygen, nitrogen and silicon. Since the formation of the Earth all, along with other elements, have been involved in continual redistribution between the Earth's various structural units which are shown in Figure 1.5, i.e. between the lithosphere, biosphere, atmosphere etc. Each element has a unique cycle known as a biogeochemical cycle, a term which expresses the linkages between the living and the non-living components of the Earth. Individual biogeochemical cycles are not discrete as there are linkages between different biogeochemical cycles. This occurs because most elements react with others to form compounds; carbon combines with oxygen, for example, to produce carbon dioxide and so the cycles of carbon and oxygen are intertwined in the same way that the cycles of carbon and hydrogen are linked through compounds like methane and various hydrocarbons. It follows that alterations in one cycle will inevitably have repercussions for other cycles. Such interrelationships make the analyses of biogeochemical components and processes difficult and their management problematic. Changes within individual cycles and linkages between cycles can occur at different spatial and temporal scales. Movement within a cycle may occur on short time scales of seconds to long term shifts associated with geological processes. For example, photosynthesis and respiration involve the uptake and output respectively of carbon dioxide on very short time scales while the formation of limestone, coal or peat takes millions or thousands of years. Spatial variations also occur in these two processes; for example, globally and on an annual basis rates of photosynthesis are greater at low latitudes than at high latitudes. Similarly more carbon is stored in forested lands than in land under agriculture at the same latitude and longitude. Most biogeochemical cycles are also subject to perturbation by human activity; indeed most forms of pollution are caused by such intervention as is discussed below in relation to the carbon cycle.

All biogeochemical cycles comprise pools or reservoirs of the element and linkages occur through flux rates. Those elements, like carbon, which have a reservoir in the atmosphere are said to have a gaseous biogeochemical cycle while those whose major reservoirs are in the Earth's crust are termed sedimentary. Gaseous and sedimentary cycles also interact at different spatial and temporal scales, an example being that of the silicon and carbon cycles which operate on geological time scales. It involves the removal of carbon dioxide from the atmosphere, the formation of carbonate rocks in the ocean basins and subsequent release of carbon dioxide consequent on tectonic processes that compress the carbonates. This relationship illustrates the fact that biogeochemical cycles can be analysed on two time scales, i.e. contemporary and long-term processes. The long-term cycle of carbon is examined in Section 4.2.

The contemporary pools and approximate annual flux rates in the carbon cycle are shown in Figure 2.17 and a carbon budget is given in Table 2.10. The unit of measure for both the Figure and the Table is 10^9 tonnes carbon per year. It must be emphasised that these data are simplistic representations of complex processes and are thus generalisations of reality. Indeed, current understanding of the carbon cycle is particularly inadequate given how important this cycle is in terms of climatic change and human dependence on fossil fuels.

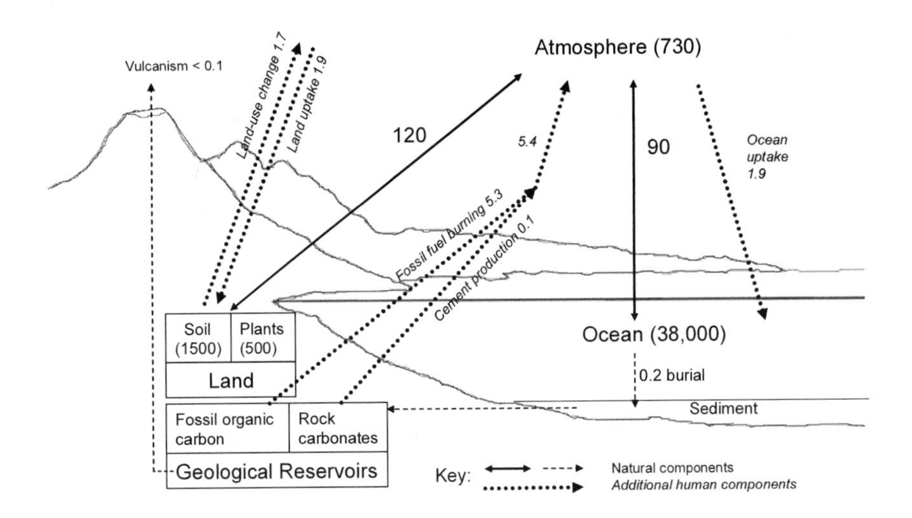

Figure 2.17. The carbon cycle (based on IPCC, 2001a)

Table 2.10. The global carbon budget for the 1980s and 1990s (based on IPCC, 2001a)

	1980s 10^9 tC yr^{-1}	**1990s** 10^9 tC yr^{-1}
SOURCES		
Emissions from fossil fuels and cement production	5.4 ± 0.1	6.3 ± 0.1
Flux from land to atmosphere due to land-use change	1.7 (0.6 to 2.5)	Not available
SINKS		
Increase in the atmosphere	3.3 ± 0.1	3.2 ± 0.1
Flux from atmosphere to ocean	1.9 ± 0.6	1.7 ± 0.5
Missing sink (possibly terrestrial vegetation)	1.9 (0.3 to 3.8)	Not available

The major pools are, in order of magnitude, the oceans, the biosphere and the atmosphere. Amongst the major flux rates are uptake from the atmosphere through photosynthesis and return to the atmosphere through respiration and decomposition of organic material, processes which reflect the significance of life in the carbon cycle. Details of these processes are given in Chapter 3. Carbon is also fluxed to the atmosphere by fire which converts the organic components of stored organic matter, i.e. biomass, into carbon dioxide, carbon monoxide and methane. Some fire occurs naturally, and has done so throughout geological time, but human activity has substantially altered the rates of occurrence, as discussed below.

In addition, diffusion occurs between the oceans and the atmosphere. This is a chemical/physical rather than biological process while photosynthesis and respiration by marine organisms also involve the exchange of carbon dioxide as occurs in terrestrial environments. Carbon dioxide is dissolved in sea water, where it can be converted to carbonate or bicarbonate. These processes are largely dependent on sea-surface temperatures and currents which also influence the distribution and populations of marine organisms. For example, cold temperatures are conducive to the uptake of carbon from the atmosphere whilst warm temperatures have the opposite effect. Currents shift carbon dioxide-rich waters within the oceans; upward or downward currents remove or return carbon-dioxide-rich water to or from the surface. Algae within the ocean are the most important consumers of carbon dioxide through photosynthesis; they are termed phytoplankton and are the basis of marine food chains. All marine organisms return carbon dioxide to the oceans through respiration and decomposition. Unlike the terrestrial environment the storage of carbon, the biomass, in the oceans is limited as there is rapid turnover and no equivalent of the terrestrial storage in soils and vegetation. However, some organisms use carbonate from the ocean water by biologically combining it with calcium to produce shells of calcium carbonate. On death this accumulates on the ocean floor and with time sedimentary rocks such as chalk and limestone are formed (see Section 4.2).

Human activity as a significant domesticating influence on the carbon cycle began when humans harnessed fire, escalated with the advent of agriculture c. 12,000 years ago and escalated again with the Industrial Revolution of the mid eighteenth century (see Section 1.5). Today the footprint of humanity on the face of the Earth grows ever deeper and broader and is manifest in alterations to the carbon cycle that are causing environmental change on local, regional and global scales. This domestication of carbon focuses on two widespread activities which underpin this alteration: agriculture which requires removal or alteration of natural vegetation, especially forest, and the burning of fossil fuels. Both accelerate carbon cycling by altering the flux rates between carbon stores and the atmosphere. Agriculture releases carbon from the store in the biosphere while fossil-fuel use releases it from the geological store in the lithosphere. Other activities also contribute to the enhanced flux of carbon to the atmosphere such as logging, mining and any land-use change which involves the loss of carbon storage such as urbanisation.

There is abundant evidence for these alterations. Measurements of carbon dioxide concentrations in polar ice cores provide data relating to the last c. 420,000 years (Barnola *et al.*, 2003) As ice accumulates, mini atmospheres are trapped in bubbles so the accumulation of ice year on year generates an environmental archive with a record of the composition of air over time. Such work from the Antarctic and Arctic confirm that there are substantial differences between ice age (cold stage) and interglacial (warm stage) atmospheres, as shown in Figure 2.18.

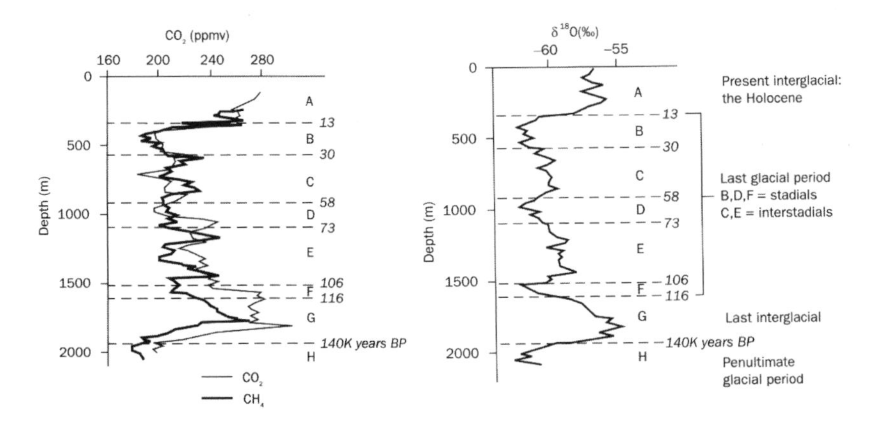

Figure 2.18. *a.* Carbon dioxide and methane data (based on Barnola *et al.* 1987 and Chapellaz *et al.* 1990)
b. Oxygen isotope data (based on Lorius *et al.* 1985)

In particular, there is at least a c. 25 percent increase in the average concentration of carbon dioxide and a doubling of that of methane in warm (interglacial) stages when compared with cold (glacial) stages. The differences are especially marked when comparing the peak values of a warm stage with those of the nadir of a glacial stage. For carbon dioxide this involves a shift from 280-300 ppmv to 180 ppmv with a shift from c. 750 to 1750 ppbv for methane. Temperature reconstruction for the same period, based on isotopes of oxygen and hydrogen, which are independent of carbon dioxide concentrations, have also been undertaken for polar ice cores and as Figure 2.18 shows, there is a high degree of correspondence between the two. This re-emphasises the role of carbon, as heat-trapping 'greenhouse' gases, in the atmosphere as a determinant of global temperatures. Analyses of the data do, however, raise further questions such as where did all the carbon go during the cold stage and how did it arrive in these unidentified sinks? Such questions are important insofar as they reflect an inadequate understanding of vital environmental processes and thus an unsound basis on which to formulate management policies. This is reconsidered below. Moreover, measurements of carbon dioxide and methane in contemporary air at various monitoring stations such as Mauna Loa in Hawaii since the 1950s demonstrate that the concentrations of both gases have been steadily increasing. As Table 2.11 shows, the mean annual concentration of carbon dioxide was almost 317 ppmv in 1960 and by 2002 it had risen to just over 373 ppmv, representing an 18 percent increase in 43 years.

Table 2.11. Past and recent concentrations of carbon dioxide in the atmosphere (data from Barnola *et al.*, 2003 and Keeling and Whorf, 2004).

YEAR	CO_2 Concentration ppmv
Glacial-interglacial transition	180 - 280 -300
Highest interglacial concentration	260 - 280
20,000 years ago	189
10,000 years ago	262
1960	317
1970	326
1980	339
1990	354
2000	369
2002	373

Similarly, the concentration of methane in the atmosphere has risen from 666 ppb in 1008 AD to 1751 ppb today (Etheridge *et al.,* 2002), an increase of 74 percent. These data testify to considerable alteration of the carbon cycle but equally they highlight the fact that the high atmospheric carbon dioxide and methane concentrations of the present have no precedent in the ice core data of the last 420,000 years. Human activity has increased this concentration from 280 ppmv at the start of the current warm stage to 373 ppmv now. This represents a substantial increase of c. 33 percent, about the same magnitude as the shift from cold to warm conditions that occurred at the end of the last and earlier cold stages. Little wonder then that the Earth is already experiencing global warming.

Should any scepticism remain concerning the role of human activity in altering the concentration of greenhouse gases in the atmosphere two further lines of evidence add weight to an already weighty argument. First, atmospheric carbon dioxide concentrations are not uniform spatially; higher concentrations occur in the northern hemisphere when compared with the southern hemisphere because this is where human activity, especially the consumption of fossil fuels, is concentrated. Second, there is a growing abundance of carbon atoms from fossil fuels in the atmosphere. Carbon atoms so liberated have lost their proportion of radioactive carbon which decayed in their lithospheric repositories to stable carbon. This contrasts with naturally- produced carbon atoms, a constant fraction of which remain radioactive. The overall measurable effect of the coalescence of the carbon from both sources is a dilution of the radioactive fraction in the atmosphere which is also apparent in tree rings since they comprise carbon of both sources from the atmosphere.

As stated above, the major causes of this accelerated flux of carbon to the atmosphere is agriculture and use of fossil fuels. The former has a much longer history than the latter (see reviews in Mannion, 1995, 1997a and 2002), as discussed in Chapters 5 and 6. Its origination and initial expansion, especially in Europe and Asia, in the early part of the present warm stage has been attributed by Ruddiman (2003) as the trigger for the start of global warming. He opines that the cycles of temperature change associated with the glacial-interglacial (cold – warm) cycles described above were initially disrupted by deforestation for agriculture c. 8,000 years ago which began the trend of increasing atmospheric carbon dioxide concentrations. He also suggests that methane concentrations began to increase c. 5,000 years ago due to the initiation of irrigation and the expansion of livestock herding. In a detailed analysis relating ice-core data to major events, Ruddiman correlates short-term reversals of carbon dioxide and methane concentrations in the historic period with the incidence of various plagues. These reduced human populations and probably caused the abandonment of agricultural land so that forests could recolonise and so sequester carbon from the atmosphere. As a result atmospheric carbon dioxide concentrations declined with the reverse occurring when populations recovered and deforestation recommenced. At the core of these change in atmospheric composition is loss of natural vegetation communities, especially forests. Their removal not only releases carbon dioxide to the atmosphere from biomass and soil organic matter but also impairs an important natural mechanism for removing it from the atmosphere. Moreover, fire has and continues to be widely used to effect land clearance for agriculture (see Pyne, 2001, for a review of fire in the environment). In chemical terms, oxidation through combustion and decomposition transport carbon from the biosphere to the atmosphere.

As Williams (2002) has discussed, a considerable proportion of the Earth's interglacial forests have already been removed through human action, a process that continues apace today especially in tropical and subtropical regions. Quantifying the amount of carbon released through fire and fuel wood collection is difficult but globally, vegetation loss is considered to emit between 1 and 1.5 x 10^9 tC to the atmosphere annually. This is approximately 30 percent of that released by fossil-fuel combustion which amounts to a massive c. 5.5 x 10^9 tC yr^{-1} (IPCC, 2001). Accelerated fossil-fuel use began in the mid-seventeenth century with the onset of industrialisation in Britain and Europe. Although peat was used where it was locally available, coal was the first fossil fuel to be widely exploited. It dominated the fuel market until the 1950s when the emphasis shifted to oil and natural gas which remain the main sources of energy today (see Chapter 5). The use of fossil fuels is increasing rapidly as industrialisation occurs in developing countries and as other, developed, nations fail to reduce consumption.

Figure 2.17 shows that these anthropogenic fluxes are relatively small numerically when compared with natural fluxes between pools but they are of sufficient magnitude to cause global climatic change, notably global warming. It is now a matter of record that the last few years have been the warmest since records began and while this can be dismissed as simply a short term trend it is hardly coincidence that they correspond with carbon dioxide and methane concentrations that are higher than ever before. Moreover, the increase could theoretically be in excess of what it actually is; this is because some of the anthropogenically-produced carbon is absorbed in sinks. One such sink is the oceans but where the remainder, approximately 30 percent (c. 1.9 x 10^9 tC) of anthropogenic emissions, is sequestered remains uncertain (see Table 2.10) though biological fixation is the most likely explanation. Grace (2004), for example, describes methods for calculating the magnitude of carbon sinks and provides evidence for sinks in boreal, temperate and tropical forests and the ocean. The significant amount of carbon for which there is no true account is another testament to the inadequate understanding of biogeochemical processes. What is clear, however, from the record of past and recent carbon emissions is that management of the carbon cycle is vital and that to be effective it requires international co-operation. All nations of the world have a vested interest because they produce carbon (coal-, oil-or gas-rich nations), use it intensively (e.g. USA, Europe), require increasing volumes for development (e.g. the newly industrialising countries) and /or will be affected by climatic change which is likely for all nations though some are more prepared than others and some are more susceptible than others. These issues are discussed in Chapter 7.

Chapter 3

3 THE BIOLOGY OF CARBON

All biology is about carbon and carbon is (almost) all about biology. This is because all living things are composed of carbon compounds and because the environment in which life exists is regulated by carbon exchanges and storage that are components of biogeochemical cycles (Section 2.3). The exceptions are those allotropes of elemental carbon, e.g. diamond and graphite, which have been generated by geological and not biological processes but which do not loom large in the carbon budget of the Earth (see Sections 1.1 and 2.1). Life on earth is highly diverse, reflecting the myriad ways in which carbon, as carbon compounds, may be present. Not only is the range of species diverse but so is their chemical composition. This diversity of life has attracted scientific comment for millennia, especially in relation to its classification; its basic subdivision into plants and animals is no longer acceptable. The classification of living organisms is in a constant state of flux as new discoveries are made and methods of identification, ranging from large scale morphology to micro-scale genetic analysis, are refined. Moreover, there are still no unequivocal criteria for what does and does not constitute a living organism. The question of how to classify viruses and prions (these are proteins associated with various diseases of the brain in cattle and humans) are cases in point.

No organisms exist in isolation as all are linked in some way through the processing and manifestation of carbon in various forms. Organisms undertake processes which are vital to their own existence but also to the existence of others. These relationships are not, however, part of a closed system because neither the components nor the processes would exist without the transfer of energy. This energy derives from the sun and is harnessed through photosynthesis, the first stage in the passage of energy through ecosystems. Photosynthesis is the domain of green plants and supports, through the provision of food, all organisms that cannot manufacture their own food, including humans. For this reason photosynthesis not only underpins all types of ecosystems but also agriculture which is itself a carbon-processing system (see Chapter 5). The counterpart of photosynthesis is respiration, whereby food is metabolised to release energy for the whole range of activities undertaken by organisms and carbon is released to the atmosphere as carbon dioxide. In contrast to the confinement of photosynthesis to green plants, all living organisms respire.

Organisms, unless consumed by other organisms, decay after death. Decay involves chemical alteration of body tissues and the actions of various micro-organisms which use the dead material as a source of energy. Again, carbon dioxide is returned to the atmosphere while other compounds are returned to the soil. Living and dead organic material is present over much of the Earth's surface as biomass. This constitutes an important store of carbon in many different forms which include the wood of trees, the plant material of marshlands and peatlands, the organic matter in soils and the bodies and shells of all other living organisms in terrestrial and aquatic environments. This is not a long-term store of carbon as respiration occurs throughout an organism's life and decay generally takes place relatively rapidly after death, but in the geological past the biological fixation of carbon through photosynthesis did result in carbon storage which was subsequently confined to geological repositories. These are the coal, oil and gas deposits that today provide society with fuel energy (see Chapter 4).

Green plants manufacture their own food through photosynthesis but the majority of other organisms have to rely on the products of photosynthesis for their food. Thus food chains and food webs have developed in ecosystems. Some involve direct grazing relationships wherein green plants are consumed by herbivorous animals and these, in turn, are consumed by other animals that feed on a range of food types or only on meat, i.e. the omnivores and carnivores. Alternatively, some organisms rely on dead organic matter for their energy and are thus important in the decay of dead organic matter. This is known as the detrital pathway of energy transfer and the organisms are known as detrivores; they include fungi and bacteria. In addition, there are links between the grazing and detrital food chains to create complex relationships as carbon compounds are transferred and transformed within ecosystems.

Many factors influence the distribution of organisms, the processes they undertake and the links between them. Climate is especially significant and in the last 10,000 years human influence has also altered the processes of carbon transfer and storage. Consequently, the biology of carbon has varied over geological time and is thus linked with the geology of carbon, the subject of Chapter 4. The variation of climate that exists globally also influences the geography of carbon as does the global variation in human pressure on the environment and its changes over time. These issues are examined in Chapters 5 and 6.

3.1 The web of life

There is much debate as to what exactly constitutes a living organism but it is generally accepted that organisms are capable of growth and of reproduction. The range of organisms on Earth is diverse, ranging from relatively simple life forms such as viruses to complex life forms such as mammals. Attempts to classify this vast range of organisms, i.e. the development of taxonomy (the naming, describing and classifying organisms) and systematics (the current and evolutionary relationships between organisms), began in classical times. Scientific developments in the last 300 years have, however, shown that a simple division into plants and animals is inadequate. Nevertheless the issue remains controversial and taxonomy and systematics are in a continuous state of flux. Today there are considered to be six kingdoms (major subdivisions) of organisms though even this is not entirely accepted, as discussed below. Each of these major groups will be briefly examined to reflect its major characteristics and its relationships.

3.1.1 The classification of living things

The Greek philosopher Aristotle (384–322 BCE) was the first to classify organisms. He is credited with introducing two groups: plants and animals and he subdivided the latter into land dwellers, water dwellers and air dwellers. His pupil Theophrastus (370–285 BCE) introduced a subdivision of plants based on stem size, notably into herbs, sub-shrubs, shrubs and trees. Further attempts to refine these classifications and to introduce new ones proliferated in the 1500s and 1600s as many 'new' organisms were discovered in the Old World and especially as 'new' organisms were introduced to Europe from the New World. This was an age of discovery that accelerated the spread of plants and animals from their centres of origin around the globe and which was the second wave of 'biological globalization', the first being the spread of domesticated organisms from their centres of origin many millennia earlier as agriculture was initially established (see Chapter 5). The importance of herbal medicines to people of the Old and New Worlds prompted published descriptions of plants, the establishment of herbaria and botanical gardens, and the publication of *Materia Medica,* i.e. lists and illustrations of herbs and their uses. Amongst these authorities were Otto Brunfels (1464-1534), Nicholas Monardes (1493-1588), Leonhard Fuchs (1501-1566), John Gerard (1545-1612) and Andrea Caesalpino (1519-1603).

However, it was Carolus Linnaeus (Carl von Linné, 1707-1778), a Swedish naturalist, who established the system of taxonomy which is the basis for the modern classification of organisms (see Frängsmyr, 1983, for a review). This was published in 1735 in a book entitled *Systems Naturae.* The Linnaean taxonomic system retains Aristotle's subdivision into plants (*Plantae*) and animals (*Animalia*) which are the broadest categories possible and are known as kingdoms. Further subdivisions are given in Figure 3.1 which shows four further levels in this hierarchical or taxonomic system.

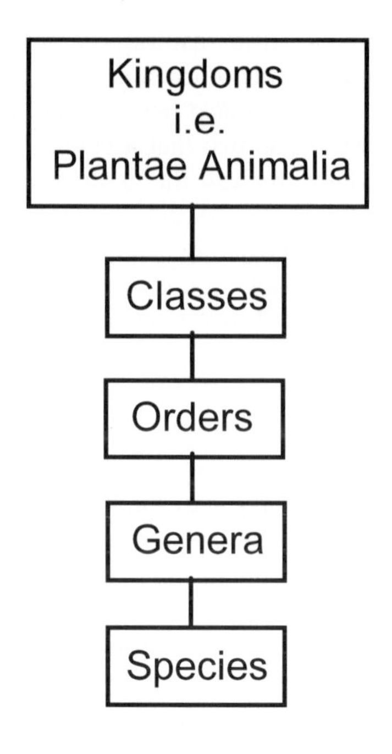

Figure 3.1. Linnaean classification of organisms

The naming of organisms in this classification requires reference to two parts; the first relates to genus and the second to species. Thus the common dandelion is termed *Taraxacum vulgare* and the wolf is *Canis lupus.* The use of two names, representing the highest levels of classification, is referred to as binomial nomenclature and remains the method by which organisms are named today. The criteria used by Linnaeus for grouping or separating organisms were based on the observable morphological/anatomical characteristics of organisms. For example, his means of differentiating between plants involved variation in the numbers of flower parts, such as numbers of pistils and stamens. Such criteria were published by Linnaeus in his book *Species Plantarum* in 1737. This approach is a so-called 'artificial' method of classification because it relies on a limited set of criteria rather than an appraisal of all the possible attributes that give an organism its 'character', i.e. its natural character. Linnaeus's system is not perfect but it has stood the test of time, especially in relation to its hierarchical relationships that, in many but not all cases, have been uncovered by recent genetic investigations. Linnaeus' original scheme comprised five levels overall (Figure 3.1) but later discoveries in the nineteenth and twentieth centuries resulted in the addition of extra levels, notably phylum and family. Sometimes further divisions are included, such as sub-orders and domains to give the nine-fold scheme shown in Figure 3.2.

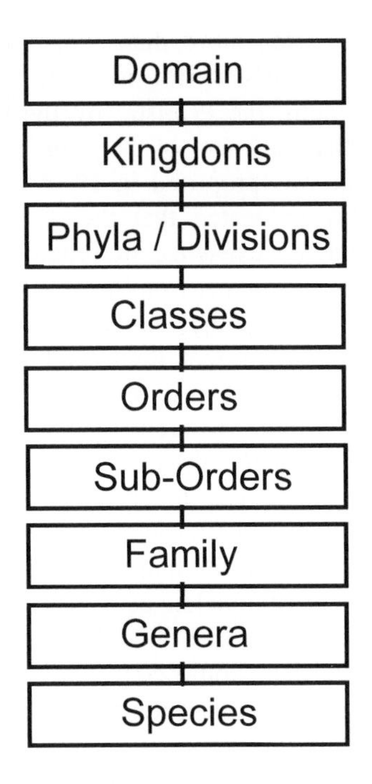

Figure 3.2. Taxonomic system used today

Today, taxonomists have many more tools at their disposal for determining the characteristics of organisms than their eighteenth and nineteenth century counterparts. As a result, and because of continuous revisions as new species are discovered and known ones reassessed, the classification of organisms is in continual state of flux. This also makes estimates of the total number of organisms in the Earth's biota difficult. As Stork (1997) has pointed out, there are two main problems. First, researchers working in different parts of the world may describe the same species but give it a different name; this leads to synonyms and exaggeration of species diversity. Second, there is usually considerable variation within a population of a species, making it difficult to determine if the individuals reflect gradation within a species or are in fact different species. Consequently, taxonomists are regarded as 'lumpers' or 'splitters', i.e. those who group individuals into genera and those who divide them into species. Indeed, some of the problem relates to the fact that there is no universally accepted definition of a species, as discussed by Groombridge and Jenkins (2002) and Lévêque and Mounolou (2003).

Moreover, as science has advanced it has become apparent that the simple subdivision of organisms by Linnaeus into the two kingdoms of plants and animals is inadequate. Some organisms do not fit into either category. This was recognised more than a century ago when there was sufficient momentum for a category for micro-organisms. It resulted in the addition of a third kingdom, the Protista, by the German biologist Ernst Haeckel, who was also the originator of the term and science of ecology, in 1866. A fourth kingdom was proposed in 1938 by Herbert Copeland, an American biologist, who advocated that there should be a kingdom for bacteria. This recognised the existence of organisms without nuclei (e.g. bacteria), the prokaryotes, as distinct from those with nuclei, the eukaryotes. In 1957 Robert H Whittaker, another American biologist, proposed a fifth kingdom for fungi on the basis that their method of obtaining food, by excreting enzymes onto food that is external to their body and then absorbing the nutrients, is quite different to that of other organisms. This five-kingdom scheme is widely though not universally accepted (e.g. Margulis and Schwartz, 1998). The advent of genetic studies officially left its mark on taxonomy when Carl Woese, an American molecular biologist, proposed the introduction of the domain, at the level above kingdom, in 1990. Based on the types of ribonucleic acid (RNA) present in an organism, Woese proposed that there should be three domains: Archaea, Bacteria and Eukarya, and that the Archaea should comprise two kingdoms, to give six kingdoms in all. The relationships reflect a common ancestry and thus reflect evolutionary trends, the theory of which had not been established during Linnaeus's lifetime. Such relationships are known as phylogeny and Figure 3.3 shows the evolutionary relationships between all living species while Table 3.1 gives details of the numbers of described species and estimates of the total numbers of species in each kingdom. Recent work by Rivera and Lake (2004) indicates that the idea of a tree of life is not a reflection of reality. This is because their genetic analysis of numerous eukaryotic and prokaryotic organisms reveals that the former arose from the fusion of two diverse prokaryotic genomes. This has occurred through lateral gene transfer so that eukaryotes have genes which originated in both bacteria (also known as eubacteria) and Archaea (also known as archaeobacteria) which are discussed in Sections 3.2 and 3.3. In this case depiction of a ring of life would appear to be more appropriate than a tree of life, as shown in Figure 3.3. Table 3.1 reflects the amazing diversity of life which in turn reflects the amazing forms adopted by carbon! Table 3.1 also indicates how imperfect the current state of knowledge is in relation to the Earth's likely total tally of living species, especially in terms of insects and arachnids.

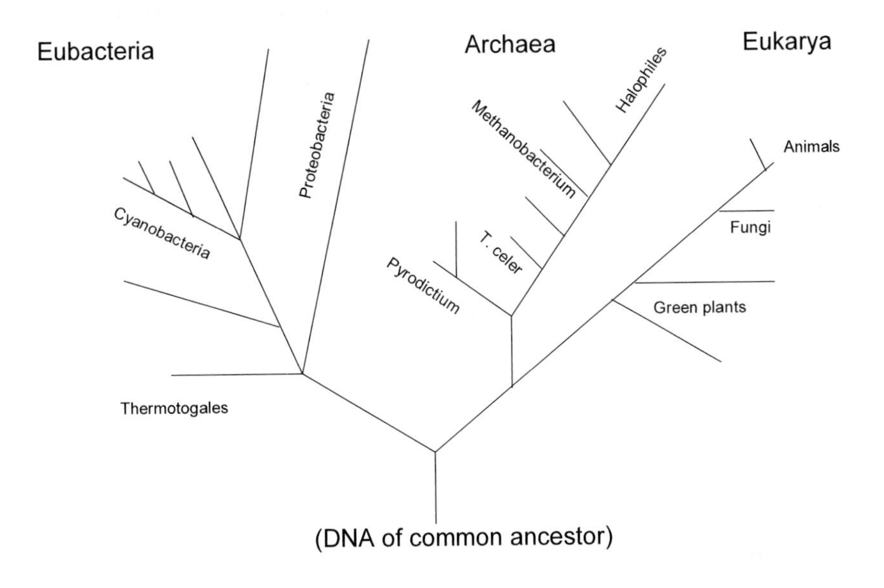

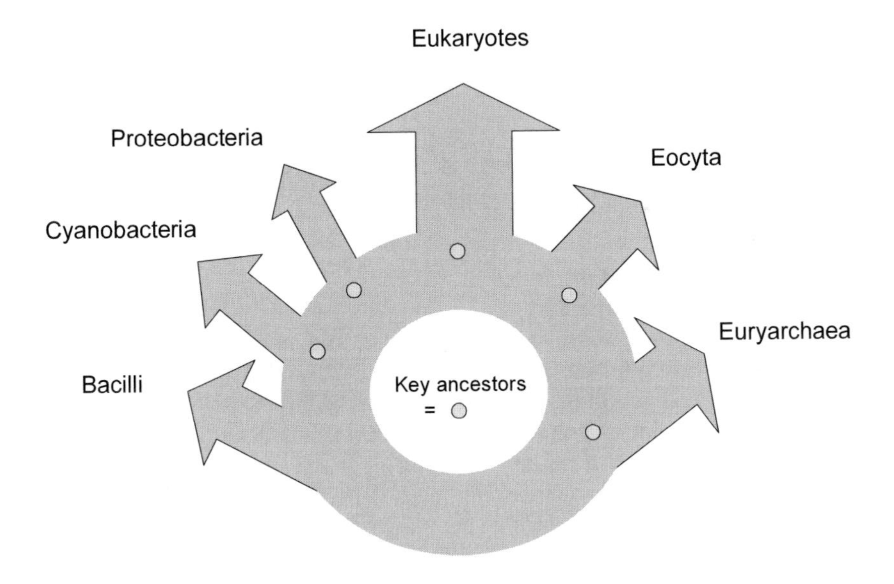

Figure 3.3. The Tree of Life – two of many possible views

Table 3.1. Estimated numbers of described species and possible totals (from Groombridge and Jenkins, 2002)

DOMAIN	**EUKARYOTE KINGDOMS**	**No. of KNOWN SPECIES**	**TOTAL (Estimate)**
Archaea		175	Unknown
Bacteria		10,000	Unknown
Eukarya			
	ANIMALIA		1,320,000
	Craniata (vertebrates), total	52,500	55,000
	Mammals	4,630	
	Birds	9,750	
	Reptiles	8.002	
	Amphibians	4,950	
	Fishes	25,000	
	Mandibulata (insects, myriapods)	963,000	8,000,000
	Chelicerata (arachnids etc.)	75,000	750,000
	Mollusca	70,000	200,000
	Crustacea	40,000	150,000
	Nematoda	25,000	400,000
	FUNGI	72,000	1,500,000
	PLANTAE	270,000	320,000
	PROTOCTISTA	80,000	600,000
TOTAL		**1,750,000**	**14,000,000**

3.2 Domain Archaea

The Archaea are similar to bacteria and until 1977 most known species were classified as bacteria (see review by Makarova and Koonin, 2003). It was the work of Carl Woese (see Woese and Fox, 1977) who discovered that some so-called bacteria were different genetically and separated the Archaea from the bacteria on the basis of particular rRNA sequences present in the cells of the organisms. The Archaea in existence today are thought to have descended from the first forms of life to have evolved on Earth some 3.5 to 4.0 x 10^9 years ago. This relates to the fact that many (but not all) members of the existing Archaea are anaerobic, i.e. they do not require oxygen for survival, and thus could inhabit a world with an oxygen-poor atmosphere as was the case during the early history of the Earth. They also survive under extreme habitats today (see Table 3.2) and so could have tolerated the harsh conditions of the newly-formed planet.

These unicellular organisms occur as individuals or as aggregates or filaments of cells; they are morphologically heterogeneous with many being rod–shaped or circular and most have cell walls. Such organisms are microscopic, ranging from 0.1 to 15 μ and they reproduce asexually, mainly through cell fission. They have adapted to extreme conditions by developing varied strategies for obtain their energy. Many are chemotrophic, meaning that they extract energy from chemicals in the environment in which they live. Some chemotrophs have developed symbiotic relationships with eukaryotic organisms, such as those which live in the digestive organs of animals. Under these circumstances methane is produced as the organisms contribute to the breakdown of food, which is of benefit to their hosts, and so obtain their energy as well as a habitat. Technically, such Archaea are chemoorganotrophs. Others are chemolithautotrophs, obtaining their energy from inorganic compounds in their habitats, such as sulphurous compounds. Others have the capacity to alter their mode of nutrition with changing circumstances.

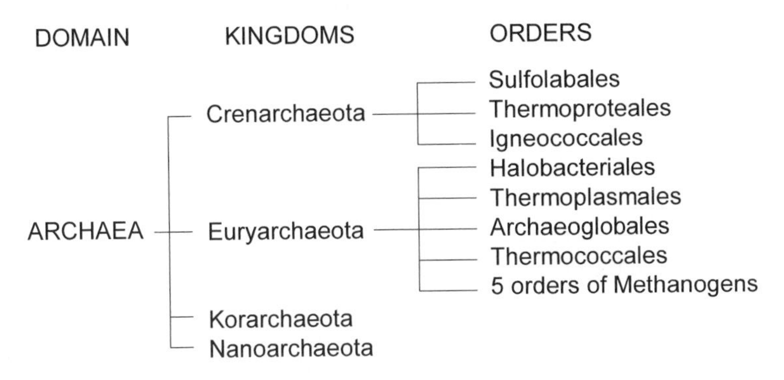

Figure 3.4. Kingdoms of the Archaea (based on Tree of Life, 2004)

Figure 3.4 shows the kingdoms of the Archaea and Table 3.2 gives details of their orders and their characteristics. They have been found in extreme environments which range from ocean hydrothermal vents to active terrestrial volcanic areas, including hot springs. Some of the crenarchaeota also occur in very cold environments. Others, like the halobacteriales colonise concentrated salt solutions such as the brines of salt lakes while the thermoplasmales enjoy warm ocean waters where they can form blooms, a process which gives the Red Sea its name. The orders of methanogens are diverse. Many species commonly occur in landfill sites where they play a major role in the decomposition of organic matter which they use as a source of energy and thus release methane gas.

Table 3.2. The orders of the Archaea and their characteristics

ORDERS	CHARACTERISTICS
Sulpholabales	Thermophilic, acidophilous, organic substrates
Thermoproteales	Thermophilic; terrestrial hot springs
Igneococcales	Terrestrial and marine active volcanic regions
Halobacteriales	Extreme halophiles: brine dwellers
Thermoplasmales	Acidophilous, anaerobic; coal residues
Archaeoglobales	Sulphate reducing, oceanic hydrothermal vents
Thermococcales	Sulphur metabolizers; hot marine sediments
5 orders of methanogens	Energy from organic matter, produce methane
Orders of kingdom korarchaeota	Little known; present in Yellowstone hot spring
Orders of kingdom nanoarchaeota	Found in hot vent off coast of Iceland

Methanogens have also been responsible for the generation in past geological eras of so-called natural gas (see Chapter 4), an important fossil fuel; the gas produced from marshes (marsh gas) is also methane generated by Archaean methanogens. Other methanogens inhabit the digestive tracts of animals and termites in a symbiotic relationship; the animals provide a habitat and nutrients while the organisms aid digestion to produce methane.

Little is known about the kingdoms korarchaeota and nanoarchaeota. Genetic characteristics indicate that they are distinct kingdoms (or phyla) from each other and from the other kingdoms of the Archaea. Both are thermophilous. The former was discovered in a terrestrial hot spring in Yellowstone National Park, USA, while the discovery of the latter in a hot vent off the coast of Iceland occurred in 2002. Huber *et al.*, (2002) report that it is a symbiont with a member of the igneococcales.

The capacity of Archaea to inhabit extreme habitats and their varied metabolic pathways may prove useful to society. For example, methane is a valuable source of energy and already its release from landfill sites is being harnessed in some countries to provide energy for domestic consumption. Other Archaea may be valuable for the bioremediation of polluted areas because they prefer highly acidic or highly alkaline media and several species degrade organic molecules such as polychloro biphenyls (PCBs) while others can extract heavy metals from waste materials.

3.3 Domain Bacteria

Little is known about the domain Bacteria in comparison with the kingdoms of the Plantae and Animalia and the evolution of its members is much disputed (e.g. Margulis and Dolan, 2002). It is possible that the bacteria will prove to be the largest group of organisms in existence. Bacteria are single-celled prokaryotic organisms, i.e. they have no nucleus; their genetic material occurs as strands (known as plasmids) and their organelles (the cell components associated with specific functions such as respiration) are also unconfined by membranes. Their cell walls comprise peptidoglycan, a carbohydrate, as is the case of the cell walls of plants. They are mainly spherical, spiral, comma- or rod-shaped organisms which obtain their food by coating it with secretions and then absorbing the break-down products. Most bacteria reproduce by binary fission (cell division) though a few species produce spores and others reproduce by receiving genes from other bacteria or from viruses and then undergoing binary fission. Some are autotrophs, obtaining food by processes similar to photosynthesis, notably the cyanobacteria which inhabit the oceans. These are described as photoautotrophs while bacteria which obtain their energy from the decomposition of substrates are known as chemoautotrophs. Many species are heterotrophs, deriving energy from the digestion of organic material produced by autotrophs. Bacteria are involved in a range of processes which are beneficial, indeed vital, to the maintenance of life, such as the biogeochemical cycles including that of carbon (see Section 2.3). Others are harmful to humans insofar as they transmit diseases, including tuberculosis, anthrax, botulism, pneumonia, salmonella and typhoid. In contrast many bacteria provide useful substances such as compounds used in industrial contexts, for bioremediation, in food preparation and for pharmaceuticals. The domain is very diverse and its members occupy almost all types of ecological niches. According to Oren (2004) there are c. 6,200 validly named species when the criterion of a 70 percent similarity of DNA between individuals is adopted, but if other methods of DNA characterisation are used the numbers could rise by two orders of magnitude.

The subdivisions of the domain bacteria are subject to alteration as revisions of existing groupings are made and new groupings are created. Some schemes promote five kingdoms as shown in Figure 3.5 while others suggest as many as 22 phyla (this term is used widely because of the recent adoption of the term domain which is almost synonymous with the term Kingdom – see Section 3.1.1). For simplicity the former scheme is adopted here.

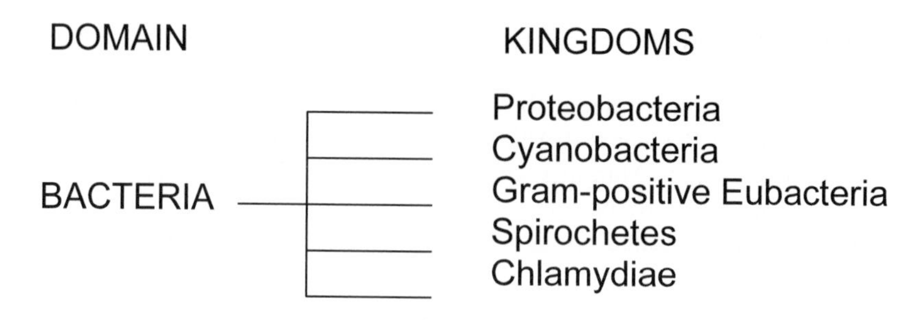

Figure 3.5. Five kingdoms of the domain Bacteria (based on Tree of Life, 2004)

The significance of bacteria as life forms and as facilitators for most other life forms cannot be overestimated. Whilst space precludes a comprehensive appraisal here, the following examples illustrate the role of bacteria in biospheric processes. Cyanobacteria play a major role in biogeochemical cycles and thus facilitate nutrient cycling and because they can photosynthesize they are important in food chains and webs, especially in the oceans. Formerly referred to as blue-green algae, cyanobacteria are not algae, which are eukaryotes, though many species inhabit aquatic environments where they, like algae, undertake photosynthesis. They are unicellular but may form filaments, sheets and colonies which are visible to the naked eye. Examples include the species *Prochloron* which is a major photosynthesizer in the oceans. The tiny organisms which comprise the cyanobacteria are important in Earth history. Their ancestors are amongst the earliest forms of life so far identified and created reef-like structures known as stromatolites. The oldest found are considered to be c. 3.5×10^9 years old (see Chapter 4). Their capacity to photosynthesize has had two further repercussions. First, it resulted in the gradual alteration of atmospheric composition to include oxygen and reduce the concentration of carbon dioxide with attendant consequences for global climate. Second, the green plants evolved through the inclusion of the cyanobacterial chloroplast in eukaryotic organisms; this promoted photosynthesis and contributed to changes in atmospheric composition through geological time.

Other species of cyanobacteria, notably *Nostoc* spp., contain the enzyme nitrogenase which confers the unique capacity for fixing nitrogen from the atmosphere and transforming it into ammonia, nitrites and nitrates. These substances can then be utilised by plants for producing proteins, amino acids etc. and so nitrogen can enter food chains and webs. Cyanobacteria with this ability have thickened walls inside which is an anaerobic environment to facilitate this chemical reaction; they are known as heterocysts. They are widespread in soils and water bodies, including paddy fields, and thus contribute considerably to ecosystem function and global food production. Some nitrogen-fixing cyanobacteria also form symbiotic relationships with other organisms that benefit from nitrogen in an available form, including algae which combine with bacteria to form lichens, as well as a wide range of eukaryotes. How this capacity to fix nitrogen arose has been investigated by Raymond *et al.* (2004) who indicate that its origins may be linked with methanogenic organisms and that it developed early in the evolution of life. In contrast, some cyanobacteria, e.g. *Lyngbye* spp., form pond scums, are toxic to animals and humans and can cause death.

The kingdom proteobacteria is particularly varied, with at least 17 orders. Many species are nitrogen-fixing bacteria which enjoy a symbiotic relationship with higher plants, notably the legumes. The bacteria inhabit the root nodules of the plants, which provide a habitat, and ensure that their hosts have a source of available nitrogen. As in the case of the nitrogen-fixing cyanobacteria discussed above, the nitrogen-fixing proteobacteria are vital components of the nitrogen cycle and facilitate the process of primary production in ecosystems and agricultural systems. Examples of root-nodule bacteria are *Nitrosomonas* spp. and *Nitrobacter* spp.; *Azotobacter* spp. are free-living soil proteobacteria that also fix nitrogen while *Escherichia* spp. (e.g. *E.coli*) and *Enterobacter* spp. are examples of proteobacteria which inhabit the digestive tracts of animals where they assist in the digestion of food by promoting fermentation. Other types of bacteria are involved with the equally vital process of decomposing organic matter which involved the release of nitrogen. These are the denitrifying bacteria which include the proteobacteria *Pseudomonas* spp.

Many of the Chlamydiae bacteria cause disease. For example, many are parasites and strip cells of vital substances such as vitamins and amino acids and so cause cell mortality while the species *Chlamydia trachomatis* causes a venereal disease known as Chlamydia. A member of the Chlamydiae also causes psittacosis, a disease of the birds that can be passed onto humans to cause respiratory problems. However, perhaps the most notorious bacterium is *Yersinia pestis* which caused the Black Death in Europe in the 14th century. Spread by fleas and rodents, this bacterium resulted in the death of some 25 million people. The drop in population resulted in abandoned settlements and possibly a reduction in the carbon dioxide concentration in the atmosphere as forest regrowth occurred in the wake of a decline in agriculture and forest clearance (see Section 2.3. and Ruddiman, 2003). This was not an isolated outbreak of this pestilence, otherwise known as bubonic plague, as is documented along with details of many other outbreaks of diseases caused by bacteria by the Bacteria Museum (2004). Today, this and many other bacterial diseases can be treated with antibiotics, some of which derive from fungi (see Section 3.4.2) while others derive from bacteria themselves, e.g. streptomycin. This is used to treat tuberculosis and originates from the soil bacteria streptomyces. Moreover, bacteria are widely used in the food industry and are, in many cases, indispensable. For example, *Lactobacillus* spp. and *Streptococcus* spp. are involved in the production of milk-related products such as buttermilk, yoghurt and cheese. Thus the world can live quite happily with some bacteria but has considerable difficulties with others!

3.4 Domain Eukaryote

This is the most well known and best documented of the three domains. It comprises a vast array of organisms, ranging from the relatively simple to the most complex. In contrast to prokaryotes, eukaryote cells have a nucleus which contains genetic material; other organelles are also contained within membranes. Almost all eukaryotes are aerobic and so require oxygen for survival while the prokaryotes may be aerobic or anaerobic. As Figure 3.3 shows, the eukaryotes evolved from the prokaryotes. The earliest evidence for the former comes from shales of the Pilbara craton in northwest Australia. Here molecular fossils, i.e. chemicals of biological origin, include steranes which are indicative of eukaryote activity. They are thought to have been produced by ancient marine algae, possibly acritarchs. Brocks *et al.* (1999) report that these fossils are 2.7×10^9 years old which means that eukaryotes were probably present on Earth prior to this date (see also Chapter 4). There are four kingdoms in this domain are outlined below.

3.4.1 Protista (also protoctista)

This is a diverse kingdom which comprises mainly micro-organisms. These are single-celled though many persist in colonies. Some protists, notably the algae, can photosynthesize and are thus autotrophic while the remainder consume bacteria or organic particles. Most protists live in aquatic environments and move using flagella. They are mainly aerobic except for some species that inhabit anaerobic mud/silt habitats and they reproduce sexually. Many organisms now classified as protists were once considered to belong to the other eukaryote kingdoms, i.e. were plants, animals or fungi. Evolution of protists from the prokaryotes involved the incorporation of membranes around cell functional units to produce discrete organelles and a process called endosymbiosis which was proposed by Margulis in 1981 (see Margulis and Sagan, 2002 for a discussion of recent advances). This involved the inclusion of prokaryote bacteria into host cells in a symbiotic relationship to produce mitochondria and chloroplasts. For example, Yoon *et al.* (2002) present evidence for the incorporation of a cyanobacterium (see Section 3.3) into an early 'protist' cell to produce an alga. This and similarly 'new' organisms apparently proved to be successful and established the eukaryotes.

The protists have been classified in different ways. This is because of the reorganization of domains and the reclassification of families and species, especially as information on genetic characteristics and relationships has become available (see discussions by Schlegel, 2003 and Finlay, 2004). For example, protists have been classified on the basis of lifestyle and/or ecology, a system which distinguishes between protozoans, slime molds, unicellular algae and multicellular algae. Some classifications have 12 phyla; others have as many as 50 phyla. Groombridge and Jenkins (2002) list 30 phyla which are given in Table 3.3.

In general, those protists which are unicellular and free-floating make up the organisms known collectively as the plankton. Those which can photosynthesize are the phytoplankton and those which consume phytoplankton, bacteria or organic particles are the zooplankton. The autotrophic protists, especially the algae (see Graham and Wilcox, 1999) and the diatoms, are very important primary producers in aquatic ecosystems and are thus influential in the cycling of carbon between the atmosphere and the biosphere. Some algae are single celled, others are multicellular while diatoms are unicellular; some algae and some diatoms can form colonies but all contain chloroplasts which facilitate photosynthesis. Many species of algae and diatoms occupy varied habitats; apart from those in the phytoplankton some species grow attached to plants and are thus epiphytic while others grow on or in sediments of the ocean or lake beds. All the primary producing protists are the basis of aquatic food chains and webs.

The slime molds (see Stephenson and Stempson, 2000) have life cycles involving unicellular and multicellular stages. The Labyrinthulata grow as colonies which form net-like structures, mainly on eel grass which occurs in littoral marine plant communities. The plasmoidal slime molds (Myxomycota) and cellular slime molds (Rhizopoda) feed on bacteria and other organisms found on decaying vegetation by surrounding and digesting them. In contrast, the water molds (Oomycota) obtain food by piercing host tissue with hyphae from which enzymes are produced. The break-down products are then absorbed. Species of this group of organisms, hitherto classified as fungi, occur in freshwater and soil and some species live parasitically on certain plants, including several of agronomic value, and fish. For example, *Phytophthora infestans*, is the organism responsible for potato blight and was the cause of the Irish potato famine in the 1840s when more than one million people died. Slime molds can, thus, have a considerable economic and social impact. Many of the remaining protists of Table 3.3 were formerly considered to be protozoa. All ingest food and are heterotrophic. Some species live in the digestive tracts of animals and some cause disease in animals and humans.

Table 3.3. The phyla of the kingdom protista (based on Groombridge and Jenkins, 2002).

PHYLUM	No. of species	Comments on main group in phylum
Actinopoda (Radiolarians)	4000	Mainly marine, siliceous
Apicomplexa(Sporozoa)	5000	Parasites in animal blood
Archaeoprotista (Amitochondriates)	?	Anaerobic, some parasitic/ anaerobic in animals
Chlorophyta (Green algae)	16000	Main primary producers
Chrysomonada (Chrysophyta)	?	Golden-yellow marine algae, autotrophic
Chytridomycola	1000	Hyphae, cause plant diseases, heterotrophic
Ciliophora (Ciliates)	10000	Consume bacteria
Cryptomonada (Cryptophyta)	?	Elliptical, free swimming, Hetero and autotrophic
Diatoma (Diatoms)	10000	Siliceous autotrophs
Dinomastigota (Dinoflagellates)	4000	Mainly planktonic, hetero and autotrophic
Discomitochondria (Flagellates, zoomastigotes)	?	Once the protozoa, hetero and autotrophic
Eustigmatophyta (Green eyespot algae)	?	Plankonic, yellow-green algae, autotrophic
Gamophyta (Conjugating green algae)	~1000	Filament forming autotrophic algae
Granuloreticulosa (Foraminifera Reticulomyxids)	4000	Calcareous tests, marine, heterotrophic
Haplospora	33	Heterotrophic, symbionts
Haptomonada (Prymnesipophytes)	? few	Mostly marine, autotrophic, calcareous
Hyphochytriomycota	23	Hyphae, heterotrophic
Labyrinthulata (Slime nets and Thaustochytrids)	?	Colonial, heterotrophic, marine
Micospora (Microsporans)	800	Anaerobic, heterotrophic
Myxomycota (Plasmoidal slime molds)	500	Heterotrophic, damp terrestrial habitats
Myxospora (Myxosporidians)	1100	Heterotrophic, symbiotic and parasitic in fish
Oomycota (Water molds)	100s	Hyphae, heterotrophic
Paramyxa	6	Multicelled spores, heterotrophic, symbionts
Phaeophyta (Brown algae)	900	Autotrophic, marine

Plasmodiomorpha	29	Heterotrophic, symbionts
Rhizopoda (Amoebas, cellular slime molds)	200	Heterotrophic, benthic and soil based
Rhodophyta (Red algae)	4000	Most autotrophic, marine
Xanthophyta (Yellow-green algae)	600	Autotrophic, freshwater, yellow-gold pigments
Xenophyophora	42	Heterotrophic, benthic
Zoomastigota	? few	Heterotrophic on bacteria

3.4.2 Fungi

Fungi are considered to be a separate kingdom of the eukaryotes because they have genetic characteristics that distinguish them from the other eukaryote kingdoms. Like these kingdoms, the classification of fungi has been revised as new advances in molecular science have been made resulting in additions, losses and the reorganization of phyla. As shown in Table 3.1, some 72,000 species have been described but this is considered to be a large underestimate of the true number of species which may be as many as 1.5 million, especially as they can grow in a wide variety of environments. In terms of general characteristics (see Carlile *et al.*, 2001, Burnett, 2003 and Kendrick, 2001/2003), fungi have cell walls comprised of chitin, a complex carbohydrate, and most consist of hyphae which are thread-like protuberances that intertwine to produce a mass known as a mycelium. They are non-motile and can reproduce sexually or asexually though neither method involves an embryonic stage in the production of single– or multi-celled species. It has been argued that fungi are less close to plants than to animals because chitin is also the material of insect exoskeletons and because the chief storage carbohydrate is glycogen as it is in animals. They are heterotrophic and obtain food by surrounding it, excreting digestive chemicals and absorbing the resulting compounds in a similar way to the slime molds (Section 3.4.1). In this respect they can be described as osmotrophic.

In terms of their geological history, fungi were present more than 500 million years ago and it is likely that they evolved from aquatic prokaryote organisms which evolved mechanisms for maintaining dryness. With few exceptions, modern fungi are terrestrial, though this depends on the classification adopted (see below) and all but the yeasts are multicellular. They can be grouped according to their means of subsistence. For example, some are saprophytes which live on dead organic matter, others are parasites and live on living organisms, and yet others are so-called mutualists involving a symbiotic association with other organisms such as vascular plants or algae. The latter association gives rise to lichens.

There is much dissension in the literature regarding the taxonomy and systematics of the fungi. For example, Groombridge and Jenkins (2002), following Margulis and Schwartz (1998), identify three phyla while Kendrick (2001/2003) identifies two kingdoms rather than one, notably the eumycota and the chromista and within the former, the most prolific group, he lists four phyla: chytridiomycota, zygomycota, glomeromycota and dikaryomycota. This exemplifies the state of flux in which fungal (and other) systematics exists at present. It also explains why some researchers believe that certain organisms, namely the chromista, are a group of fungi which include species in aquatic habitats (see above). For present purposes the scheme used by Groombridge and Jenkins (2002) will be followed and is presented in Table 3.4.

Table 3.4. The phyla of the kingdom fungi (based on Groombridge and Jenkins, 2002).

PHYLUM	**No. of species**	**Comments**
Ascomycota	30000	Diverse, including edible species e.g. truffles, yeasts, others cause disease, some are valuable medicines; many form lichens
Basidiomycota	22250	Diverse, including edible species e.g. mushrooms; many associate with vascular plants.
Zygomycota	1100	Most are saprophytes or parasites on protists; some associate with vascular plants

The role of fungi in the carbon cycle is primarily the breakdown of dead organic matter (see Section 3.6.3). This releases carbon dioxide and other compounds which can then be recycled. The biomass of fungi and lichens is also a significant store of carbon in the biosphere. Some types of fungi are sought after for food; indeed many species are cultivated while others cause plant and animal diseases and yet others produce beneficial substances such as antibiotics. The most well known species used as food are the mushrooms, though in terms of value the truffles are most prized. There are many species of edible mushrooms and some species which are poisonous (see www.mykoweb.com). There are also several species of truffle but only three are edible and some are poisonous, as explained by the Mycological Society of San Francisco (2004). The former include *Tuber melanospermum,* the black truffle from the Perigord region of France, and *T. magnatum,* a white truffle from Italy. Both occur as part of a mycorrhizal association with trees and are retrieved traditionally by pigs or dogs which recognise their distinctive aromas. Many fungi can cause diseases in crops (Prell, 2001); for example the rust diseases in wheat include brown rust which is caused by *Puccina recondite*. Fungal diseases in humans (see Hudler, 2001; San-Blas and Calderone, 2004) include *aspergillosis*, also known as brooder pneumonia, which is caused by *Aspergillus fumigatus* while *A. flavus* can contaminate food and produces toxins, known as aflatoxins, which are poisonous to humans and can cause death. In contrast, some fungi produce useful substances which can combat disease, the classic case being penicillin which derives from *Penicillium notatum* and which is now produced synthetically. It is an antibiotic and is used to treat infections in humans caused by bacteria. There are also increasing applications of fungal use in biotechnology (see Arora *et al.*, 2003). For example, fungi are used to breakdown the lignin of wood for the paper industry, some species, e.g. *Phanerochaete sordida* and *P. chrysosporium* can degrade creosote in soil and are thus effective bioremediation agents while others, as mycopesticides, are used for pest control in selected crops. Another group of fungi, the yeasts which are single celled, are very important in the food industry for promoting fermentation. Examples of their use include bread production and the brewing of beer.

3.4.3 Plantae

Plants have been classified in many different ways and like the bacteria, fungi etc. discussed above there are continual additions, revisions and regrouping, as discussed in Stuessy *et al.* (2001) and Judd *et al.* (2002). As shown in Table 3.1, the plants rank second, after insects, in terms of total number of described species which are thought to constitute c. 84 percent of the total number of plants. Groombridge and Jenkins (2002), following Margulis and Schwartz (1998), list 12 phyla which are given in Table 3.5. All plants are multicellular and contain chloroplasts which facilitate photosynthesis (see Section 3.6.1) and so they are autotrophs. This function also captures energy from the sun and it is this energy which underpins processes in the biosphere. Plants thus provide a source of food for other living organisms as living or dead matter. Plants are either non-vascular or vascular. Most are vascular which means that they possess dedicated systems for transporting water and nutrients. Non-vascular plants include the bryophytes (mosses) in which food and nutrients are absorbed on the surface and transported from cell to cell by osmosis.

Plants are diverse, especially the angiosperms (anthophyta), and inhabit all but the most extreme environments on the Earth's surface. They are vital components of the global carbon cycle (see Section 2.3 and Figure 2.17) as they flux vast quantities of carbon dioxide from the atmosphere to the biosphere, and they constitute a large carbon store. In addition, oxygen is released by photosynthesis (see Section 3.6.1). This maintains oxygen concentrations in the atmosphere which favours aerobic life forms and processes.

Table 3.5. The phyla of the kingdom plantae (based on Groombridge and Jenkins, 2002).

PHYLA	**No. of species described**	**Comments**
Anthocerophyta (Horned liverworts)	100	Non-vascular, pioneer species, moist habitats
Anthophyta (angiosperms)	270000	Vascular, diverse, most terrestrial habitats, large biomass
Bryophyta (mosses)	10000	Non-vascular, diverse habitats
Coniferophyta (conifers)	630	Cone-bearing, evergreen, widespread, soft wood.
Cycadophyta (cycads)	145	Seed-bearing, low latitudes
Filicinophyta (ferns)	12000	Diverse, moist habitats
Ginkophyta (ginko)	1	Vascular, seed-bearing tree
Gnetophyta (gnetophytes)	70	Vascular, 3 genera
Hepatophyta (liverworts)	6000	Non-vascular, moist habitats, widespread, pioneer species
Lycophyta (club mosses)	1000	Vascular, evergreen, small, widespread, some epiphytes
Psilophyta (whisk fern)	10	Vascular, no roots or leaves
Sphenophyta (horsetails)	15	Vascular, seedless, moist and disturbed habitats

Plants have varied reproductive strategies including vegetative reproduction which involves subdivision of bulbs, rhizomes or tubers to produce independent plants but which are identical genetically. In contrast, sexual reproduction involves the fusion of reproductive units (gametes) from two different individuals to produce offspring with a new and unique genetic signature. For example, pollen is produced by many plants and is disseminated into the environment; some will reach the sporangia produced by the ovary in other plants of the same species and fertilisation will take place. The resulting seeds have the potential to produce new plants.

The divergence of plants from their prokaryote ancestors occurred c. 1000 x 10^6 years ago at much the same time as the other eukaryotes emerged. However, the first terrestrial plants appear in the fossil record at about 425 x 10^6 years ago following the development of strategies that allowed survival without submergence in water. Nevertheless, the earliest land plants, notably lichens (a symbiotic association of a fungus and an alga – see Section 3.4.2), mosses and liverworts remained dependent on water to disseminate their propagules. The mosses and liverworts are non-vascular, reflecting their early origins. By 200 x 10^6 years ago pollen-producing plants had evolved, providing opportunities for heightened dispersal by insects and wind. The angiosperms were in evidence 135 x 10^6 years ago and thereafter underwent substantial radiation. As Table 3.1 shows, this phylum contains the most species. The issue of plant evolution has been discussed by Bell and Hemsley (2000) and is considered again in Chapter 4. Today, one of the major environmental issues of the 21st century is the loss of vegetation cover and associated extinction of all types of organisms.

As illustrated throughout this book, plants have myriad uses. The provision of food and shelter is basic. Wild plants sustained hunter-gatherers and then, beginning c. 10,000 years ago, specific plants (and animals) were selected for domestication; they were too vital a resource to leave their chances of survival etc. entirely to nature. The cereals and legumes were amongst the first to be domesticated and they remain the most important agricultural crops today. Plants also support those agricultural systems which produce animal products. Wood is eminently versatile; it provides a flexible building material, a source of energy when burnt and raw material for paper making. Plants provide decoration, medicines and perfumes as well as being major sources of fibre, especially cotton. They assist in land reclamation where plants tolerant of heavy metals are required and they provide protection from wind and water erosion. Increasingly important is their role in landscape aesthetics which is important for tourism and of course plants of the past have left a geological and wealth-generating legacy in the form of coal, oil and natural gas (Chapter 4).

3.4.4 Animalia

The animal kingdom is diverse, ranging from organisms with only a few cells, such as trichoplax (see Table 3.6) to the largest organisms such as the blue whale and elephant. All animals are multicellular and heterotrophic, and most are macroscopic. Almost all develop from an embryo produced by the fertilisation of an egg by sperm. Another feature that distinguishes animals from the other kingdoms is the fact that most ingest their food into a body cavity (coelum) where it is digested. The exceptions are the phyla porifera and cnidaria which do not have specialised tissues. Other criteria employed to classify animals include body symmetry and presence or absence of segmentation Some phyla have been almost completely described, notably the vertebrates but as Table 3.1 shows, the most prolific phylum, the mandibulata (insects and myriapods), is not as well documented as those described are considered to be only c. 12 percent of the total. Moreover, the beetles (coleoptera) are the most numerous insects, comprising some 300,000 species which makes them the single most prolific life form. As in the case of the kingdoms described above, the animalia are continually being reclassified and there is much debate as to how many phyla there should be. Groombridge and Jenkins (2002), following Margulis and Schwartz (1998), list 37 phyla (see Table 3.6). However, it is generally agreed that most of the Earth's animals are contained within nine phyla, as highlighted in Table 3.6. Most phyla are likely to contain many more species than those so far described, especially as the marine environment is investigated further and as the less accessible parts of the world are explored.

Table 3.6. The phyla of the kingdom Animalia (based on Groombridge and Jenkins, 2002).

PHYLUM	**No. of described species**	**COMMENTS**
Acanthocephala (Thorny-headed worms)	>1000	Parasitic, live in guts of vertebrates
Annelida (Annelids) *	16000	Segmented worms, soils, sediments, scavengers/predators
Brachiopoda (Lampshells)	350	Marine, benthic , sessile
Bryozoa (Ectoprocts)	4000	Sessile, colonial, filter feeders
Cephalochordata (Lancelets)	23	Marine, filter-feeders, shallow water sediments
Chaetognatha (Arrow worms)	70	Zooplankton, marine, widespread, warm seas
Chelicerata (Chelicerates)*	75000	Arthropods, diverse, widespread
Cnidaria (Cnidarians, hydras) *	9000	Aquatic, marine, varied, symmetrical and widespread
Craniata (Vertebrates)*	52,500	Free-living, widespread
Crustacea (Crustaceans)*	40000	Free-living, diverse, mostly aquatic, widespread
Ctenophora (Comb jellies)	100	Free-swimming, marine
Echinodermata (Echinoderms)	7000	Marine benthic free-living invertebrates, calcium carbonate
Echiura (Spoon worms)	140	Marine free-living worms
Entoprocta (Entoprocts)	150	Marine colonial organisms
Gastrotricha (Gastrotrichs)	400	Marine free-living worms
Gnathostomulids (Jaw worms)	80	Marine free-living worms, benthic, low oxygen sediments
Hemichordata (Acorn worms)	90	Marine, benthic, sedentary, soft bodied

Kinorhyncha (Kinorhyncha)	150	Free-living marine benthic especially mud
Loricifera (Locificerans)	100	Marine, benthic, microscopic
Mandibulata (Mandibulates) *	950000	Free-living terrestrial, diverse, chitin exoskeleton, inc. insects
Mollusca (Molluscs)*	At least 70000	Widespread, marine, freshwater
Nematoda (Nematodes)*	20000	Free-living, parasitic,widespread
Nematomorpha (nematomorphs)	240	Aquatic, free-living, unsegmented worms
Nemertina (Ribbon worms)	900	Predatory marine worms, intertidal zone
Onychophora(Velvet worms)	100	Terrestrial worm-like, prefer high humidity, chitin cuticle
Orthonectida (Orthonectida)	20	Symbionts, parasites, worm-like marine organisms
Phoronida (Phoronids)	14	Marine worms, filter feeders
Placozoa (Trichoplax)	1	Least complex animal
Platyhelminthes (Flatworms) *	20000	Symbionts, parasites e.g. tape worms, flukes
Pogonophora (Beard worms)	>120	Marine benthic worms, chitin tubes, cold deep water
Porifera (Sponges) *	5-10000	Mostly marine filter feeders
Priapulida (Priapulids)	17	Marine sand/mud
Rhombozoa (Rhombozoans)	70	Kidneys of squid and octopi
Rotifera (Rotifers)	2000	Freshwater
Sipuncula (Peanut worms)	150	Marine benthic
Tardigrada (Water bears)	750	Widespread. aquatic
Urochordata (Sea squirts)	1400	Marine, benthic, filter feeders

* Designates the most diverse phyla

It is generally accepted that animals originated in the ocean though when this occurred and how evolution proceeded is the source of much debate (see Carroll *et al.*, 2001). Until recently, evidence for a major explosion of animal species during the Cambrian period, which began 543 x 10^6 years ago, was considered to reflect the origins of most of the animal groups known today. This is certainly indicated by the abundant fossil evidence from rocks of this period but for it to have occurred it seems likely that the ancestors of this Cambrian fauna were in existence some considerable time before. Fossil evidence is not yet available to corroborate this. Moreover, animal evolution is linked with that of the plants and the changes in atmospheric composition, especially the increase in oxygen, to which plants contributed (see Section 3.6.1 and Chapter 4). Wang *et al.* (1999), on the basis of molecular clocks, have suggested that the divergence of animals from the plants and fungi occurred as long as 1,500 x 10^6 years ago, with sponges and coelenterates (primitive animals) developing between 1,500 x 10^6 and 1,200 x 10^6 years ago. Nematodes evolved c. 1,200 x 10^6 years ago while birds and mammals diverged c. 320 x 10^6 years ago. The earliest hominid species, i.e. those primates with human-like traits, diverged from the apes some 5 x 10^6 years ago, and modern humans evolved c. 250,000 years ago. Thus humans are a recent addition to the Earth's fauna.

Animals feed on plants, respire and are stores of carbon and thus part of the global carbon cycle (Section 2.3 and Figure 2.17). They have been extensively exploited by humans from the days of the hunter-gatherers of prehistory to modern-day use of animal labour and transport and sport facilitation such as horse racing and hunting though by far the largest and economically valuable use is in animal agriculture. This relies on a few types of animal selected millennia ago for their meat and milk, especially in the centre of agricultural innovation in the Near East. These include cattle, sheep, goat and pig. Other domesticated species include the chicken, turkey, camel, horse and llama. Fish and shellfish are also important food sources while some animals are reared for leather, wool and fur.

3.5 Viruses etc.

There are some entities that are at the margin between the living and non-living worlds. The most common are the viruses which are present throughout the environment and which were first identified in the late 1890s. The many different types of viruses all comprise organic matter, including genetic material, and are thus made of carbon compounds. The genetic material, DNA or RNA, is contained within an outer 'shell' of protein known as a capsid. When outside living cells, viruses are inert but become viable when they infect living cells which they effectively hijack. Once inside a cell a virus inserts a strand of its genetic material into the host's nucleus where it replicates and damages or destroys the host's DNA/RNA. The host then replicates the viral DNA which accumulates until the host cell ruptures to release the newly generated viruses. These may not be entirely identical to the parent as genetic material from the host can be incorporated and transmitted to a new host. This is transduction which may be one process whereby evolution occurs. In general, viruses replicate rapidly so their genetic constitution also has the potential to alter rapidly, as occurs with the virus that causes influenza. Some viruses have the capacity to remain dormant for long periods until they reach a suitable host or even within the host.

Plants, animals, fungi and bacteria are susceptible to infection by viruses and the disease caused can be fatal. The most common viruses which affect humans include those which cause the common cold, measles, mumps and influenza. More serious diseases caused by viruses include rabies, yellow fever, hepatitis, ebola, dengue fever, Japanese encephalitis, west Nile fever and poliomyelitis while HIV, technically a retrovirus, causes autoimmune deficiency disease (AIDS). These and other viral diseases are examined by Oldstone (2000) and a detailed list has been compiled by the Garry Laboratory (2004). Viral diseases of plants are also numerous, as discussed by Hull (2002). Examples include a range of mosaic viruses which impair many leaf vegetable crops. Other viruses attack root vegetables, e.g. potato Y potyvirus and many other potato viruses; or fruits e.g. strawberry latent ring spot, various grapevine viruses and apple leaf spot; or cereals e.g. barley mosaic virus and wheat yellow mosaic *bymovirus* (see Brunt *et al.*, 1996 onwards). Diseases caused by viruses in animals include cowpox and influenza in horses, pigs, ducks and chickens. Many other viruses attack wild plants and animals. Those affecting agricultural plants and animals cause considerable crop losses and financial losses every year.

Another group of 'organisms' that straddle the life - non-life border are the prions. These have come to public attention recently because of their controversial association with Creutzfeld-Jacob disease (CJD), a fatal infection of the brain in humans, so-called scrapie in sheep, and bovine spongiform encephalopathy (BSE), a related disease in cattle, as discussed by Heaphy (2004) and in Prusiner (2004). Several other neurological diseases in animals and humans are also linked with prions which are proteinaceous particles and thus carbon-based. Unlike other disease-causing organisms, prions do not contain nucleic acids and consist of modified forms of normal cellular protein. They can be incubated in brain tissue for long periods before they have an effect which involves impairment of tissue function and ultimately impairment of human/animal function prior to death. How prions enter the body and infect brain tissue has not yet been ascertained in detail.

3.6 Vital processes

The cycling of carbon at short timescales involves a number of processes that are vital to the operation and renewal of the biosphere. These are photosynthesis, respiration and decomposition. All are involved in the life cycles of organisms and are related to reproduction, survival, death and subsequent decay. They are involved in inter-organism exchanges, such as the provision of food by photosynthesizing organisms, the autotrophs, for non-photosynthesizing organisms, the heterotrophs (see Section 3.7 below), and link the organic and inorganic components of the environment. Thus they are significant components of the carbon cycle (see Figure 2.17).

3.6.1 Photosynthesis

Photosynthesis is a crucial and complicated chemical process which is carried out by organisms containing chlorophyll. As discussed above, those organisms with the ability to photosynthesize are eukaryotes, e.g. plants, and protists e.g. algae. The process is crucial because photosynthesis harnesses the Sun's energy for use within the biosphere and thus provides the 'fuel' for the reproduction and survival of almost all organisms. It is a complicated process involving the conversion of light energy to chemical energy, and the chemical combination of carbon dioxide from the atmosphere with water from within the soil (see descriptions in Smith and Smith, 2001; Bonar, 2002). The process is represented by the following chemical equation:

$$6\,CO_2 + 6\,H_2O \rightarrow C_6H_{12}O_6 + 6\,O_2$$

Carbon dioxide is absorbed from the atmosphere and combined, using energy from sunlight, with water taken up from the soil. The resulting sugars (types of carbohydrates, see Section 2.2.3) are stored within the tissues of the autotrophs and oxygen is released to the atmosphere. This release of oxygen has been occurring since the process of photosynthesis evolved many millions of years ago (see below) and is the reason why aerobic organisms, i.e. organisms which respire by taking in oxygen, have evolved (see Chapter 4). Photosynthesis takes place within the cells of those organisms which contain chloroplasts. These are cell units, or organelles, contain chlorophyll, a complex substance which is a photoreceptor and thus has the capacity to absorb light energy and convert it to chemical energy. The chemical structure of chlorophyll *a* (a carboxylic acid in Beilstein's classification system for organic compounds given in Table 2.3) is given in Figure 3.6, which shows that it comprises four heterocyclic carbon rings, each with one atom of nitrogen, arranged around a central atom of magnesium. This structure is similar to that of haem in the haemoglobin of blood which has a central atom of iron, rather than magnesium, and which is vital for the transport of oxygen through the bodies of animals. Other types of chlorophyll exist, with minor chemical differences. All absorb light energy effectively but differ in the wavelength that they capture. Photosynthesizing organisms obtain their energy from light in the wavelengths which correspond with the blue and red spectra. Little light is absorbed from the green spectrum and so is reflected giving plants their green colour.

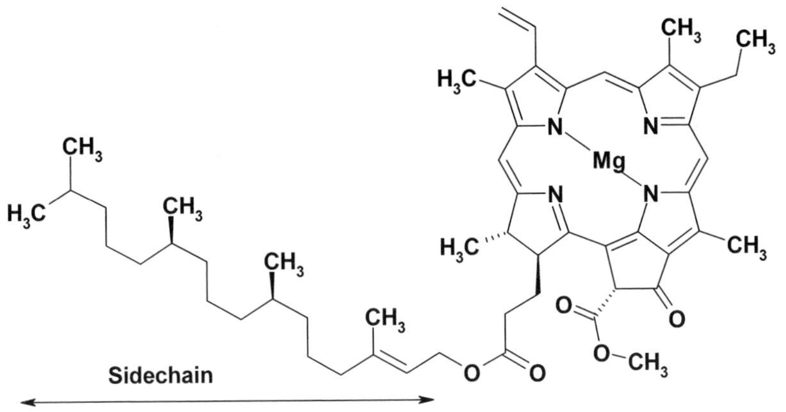

Figure 3.6. Chemical structure of chlorophyll *a*

Chloroplasts comprise inner and outer membranes which surround thylakoids. These are pillow-shaped structures arranged in stacks known as grana and they have membranes in which the chlorophyll resides. The first stage of photosynthesis is called the light reaction and involves the capture of light energy. The units within the chlorophyll essential for this are two membrane protein complexes referred to as photosystems; the crystal structure of one of these subunits has recently been ascertained (Ben-Shem *et al.*, 2003). The capture of light raises the energy level of the chlorophyll and it becomes unstable. Stability is restored when the energy is released as heat or fluorescence (a low energy light), or is transferred to another molecule from the chlorophyll. The latter involves the formation of adenosine triphosphate (ATP) and nicotine adenine dinucleotide phosphate (NADPH). Both are required for the next stage of photosynthesis known as the dark reaction (since further light is not needed) in which carbon dioxide is converted to carbohydrate. It takes place in the stroma of the cell, which are the tubular connections between the grana housing the chlorophyll-containing thylakoid membranes. This part of the process is complex and is known as the Calvin – Benson cycle, after its discoverers Melvin Calvin and Andrew Benson, at the University of California, Berkeley, 1946-1953. The four stages of this cycle are as follows:

1. Carbon dioxide combines with ribulose 1,5-biphosphate catalysed by an enzyme called rubisco, to produce two molecules of 3-phosphoglycerate (3-PGA)
2. ATP and NADPH react with the 3-PGA to produce glyceraldehyde-3-phosphate (G3P), an energy-rich sugar molecule
3. One-sixth of the G3P is diverted to ultimately produce glucose

4. The remaining G3P molecules are converted back to ribulose biphosphate, with ATP supplying the necessary energy.
   The stages are shown in diagrammatic form in Figure 3.7.

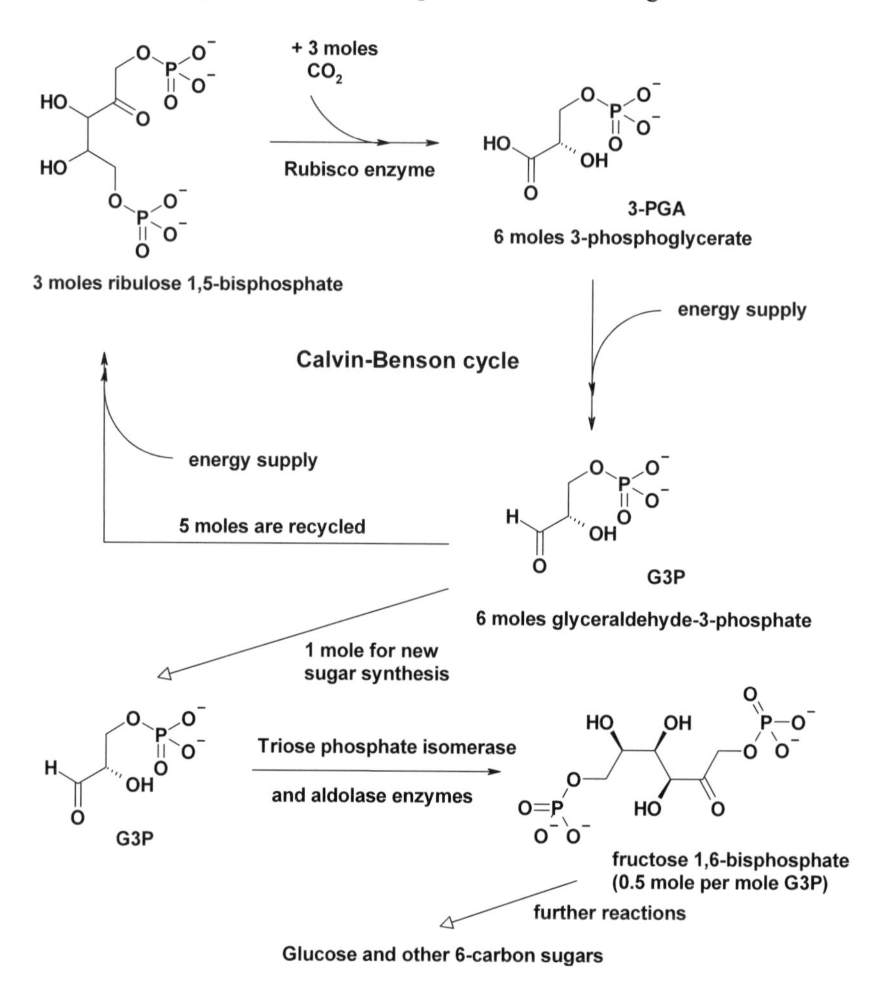

Figure 3.7 Calvin Benson reactions

Plants utilize carbon dioxide in two ways. Most allow the carbon dioxide they absorb to enter the Calvin-Benson cycle immediately and in this case the first stable organic compound produced is glyceraldehyde-3-phosphate. This contains three carbon atoms so plants operating in this away are known as C_3 plants. A few plants, including maize, sugar cane and other tropical and sub-tropical grasses, operate differently in order to conserve water. Prior to the Calvin-Benson cycle, these plants have a special enzyme, known as phosphoenolpyruvate (PEP) to trap carbon dioxide and turn it into oxaloacetate. As this contains four carbon atoms, plants exhibiting this mechanism are known as C_4 plants. Carbon dioxide is then released from the oxaloacetate and enters the Calvin-Benson cycle. C_4 plants have different anatomical leaf characteristics than C_3 plants in order to carry out this process. They contain two different groups of cells capable of photosynthesis, one group of which concentrates carbon dioxide as it enters the plant. Overall, a higher rate of photosynthesis occurs than in C_3 plants and water is conserved because the stomata, the 'pores' through which gases and water enter leaves, can close or partially close to reduce water loss. A third photosynthetic mechanism also exists. This is referred to as the crassulacean acid metabolism (CAM) which characterises some 15 families including the *Euphorbiaceae*, *Cactaceae* and *Crassulaceae*. The latter gives the mechanism its name and all such plants are succulents and/or semi-desert species. The CAM pathway is similar to that for C_4 plants insofar as the first step involves PEP which fixes carbon dioxide before the rubisco reaction occurs. However, CAM plants do not possess the specialized photosynthetic cells which concentrate carbon dioxide. Instead they open their stomata at night when cooler temperatures prevail and so they reduce water loss during the hot days. The reaction with PEP takes place at night to produce malic acid; this in turn produces carbon dioxide for entry into the Calvin-Benson cycle. The carbon dioxide concentration remains high during the day because the stomata are closed to the exterior and so facilitates the reaction with rubisco. Plants with this mechanism can survive in very dry environments.

In evolutionary terms, it is considered that photosynthesis evolved early in the history of life on Earth, probably originating c. $2.5 - 2.6 \times 10^9$ years ago in bacteria (Olson and Blankenship, 2004). Its operation was very important for the subsequent evolution of protists, eukaryotes etc., for the alteration of the chemical composition of the atmosphere and for the development of biogeochemical cycles. These issues will be discussed further in Chapter 4. Today, photosynthesis makes the 'world tick'! Photosynthesis (on land and in the oceans) is responsible for the drawdown from the atmosphere of c. 100-120 Gt of carbon per year, about half of which is returned to the atmosphere through respiration leaving c. 50-60 Gt to be stored in the biosphere. Although some of this is counteracted by respiration from heterotrophic organisms, there is a net storage of carbon within the biosphere. This means that the carbon dioxide from fossil-fuel use is not all accumulating in the atmosphere as some is absorbed by photosynthesizing organisms. These exchanges are depicted in Figure 2.17. This fixation of carbon is not uniform globally as environmental conditions set limits on the ability of plants etc. to photosynthesize. As indicated above, air temperature and water availability are important but so too is the nutrient content of the soil, in the case of terrestrial plants, and the nutrient content of water in lakes and the oceans. Globally, most carbon dioxide is fixed in the humid tropics and least is fixed in the high polar lands and seas. Moreover, and although it may be an obvious statement, photosynthesis keeps all heterotrophs (including humans) alive by maintaining oxygen concentrations in the atmosphere and by producing food through agriculture and providing many other commodities as well as supporting many other biogeochemical cycles.

3.6.2 Respiration

Respiration is the breakdown of complex molecules to produce chemical energy for all the activities an organism needs to undertake to survive and reproduce. Like photosynthesis, respiration is a complex process and occurs in all living cells; the various components are presented in Wikibooks (2004). It involves the breakdown of sugars, such as those made during photosynthesis (see Section 3.6.2), to produce carbon dioxide and water, as represented by the following equation:

$$C_6H_{12}O_6 + 6O_2 \rightarrow 6CO_2 + 6H_2O + \text{energy}$$

Respiration can take two forms: aerobic respiration requires oxygen and is characteristic of most organisms while anaerobic respiration takes place in an environment where there is no oxygen and is characteristic of some types of bacteria which obtain oxygen from nitrate or sulphate. The former produces more energy than the latter.

Aerobic respiration occurs in prokaryotes in the cytoplasm of the cell but in the eukaryotes it takes place in the cytoplasm and in cell organelles known as mitochondria. These oval- or rod-shaped structures consist of a matrix surrounded by two membranes between which there is a compartment, as shown in Figure 3.8. The inner membrane is convoluted to increase the surface area and the folds so formed are called cristae which project into the fluid-filled matrix. Respiration takes place in two stages, the first of which occurs in the cytoplasm of the cell and is called glycolysis. The second stage occurs in the mitochondria and involves the citric acid cycle (also known as the Krebs Cycle after its discoverer Hans Krebs) and electron transport. This is discussed below (Figure 3.10).

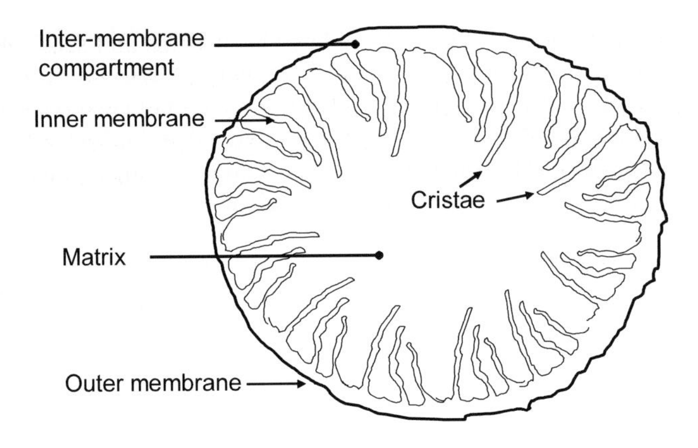

Figure 3.8. A mitochondrion.

Glycolysis (see Maber, 2004 for detailed explanations) involves nine reactions and each is catalysed by a particular enzyme: a summary is given in Figure 3.9.

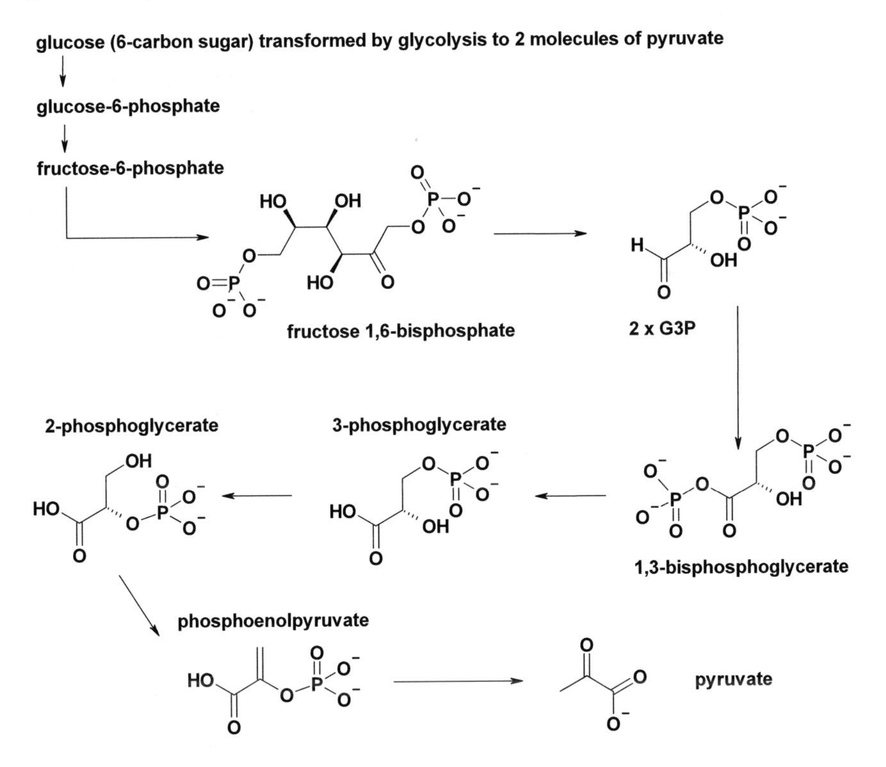

Figure 3.9. Summary of glycolysis

Many of the reactions are analogous to the Calvin-Benson cycle (Figure 3.7) but run "in reverse". Overall, a glucose molecule is cleaved into two molecules, each with three carbon atoms; and two molecules of adenosine triphosphate (ATP), as well as two other high-energy electron carrying molecules, are produced. After the final stage of glycolysis, pyruvic acid, ATP and nicotinamide adenine dinucleotide (NADH) are available. Pyruvate is then metabolized and carbon dioxide is produced along with a two carbon molecule (an acetyl unit) in the form of its coenzyme A derivative. This chemical, acetyl-CoA is the basis for the citric acid cycle, as shown in Figure 3.10.

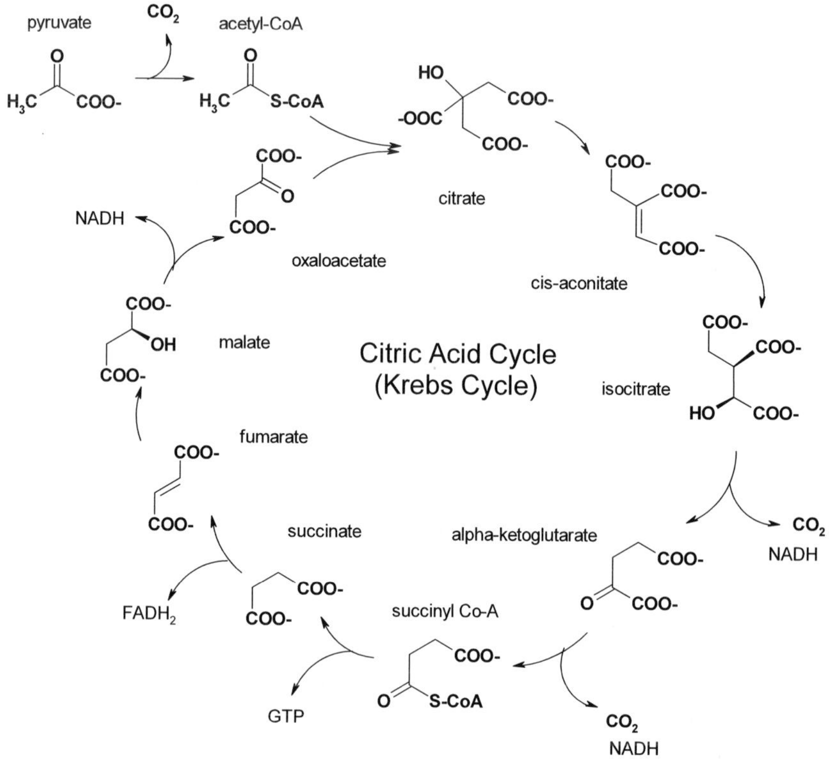

Figure 3.10. The Citric Acid Cycle

Another two molecules of carbon dioxide are produced from the acetyl-CoA, but more important is the production of the energy carriers GTP, NADH and flavin adenine dinucleotide ($FADH_2$) which are crucial to fuel the cell. This cycle occurs in the matrix of the mitochondria. The final stage occurs in the cristae of the mitochondria and requires oxygen. The high energy NADH and $FADH_2$ have their energy levels lowered in a series of reactions known as chemiosmosis to produce ATP. This is a key source of energy for all cells. Molecular oxygen is the final electron acceptor. According to Calow (1998), each molecule of glucose produces 36 molecules of ATP and is oxidized to carbon dioxide and water, as well as nitric and phosphoric acids which are recycled. The rate at which respiration occurs varies with environmental conditions, including oxygen availability and temperature; higher temperatures promote respiration.

Anaerobic respiration occurs when oxygen is absent. It is not as efficient as aerobic respiration as less ATP is produced overall. It begins in the same way as aerobic respiration with glycolysis in the cytoplasm to produce pyruvic acid. The citric acid cycle cannot take place because the presence of oxygen is necessary and so the pyruvic acid undergoes fermentation. This may be alcohol fermentation as occurs when yeast is grown in anaerobic conditions and ethanol is produced from pyruvic acid. Alternatively, pyruvic acid may be converted to lactic acid as occurs when oxygen is in short supply in the human body and anaerobic respiration takes place.

Some organisms have the ability to extract oxygen from inorganic substrates such as nitrates, sulphates, manganese compounds and iron compounds. This capacity is referred to as anaerobic metabolism rather than anaerobic respiration but it is a mechanism whereby some organisms obtain their energy. Some species of the archaea are methanogens (Section 3.2) which means that they breakdown organic matter in the absence of oxygen to obtain energy and in so doing produce methane.

The evolution of aerobic respiration was predated by the development of photosynthesis as this increased the concentration of carbon dioxide in the atmosphere (see Section 3.6.1). Indeed, respiration of some sort must have been characteristic of the earliest life forms because they required energy to survive and reproduce. The evolution of the eukaryotes also appears to have been dependent on earlier life forms such as the archaea and bacteria. As discussed in Section 3.4.1 cell organelles in bacteria, such as chloroplasts and mitochondria, became isolated as they developed membranes around them and subsequently the bacteria were incorporated into organisms to produce eukaryotes. This will be considered in Chapter 4.

3.6.3 Decomposition

When organisms die, the organic compounds of which they are made undergo the processes of decomposition. This is as complex a process as the accumulation of organic material through photosynthesis or metabolic activities such as respiration. The various components and processes have been examined by Aber and Melillo, 2001, and Smith and Smith, 2001. Decomposition occurs throughout the biosphere and comprises many chemical reactions. A host of different organisms may be involved as they break down the organic matter to obtain energy and nutrients. Organisms that rely on dead organic matter are consumers as they cannot manufacture their own food. Most fungi and many bacteria are decomposers and are sometimes referred to as microflora. Other organisms involved in decomposition include invertebrates such as earthworms and millipedes; depending on size these are classified as microfauna, mesofauna, macrofauna and megafauna. Such organisms are also known as detrivores because they feed on dead matter known as detritus. Those detrivores feeding on dead bacteria and fungi are the microbivores which include protozoa, e.g. amoebas, nematodes and some larval forms of beetles. The decomposition of organic matter is essential because it releases carbon compounds and other nutrients which can then be used by other organisms. The release of carbon, usually as carbon dioxide, means that decomposition is an essential process in the global carbon cycle (Figure 2.17).

Decomposition can occur in two ways: aerobic and anaerobic decomposition. The former occurs in the presence of oxygen while the latter occurs in an environment from which oxygen is absent and is also referred to as fermentation. Examples of anaerobic conditions include waterlogged sediments and peats. There are parallels with the various types of respiration described in Section 3.6.2. Some detrivores, notably bacteria, can operate in oxygen-rich and oxygen-poor environments as they have the capacity to switch depending on environmental conditions. These are called facultative anaerobes and are capable of using nitrates, sulphates etc. (see Section 3.6.2) within the dead matter to obtain oxygen. Some species of bacteria cannot survive in an oxygen-rich environment and are called obligate anaerobes. Generally, the sugars in dead matter are converted to organic acids and alcohols during anaerobic digestion. During aerobic decomposition carbon dioxide and water are produced. As discussed in Sections 3.3 and 3.4.2, neither bacteria nor fungi use food in the same way as animals. They secrete enzymes onto the dead material rather than ingesting it so the digestion process begins outside their bodies. The enzymes begin the breakdown process and the less complex compounds produced can then be absorbed. Different organisms utilize the dead material in different ways and at different stages until only inorganic compounds remain.

Most decomposition in the biosphere takes place in the soil. This medium provides a milieu for organisms and for the chemical alteration of its store of organic and inorganic matter which make it a vital component of many biogeochemical cycles as well as the carbon cycle. Organisms, e.g. the various faunal groups referred to above, contribute to the alteration of organic material by breaking it up into small fragments. For example, ants and beetles will fragment leaf matter while earthworms digest organic matter and pulverize it further. The resulting detritus is incorporated into the soil matrix along with sediment. It has a much larger surface area than the original matter and is thus more susceptible to further modification by physical/chemical processes and by bacteria, fungi etc. Physical processes include leaching whereby rainwater percolates through the soil taking with it any substances capable of being dissolved and it carries detritus physically down into the depths of the soil. Digestion by bacteria etc. alters the detritus further, breaking down complex compounds to simple ones in a process known as mineralization. Some of the resulting nutrients become incorporated into the detrivore biomass while a proportion resides in the soil which is thus a reservoir of nutrients. These can be utilized by plants as they begin to grow, and so are recycled.

Complete mineralization rarely occurs as almost all soils contain a proportion of organic matter derived from the decomposition of living organisms. In its amorphous form, this is called humus. Not only is this organic matter an important source of nutrients such as magnesium, calcium, etc. but it also has a vital role to play in determining the structure and stability of the soil and its water content. High mineral content and the capacity of humus to bind organic and inorganic components and to retain water gives rise to the most productive soils in terms of the flora and fauna that they support. In turn, a good plant cover provides protection from wind and water erosion.

Environmental conditions influence the nature and rates of decomposition. In terrestrial environments water availability and annual temperature ranges, for example, may promote or constrain decomposition. Waterlogging inhibits decomposition and further inhibition occurs if the water is acidic. This is why decomposition is very slow in peat and marsh environments. Where annual temperatures are low, as in Arctic and alpine environments where growing seasons are short, decomposition is inhibited as is the case where very dry and then very wet conditions occur. In contrast, warm and moist conditions all year round constitute optimum conditions. The type of organic matter also influences the rate of decay. Lignin, a group of complex organic molecules that make up wood, is particularly difficult and requires a substantial energy input to break down the many bonds that exist within the molecules, resulting in little net gain of energy for the microorganisms involved. At the other end of the spectrum are the sugars generated in photosynthesis which are small but energy-rich molecules. In between is cellulose, a major component of dead plant matter and the tannins which give the dark brown colour to water within dead organic matter. The proteins in dead animals also fall into this middle category while bone is less amenable to decomposition. In lakes and oceans particulate waste from aquatic organisms and dead organisms sink through the water body. Some of this material will be digested repeatedly by scavenger organisms, e.g. filter feeders such as sea cucumbers, sponges and barnacles consume organic particulates, and their waste products eventually accumulate on the lake or ocean floor. Various bacteria help to decompose this material and aerobic or anaerobic digestion occurs depending on whether or not oxygen is present, which in turn reflects water depth and frequency or permanency of inundation. Recycling occurs because the bacteria die and are consumed by filter feeders and other organisms in the zooplankton, i.e. small free-swimming invertebrate consumers such as copepods, krill and rotifers, and so the material of which they are made re-enters the food chains/webs (see below).

3.7 Food chains and food webs

No organisms live in isolation. There is an inter-dependence in the living world which has evolved just as organisms themselves have evolved. Moreover, the living world survives because it is inextricably linked with the non-living world (this is the ecosystem) and there are continuous exchanges between the two, not least the exchange of carbon (see carbon cycle in Figure 2.17). Organisms rely on each other for a variety of purposes. There are many reasons why such inter-dependence occurs, including the provision of shelter and various mechanisms to facilitate reproduction, such as insect pollination, but the most obvious is the reliance for food of non-photosynthesizing organisms on those which can photosynthesize; the former are the heterotrophs and the latter are the autotrophs. The relationships between these groups are referred to as trophic relationships and are described as food chains and food webs They also include the detrivores which are important in energy flow and in the recycling of carbon and other nutrients (Section 3.6.3).

Food chains and food webs are essentially hypothetical representations of a complex reality (see Calow, 1998; Smith and Smith, 2001). A food chain comprises a series of organisms between which food energy passes linearly from the base of the chain to the top. Food chains can be divided into two types: the grazing food chain and the detrital or decomposer food chain. The grazing food chains in agricultural systems are the most simple of those which occur in the biosphere because human intervention has striven to eliminate competitors and direct the end product to humans. Examples are given in Figure 3.11. The first stage involves autotrophic capture of solar energy, some of which is used by the primary producers themselves for their metabolic processes etc. Thus the amount of energy available for the consumers, the net primary productivity, is less than the original amount of energy captured and which is known as the gross primary productivity. In the first food chain given in Figure 3.11, the sheep consume a proportion of the net primary productivity generated by the pasture plants, the remainder is stored in the plants as biomass and/or enters the decomposer food chain thereby providing energy for bacteria, fungi etc. The sheep use some of the energy they consume, i.e. the gross secondary productivity, while a proportion of the remaining net secondary productivity is consumed by humans.

1. **PASTURE (GRASS) → SHEEP → HUMANS**
2. **MAIZE → CATTLE → HUMANS**
3. **CABBAGE/LETTUCE/WHEAT → HUMANS**
4. **GRASS → DAIRY CATTLE → MILK → HUMANS**

Figure 3.11. Examples of grazing food chains in agriculture.

Although much of the energy in these food chains is harnessed by humans, a proportion still enters the detrital food chain as organisms decompose leaves, stems etc. In fact in most ecosystems, the detrital pathway is dominant in terms of energy transfer. Smith and Smith (2001) state that in intertidal marshes, only c. 10 percent of the living plant material is consumed in the grazing food chain while the remainder is consumed by detrital feeders and decomposers. In contrast energy-transfer data for a plantation of Scots pine show that c. 50 percent of the energy fixed annually enters the detrital food chain while much of the rest is stored as woody biomass or removed as wood cuts. The detrital pathway also dominates in most aquatic ecosystems. In general, food chains are short and rarely comprise more than five stages. Not only is energy transferred between the organisms in a food chain but so too are nutrients. Consequently trophic relationships are components of biogeochemical cycles but it is important to note that the amount of energy transferred diminishes between trophic stages while the volume of nutrients transferred does not. This is because energy is used at each stage in respiration and is dissipated as heat into the atmosphere whereas nutrients, e.g. calcium, magnesium etc., may be shifted from repository to repository but are not lost to the system. Carbon is also a nutrient and is involved with energy transfer; its mass remains undiminished but the energy carried in organic compounds alters as it is transferred along a food chain, so there is less energy overall at the end of the food chain than was present at the start. Similarly, the number of organisms and the amount of biomass at the top of a food chain are much less than at the primary producer level.

In reality, food chains are rarely isolated in this linear fashion. They are more likely to be interlinked, especially with crosslinks involving the detrital organisms. Thus the resulting food webs are a more accurate representation of the trophic relationships in ecosystems and agricultural systems than food chains. There are many reasons why food chains and webs are a simplification of reality. First, some organisms may be herbivores in the early stages of their lives and then become carnivores, e.g. tadpoles and frogs. Some organisms are omnivores, meaning that they consume a wide variety of foods, and so can occupy various stages in a food chain/web. In addition, organisms primarily involved with decomposition e.g. certain types of bacteria and fungi, may be consumed directly by grazing organisms. These relationships illustrate the complexity of the real world. An example of a food web is given in Figure 3.12. This represents the energy flows in the Arctic environment of Bear Island, Spitzbergen, where Charles Elton, a British zoologist, collected data that pioneered the concept of trophic relationships.

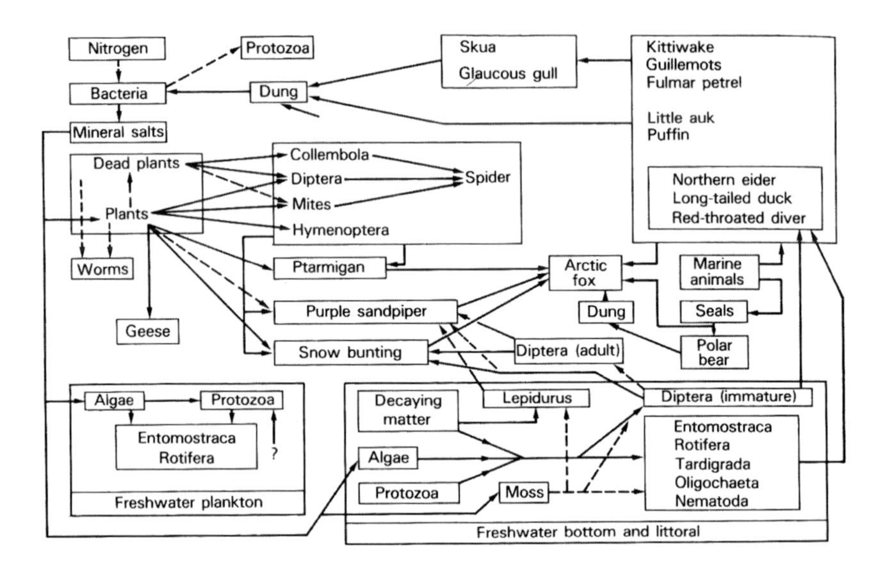

Figure 3.12. Elton's Spitzbergen food webs (based on Summerhayes and Elton, 1923)

This involves a grazing food chain comprising plants which are consumed by various birds which in turn are preyed upon by Arctic fox, but the plants are also consumed by a variety of other organisms including mites etc. which are also consumed by various birds. As Figure 3.12 shows, many additional trophic relationships also exist and all are interlinked.

The characteristics of food webs are important in determining the stability of an ecosystem, i.e. its resistance to change, which is linked to five factors (Calow, 1998). First is the diversity of species within the trophic levels and second is the degree of connection between the various food chains. Third is the length of food chains, fourth is the extent of omnivory between trophic levels, and fifth is the degree of compartmentalisation i.e. the presence of recognizable subdivisions within the food web. The more developed these characteristics are, the more resilient to change is the ecosystem. Some ecologists also point out that certain organisms are more important than others within an ecosystem. These are called keystone species without which the ecosystem would change completely. Such species are also index species because their population and its condition are surrogates for overall ecosystem health. Examples include the whale in marine ecosystems, the grizzly bears in the forests of North America, prairie dogs in the Great Plains of the USA, sea otters in coastal kelp forests, and corals in marine ecosystems. Trophic relationships are influenced by environmental characteristics, not least of which is the prevailing climate. Annual and diurnal temperature ranges and water availability are particularly important; so too are soil type and water quality as well as human impact. This is why ecosystems vary between latitude on a global basis and thus constitute one aspect of the geography of carbon, as discussed in Chapter 5.

Chapter 4

4 THE GEOLOGY OF CARBON

There is no doubt that carbon was abundant when the Earth was formed some 5 x 10^9 years ago but how it formed or why it was abundant remains obscure. Whilst it is difficult to imagine what environmental conditions on Earth at the time of origin were like, carbon in various combinations was present and especially as carbon dioxide in the atmosphere. Moreover, the Earth's physics and chemistry were constantly altering as the planet cooled and it was not until c. 3.9 x 10^9 years ago that a solid crust was in place. Generally, all the elements including carbon were present when planet Earth was first formed but they have been continually redistributed as erosion, deposition, volcanism and seismic activities have brought about change from its core to its surface and its atmosphere. Such geochemical exchanges, which characterized exchanges between inorganic components, were soon joined by biogeochemical exchanges as life emerged. Thus carbon, the most common element in life forms (see Chapter 3), became increasingly mobile. That carbon became the most important raw material for life is a function of its versatility, not least its capacity to join with other elements (Sections 1.1 and 1.2). Precisely when the first life forms evolved is not known, and may never be known for certain but recent research suggests that life was present by at least 3.8 x 10^9 years ago since when it has exerted a significant influence on the environment.

A major aspect of this influence has been the role of life in relation to atmospheric composition, especially the accumulation of oxygen and the proportion of heat-trapping gases which include carbon dioxide and methane. The latter in turn influence climate which, to complete the circle, influences life. This is the essence of the Gaia hypothesis, first formulated by James Lovelock in the 1970s (see Section 1.3 and Lovelock, 2004). Carbon distribution (as carbon compounds), life and climate have thus been intimately linked through most of geological time, not least in the maintenance of global temperatures which favour the continuance of life. This is an important aspect of the carbon biogeochemical cycle (Figure 2.17) though other factors have also been significant. Plate tectonics have influenced this relationship by altering the geography and topography of the continents and the course of evolution by separating or isolating groups of organisms, while astronomical factors such as the periodicity of Earth's orbit around the Sun have affected the amount of solar radiation received at the Earth's surface.

Overall, geological and biological processes have conspired to remove huge volumes of carbon from active circulation within and between the lithosphere, biosphere and atmosphere (see Figure 1.4). Consequently, the biogeochemical cycle of carbon has two temporal scales: one involves geological time and the other involves the recent/present period. Geologically, carbon has been sequestered in various repositories. The most important of these are the vast deposits of sedimentary rocks formed during various geological ages, the deposits of coal, mainly of the Carboniferous period c. 360 to 286 million years ago, oil and natural gas of varying geological ages and the more recent accumulations of peat in sub-arctic environments in the last 10,000 years. As discussed below, the conditions of formation were very different for all three. Sedimentary rocks form under a variety of conditions but especially in water in which material eroded from the land settles out and becomes compressed; chemical precipitation and/or the remains of marine organisms may contribute to such sediments. Oil and natural gas originated from high concentrations of marine micro-organisms which were buried and subject to pressure while coal is the product of land-based photosynthesis by plants in tropical conditions and comprises plant biomass which became incarcerated in subsiding sedimentary basins. Carbon accumulated in terrestrial plant biomass is also stored in the world's wetlands, especially in peatlands. The most extensive deposits occur in sub-Arctic to temperate regions, notably in Eurasia and North America, and have formed since the end of the last ice age to produce a significant store of carbon on the Earth's surface.

All of these deposits are important resources and their exploitation by humans is a major cause of carbon release to the atmosphere, i.e. humans are essentially accelerating the fluxes in the geological carbon cycle. This is especially so in the case of coal, oil and natural gas, the exploitation of which is closely linked to wealth generation and levels of development (see Mannion, 2000 and 2002). This is one of several aspects of the domestication of carbon. Moreover, this exploitation is giving rise to many environmental problems, not least of which is global climatic change.

4.1 Carbon, life and climate: the last 5×10^9 years

Planet Earth formed from the vapours produced during the big bang and it took a further c. 7×10^9 to 10×10^9 years before life emerged about 3.8×10^9 years ago. Although it is far from clear how this momentous development occurred, evidence suggests that organisms similar to bacteria were the earliest forms of life. Moreover, once life had originated a mutually reinforcing relationship developed between it and the environment. This central tenet of the Gaia hypothesis is illustrated by the subsequent and continued operation of biogeochemical cycles. A major way in which this influence has been manifested is through the alteration of atmospheric composition and as this has changed the course of evolution has itself been affected. The influence of life has resulted in the sequestration of carbon in the Earth's crust, as limestone, fossil fuels and peat to provide valuable resources, and a store of carbon in living organisms on the Earth's surface. This mutual relationship has also led to the emergence of *Homo sapiens,* a particularly powerful organism with unprecedented impact on the environment and other forms of life.

4.1.1 Prologue: the big bang to 3.9 x 10^9 years ago

The element carbon was created following the so-called 'big bang' between 12 x 10^9 and 15 x 10^9 years ago when the universe in which Earth and other planets came into existence was created. Indeed recent research by the National Aeronautics and Space Administration (NASA), based on a new space probe (the Wilkinson Microwave Anisotropy Probe – WMAP), has placed this event at 13.7 x 10^9 years ago with a margin of error of 1 percent (see NASA, 2004 and Seife, 2003). This expansion of space and matter caused the break up, indeed vaporization, of a dense mass of matter, into clouds of sub-atomic particles, i.e. protons, neutrons, electrons, positrons, photons and neutrinos, and their dispersion under extremely high temperatures and intense radioactivity. This was the origin of the expanding universe. As cooling ensued the relatively light elements of hydrogen (plus deuterium and tritium, its isotopes), helium and lithium were created as sub-atomic particles coalesced in a process known as nucleosynthesis. This occurred very soon after the big bang itself. The Sun and the planets emerged as clouds of particles cooled. The gases were attracted to the Sun and the larger planets of Jupiter and Saturn because of their superior gravitational pulls. Many stars were also produced in which these lighter elements coalesced via nuclear fusion to produce heavier elements which were then dispersed as the stars came to an end. The processes involved in the nuclear fusion by which carbon was created have been addressed by astrophysicists for decades (see El Eid, 2005, for a review). It involves the nucleus, an α-particle, of the element helium which combines with another α-particle to produce an unstable form of the element beryllium and then the capture of a third α-particle to produce ^{12}C. Although doubts exist as to the possibility of such a process occurring sufficiently rapidly to produce the abundant carbon found on Earth, new work by Fynbo *et al.,* (2005) indicates that within a certain temperature range the triple α reaction is fast enough to produce abundant ^{12}C. Other, heavier elements, e.g. oxygen, were then generated from ^{12}C through the addition of further α particles. Thus new stars and the smaller planets of Earth, Venus and Mars came into existence. The latter comprised the heavier elements such as carbon, nitrogen etc. In Lovelock's (1991) words "....the Earth began as a planet-sized chunk of radioactive fallout from a very large thermonuclear reaction". A later collision with another planet caused melting of Earth, with the detachment of molten material which became the Moon, and a reassembly of Earth's components with the heavier elements of iron and nickel at the core and lighter elements, such as aluminium, at the surface (Nisbet and Sleep, 2001). This occurred c. 4.5 x 10^9 years ago and caused the Earth to spin and tilt which led to day – night and seasonal cycles. Conditions during this early part of Earth's history cannot be deduced as no rocks from this period of molten substrates survive and therefore it is described as a pre-geological period. Bombardment by asteroids, planetesimals, aggregations of solid particles from the atmosphere, etc. was commonplace. There is little agreement about even the general conditions on the Earth during this period; cases have been made for an ice-house Earth, due to a weak luminosity from the Sun, and a greenhouse Earth with surface temperatures of c. 80 °C. By c. 3.9 x10^9 years ago, this so-called Hadean period (see Figure 4.1) of volcanic upheaval and general volatility was at an end. The major components of the Earth which are recognized today were in place, namely an inner core, a surface crust and an atmosphere.

4.1.2 The Archaean: 3.9 x 10^9 years to 2.5 x 10^9 years ago

Thus began the Archaean, the first true geological period (Figure 4.1). It is considered that the relative stability of the crust provided conditions in which life could evolve, though precisely when this occurred has not been ascertained. The chemical characteristics of rocks from this period indicate that the Earth's atmosphere was very different to that of today. An atmosphere dominated by carbon dioxide (from continued outgassing through volcanic emissions and constituting c. 30 percent as compared with c. 0.03 percent in today's enhanced greenhouse world), nitrogen and water vapour with traces of hydrogen and carbon monoxide (Wayne, 2003) would not have suited modern life forms and the absence of oxygen, except at trace levels, means that the atmosphere was reducing and not oxidising. Lovelock (1991) points out that there was a continuous loss of hydrogen to the atmosphere as water in the oceans reacted with carbon dioxide and the basalt rocks produced from the Earth's molten constituents. This resulted in the containment of oxygen in carbonates and ensured the relative absence of oxygen in the atmosphere. The accumulation of hydrogen at the ocean floor may have been a vital source of energy for the early Archaean organisms. These processes are shown in Figure 4.2.

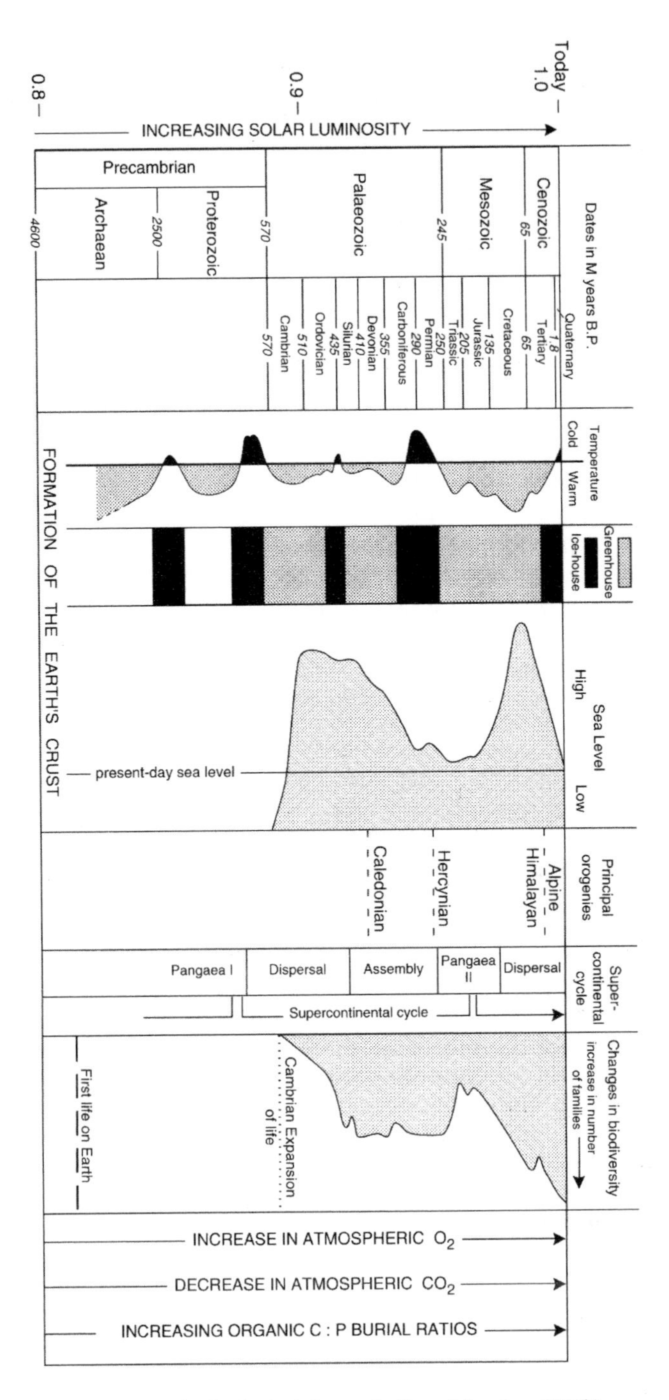

Figure 4.1. Geological timescale (from Mannion, 1997b)

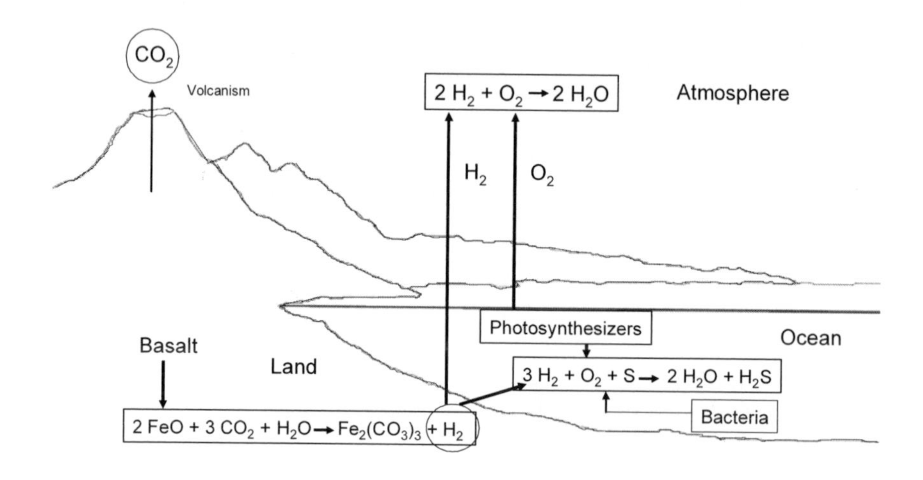

Figure 4.2. Early Earth chemistry (based on Lovelock, 1991)

The evolution of life curtailed the loss of water; otherwise Earth could have lost the moisture which became a positive factor for the maintenance of life and may distinguish the Earth from other planets. Other important processes occurring during this early period which are particularly relevant to the Earth's carbon history include the removal of carbon from active circulation by weathering processes to produce additional carbonates in the oceans, as reflected in the following equations:

$$CaSiO_3 + CO_2 \rightarrow CaCO_3 + SiO_2$$

$$MgSiO_3 + CO_2 \rightarrow MgCO_3 + SiO_2$$

When carbon dioxide dissolves in water, carbonic acid (as occurs in rain) is produced. It reacts with silicate rocks in the process of weathering to produce calcium and magnesium carbonates which are the components of limestones. According to Worsley *et al.* (1991), this process has been an important mechanism for the removal of carbon dioxide from the atmosphere throughout geological time, a process which has helped to prevent heating of the Earth's atmosphere. Moreover, even at this early stage in the Earth's history, life was beginning to emerge and it too began the process of sequestering carbon through its assimilation from the atmosphere as photosynthesis developed. Thus the advent of life, the second pillar of Gaia, became an early factor in Earth's evolution.

What were the earliest organisms and how did photosynthesis develop? These questions are fundamental in Earth history and both are inevitably the subject of a broad and controversial literature which can only be touched upon here. According to the review by Olson and Blankenship (2004) of the origins of photosynthesis, there is well-accepted evidence for the biological fixation of carbon from c. $3.8 \times 10^9 - 3.5 \times 10^9$ years ago. This relates to the ratio of two carbon isotopes (see Section 1.1), carbon-13 and carbon-12, in the organic carbon (kerogen) in sedimentary rocks; organically-fixed carbon is depleted in carbon-13 while inorganically-fixed carbon is not. However, there is no way of determining how this organic carbon was fixed. By 2.32×10^9 years ago there was a substantial increase in atmospheric oxygen (Bekker *et al.,* 2004), and at this time rocks containing iron are reddish in colour, rather than grey or blue, reflecting oxidation. Thus within the first c. 1.5×10^9 years of the Archaean, biological activity was substantially affecting the atmosphere. Indeed, the advent of oxygen was so important for life that Margulis and Sagan (2002) have described this as the 'Oxygen Holocaust'. Olson and Blankenship (2004) suggest that there were stromatolites, types of cyanobacteria, in existence by 2.8×10^9 years ago and that these organisms (see Section 3.3) contained chlorophyll and could photosynthesis. It must also be noted that there is also evidence for the presence of stromatolites about 3.5×10^9 years ago in some of the Earth's ancient rocks in Australia, South Africa and Greenland. For example, Van Krankendonk *et al.* (2003) report on bedded carbonates from the Warrawoona Group, Pilbara Craton, Western Australia, which are 3.45×10^9 years old. Their evidence supports a biogenic origin for the stromatolite carbonates, the presence of an anaerobic Archaean atmosphere and extensive erosion during this early phase of Earth history. However, Brasier *et al.* (2002 and 2004) have recently disputed the organic nature of the carbon at these sites, believing the characteristics of the contained carbon to be more in keeping with an inorganic origin. This dispute remains to be resolved though Tice and Lowe (2004) have reported carbonaceous material from the Buck Reef Chert in South Africa which they believe was produced organically by photosynthetic organisms. These organisms formed microbial mats in the ocean of c. 3.42×10^9 years ago. Colonies of these cyanobacteria, which formed reef-like structures, were amongst the earliest forms of life.

It is also likely that organisms similar to the archaea (Section 3.2), bacteria-like organisms but with distinct rRNA characteristics, were ancestors of the bacteria and were the first organisms to fix carbon biologically. Members of the archaea are today found in extreme environments, such as deep-ocean vents and hydrothermal environments, which would have been more widespread in the early days of planet Earth and they persist reducing conditions with no oxygen. The biochemical reactions developed by such organisms may eventually have led to photosynthesis. Olson and Blankenship (2004) reviewed this issue and state: "photosynthetic apparatus is best viewed as a mosaic made up of a number of substructures each with its own evolutionary history". In terms of the origin of reaction centres, two opinions prevail: the first involves energy capture in the earliest life forms and the second focuses on an origin in bacteria. Whatever the reality, it is clear that one of the most complex processes, as illustrated in Section 3.6.1, was in operation early in Earth history and was a major factor influencing the evolution of life and the environment.

4.1.3 The Proterozoic era: 2.5 x 10^9 to 544 x 10^6 years ago

As Figure 4.1 shows, the Proterozoic began and ended with Earth in an ice-house condition (see also the discussion by Tajika, 2003). The ice age toward the close of the era is known as the Varangian when it is possible that ice covered most of the Earth, not just the high latitudes, and that the oceans were frozen over (Hoffman and Schragg, 2000). The low luminosity of the Sun probably contributed to the low temperatures but the most important cause is attributed to a reduction of carbon dioxide in the atmosphere which curtailed the considerable greenhouse effect of this gas. This latter involves the trapping of energy which originates from the Sun but which is reradiated back into the atmosphere from the Earth's surface. (The enhanced greenhouse effect of modern times due to fossil-fuel use is the major cause of global warming and is examined in Chapter 5). Although atmospheric carbon dioxide concentrations were much higher than those of the present day, and have been estimated to be between 10 and 200 times higher (Kaufman and Xiao, 2003), any reduction would have impaired the greenhouse effect to cause cooling. Why should this have occurred? Or indeed why should it also have occurred in later geological periods such as the Palaeozoic and Cenozoic (see Figure 4.2)? The two major processes for removing carbon dioxide from the atmosphere are photosynthesis and the silicate-carbonate cycle. Photosynthesis (Section 3.6.1) was only just becoming a significant influence on atmospheric composition and although it now accounts for c. 20 percent of carbon dioxide removal (Worsley *et al.*, 1991) its influence in Gaia's early days was probably considerably less. Some carbon accumulation as a result of life form presence was occurring but an explanation for icehouse Earth is more likely to rest with the silicate-carbonate cycle (see equations in Section 4.1.2); too much drawdown of carbon dioxide would have impaired the greenhouse effect. A geochemical imbalance due to crustal readjustments and possibly some influence of emerging archaean and bacterial life forms accelerated weathering, as a biotic component became integrated into an essentially inorganic system.

Other characteristics of the Proterozoic include oceans with little oxygen and abundant sulphur which influenced the development and characteristics of early life forms. In addition, a supercontinent came into existence c. 1 x 10^9 years ago. This is known as Rodinia and involved the coalescence of three or four existing continents during the Grenville orogeny. By c. 700 x 10^6 years ago this supercontinent was beginning to break up to produce West Gondwana, East Gondwana and Laurasia. There was also a continued rise of oxygen due to increasing photosynthesis. This in turn reflects the origin of chloroplasts and the significant presence of life. This change in atmospheric composition may have caused the demise of many early life forms that thrived in a reducing atmosphere. When eukaryotes (see Section 3.4) evolved is enigmatic; they may have evolved as early as c. 2.7 x 10^9 years ago (see Brocks *et al.,* 1999) or as late as 1.8 x 10^9 years ago. Eukaryotes developed from a symbiotic relationship between once-independent prokaryotes (see Section 3.4) and are distinct because they have organelles (e.g. chloroplasts, a nucleus) contained within a membrane. DNA evolved as a characteristic of the nucleus and sexual reproduction developed. According to Martin *et al.* (2003) fungi may have been amongst the earliest eukaryotes to emerge; evidence for this comes from some fungal metabolic pathways and the mode of food acquisition involving osmosis (see Section 3.4.2).

There is chemical evidence for the evolution of animals c. 1 x 10^9 years ago, the earliest life forms such as stomatolites declined in importance by 0.7 x 10^9 years ago, and the first true fossils of animals are estimated at being 0.65 x 10^9 years old though animal evolution may date back to 1 x 10^9 years. At the end of the Proterozoic era (a subdivision of this era now called the Ediacaran rather than the Vendian), there is widely accepted evidence for the presence of new life forms. The life forms of the Proterozoic were marine and mostly soft-bodied quilt-like blobs or disks which occupied the seabed. This makes identification and classification difficult and the fossils of Ediacaran age have generated much controversy though they are generally considered to represent early metazoans, i.e. multi-cellular animals, possibly of early cnidarians (modern examples are corals, sea anemones, jelly fish) or sponges, or possibly of a completely different group (see Clowes, 2004, for a summary of the debate) sometimes referred to as the vendobionta. Fossils of this age are found on all continents. Other evidence of faunal life includes trace fossils described as ‘worm burrows’. The major shifts in climate and the emergence of heterotrophic animals probably influenced evolution in many ways, including the extinction of some life forms, though little can be stated with certainty. Some of the Earth’s earliest petroleum deposits accumulated during this era, reflecting the removal of carbon dioxide from the atmosphere. By the end of the Proterozoic yet more new life forms were emerging. These are known as the small shelly fauna and characterize the opening of the Cambrian.

4.1.4 The Palaeozoic era: 544 x 10^6 to 245 x 10^6 years ago

The geological history and record of life for this geological era is much better understood than the earlier eras which constitute Precambrian time. It comprises six periods beginning with the Cambrian and ending with the Permian, as shown in Figures 4.1 and 4.3. During the 300 x 10^6 years of the Palaeozoic, immense changes in the Earth's environment occurred and the evolution of life proceeded apace.

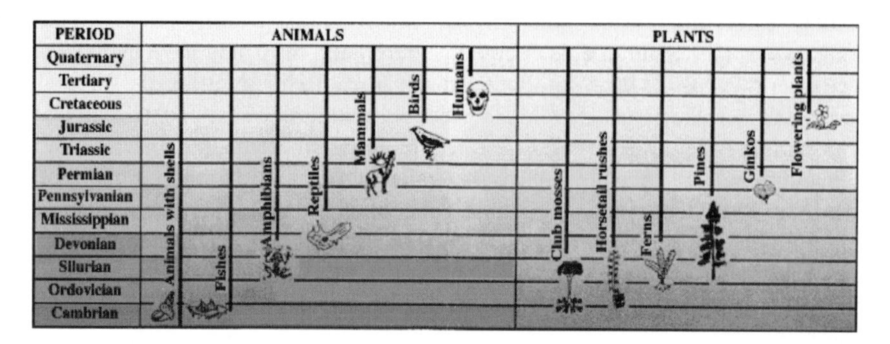

Figure 4.3. The stratigraphic ranges and origins of some major groups of animals and plants (from Edwards and Pojeta Jr., 1997; reproduced with permission from the United States Geological Survey).

A broad sketch of this is given in Figure 4.3. Despite or because of several significant extinction episodes, invertebrates, fish, reptiles and vascular plants came into existence and flora and fauna began to colonise the land surfaces as species emerged from the sea. There was a continued addition of oxygen to the atmosphere and the coalescence of continents, such as Gondwana, to produce the supercontinent Pangaea and climate changed from icehouse to greenhouse and back to icehouse. In the second half of the Palaeozoic, there was a major shift in the carbon biogeochemical cycle as vast stores of carbon were produced on the Earth's surface as tropical forests proliferated. This biomass was subsequently buried in sedimentary basins and eventually formed coal, effectively taking a huge store of carbon out of active circulation. These processes are examined in Section 4.3. The following summaries are based on information provided by the United States Geological Survey (2003), the Virtual Fossil Museum (2002) and Palaeos (2002); where possible specific authors are cited in the text.

4.1.4.1 The Cambrian period: 544 x 10^6 to 500 x 10^6 years ago

At the start of the Palaeozoic (Figure 4.1) the Earth's continents were scattered following the fracturing of the supercontinent Rodinia and metazoan life (the Vendobiota) had begun to emerge following the Varanger ice age of the Precambrian (Hoffman, 1999). Most of the landmasses were located in the southern hemisphere and most, apart from Gondwana, were submerged by shallow seas whose levels were continually changing. Climate appears to have been generally warm, wet and mild because the absence of landmasses at the poles was not conducive to ice cap formation. Consequently, the circulation of ocean currents had few barriers and so facilitated heat distribution globally.

What is quite remarkable about the Cambrian is the veritable 'explosion' of life that occurred. In its c. 45 x 10^6 years, a comparatively short time in comparison with the whole of geological history, many different major groups (phyla – see Section 3.1.1) of animal life had emerged. Amongst the most well known of these are the trilobites, brachiopods, molluscs and chordates. The trilobites, as illustrated in Figure 4.4, were primitive arthropods and thus the ancestors of the insects while the latter were the precursors of the vertebrates. Other life forms which proliferated were the sponges, echinoderms and gastropods. Fossils of this period are known from a variety of localities including sites in Wales after which the era is named. The most famous site is, however, the Burgess shale in British Columbia, Canada, where the fossils are very well preserved and which has been described by Gould (1989). What is not clear is why this explosion of life should have occurred at this time. Mild climate, abundant ecological niches, environmental change such as sea-level fluctuations and increasing oxygen in the atmosphere probably all played a part. Notwithstanding such Gaian innovations, there were also several periods of mass extinction though overall the diversity of life was considerably enhanced. For details of stratigraphy, fossils and sites see the International Sub-commission on Cambrian Stratigraphy (2003).

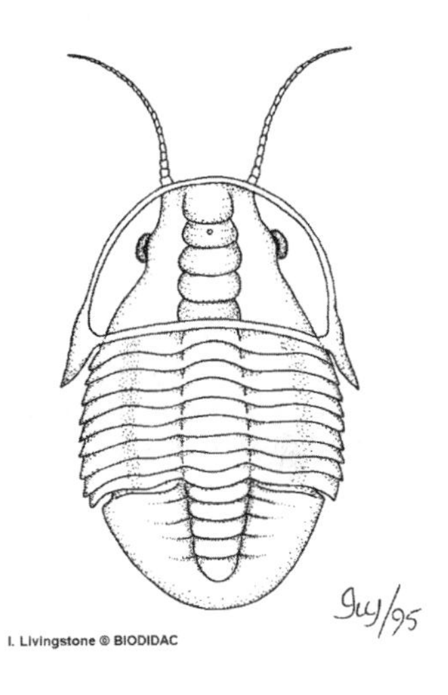

Figure 4.4. A trilobite, reproduced with permission from BIODIDAC. 2005.

4.1.4.2 The Ordovician period: 500 x 10^6 to 440 x 10^6 years ago

At the start of the Ordovician, Gondwana was a supercontinent; it comprised southern Europe, Africa, South America, Australasia and Antarctica and had drifted over the South Pole. This contributed to the development of a major ice age though unlike the Proterozoic Varanger ice age this did not encapsulate the Earth in ice but left the tropics with warm, mild conditions. Some mountain building occurred as land masses coalesced and terrestrial environments accrued at the expense of shallow marine environments until a further transgression in the mid-Ordovician which gave rise to widespread shallow seas. Marine environments and marine organisms continued to dominate though plant life, notably mosses and lichens, began to specialize and to move onto the land. This was a major threshold in evolution of life and its colonization of the Earth.

The adaptive radiation of marine organisms was remarkable; the number of families of marine invertebrates more than doubled. These heterotrophs must have been sustained by growing populations of autotrophs, notably cyanobacteria such as the stromatolites. According to Kazlev (2002a), there was a substantial and sudden increase in marine filter feeders which added to the existing variety of marine burrowers, detrivores and mud-dwellers. Other marine organisms which flourished at this time include the acrtiarchs (a type of algae), corals, bryozoa (colonial marine organisms), crinoids and brachiopods. Of particular interest are the colonial graptolites (hemichordates) fossils of which are indicative of sub-divisions within the Ordovician.

The Ordovician ended with a major extinction event which reduced the number of marine genera by c. (60 percent)

4.1.4.3 The Silurian period: 440 x 10^6 to 410 x 10^6 years ago

In geological terms this was a short-lived period but life was developing apace. Climatically, there was a shift to a greenhouse phase causing ice caps to melt and sea-levels to rise. Gondwana continued to drift south while the other continents converged around the equator. The Caledonian orogeny (mountain uplift) occurred c 430 x 10^6 years ago and created mountain belts in eastern North America and northwest Europe. There were widespread shallow seas in which coral reefs developed for the first time and carbonates were deposited (see Section 4.2). Evaporite deposits were also deposited in shallow basins.

According to Kazlev (2002b) the marine fauna recovered from the mass extinction event of the Ordovician. Brachiopods and graptolites were particularly abundant while trilobites were declining. Two important developments characterize the Silurian: the evolution of fish and the shift of life onto the land. The first fish to evolve were jawless while fish with jaws emerged later in the Silurian as did the first freshwater fish. The earliest fossils of vascular plants are found in the Silurian; although they were simple in structure, comprising mainly stems, they reflect the continued colonization of the land. There is also evidence for spiders and centipedes on land and the first fossil terrestrial animals are found in the mid Silurian. A recent find of a fossil millipede at Stonehaven, Aberdeenshire, Scotland is dated at 420 x 10^6 years ago (see Wilson and Anderson, 2004).

4.1.4.4 The Devonian period: 410 x 10^6 to 360 x 10^6 years ago

During the Devonian period two supercontinents, Gondwana and Eurasia, existed. They were moving close together (to collide during the

Permian, see Section 4.1.4.6) and were surrounded by ocean. Some mountain-building activity occurred. Known as the Arcadian orogeny, this produced the Appalachian Mountains of North America. Climate remained warm, wet and mild. Reef building by corals occurred in the oceans where life remained abundant and carbonates and evaporites were deposited.

The Devonian was a notable time of both plant and animal evolution. The small, simple plants of the Silurian diversified and radiated so that the two main lines of vascular plants, the zosterophylls and trimerophytes (both now extinct) were already in existence at the start of the Devonian. Many new families emerged and by the end of the Devonian ferns and horsetails as well as the first seed plants, the gymnosperms, were abundant. These plants formed the first forests and the proliferation of plant life encouraged diversification in the arthropods. In the oceans brachiopods continued to proliferate but the Devonian is also known as the age of the fish because of the many new species that evolved. These include the coelacanth and pteraspis; the latter, shown in Figure 4.5, was a jawless fish with a bony shield and scales. Many species of ammonoids also emerged. The first vertebrates appeared on land towards the end of the Devonian. These are known as tetrapods and are thought to have evolved from fish which developed limbs.

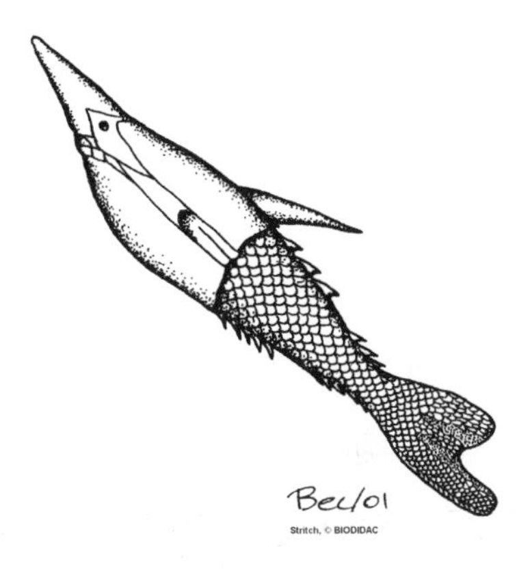

Figure 4.5. A pteraspis, reproduced with permission from BIODIDAC, 2005

At the end of the Devonian an extinction event occurred. This is known as the Frasnain – Fammenian event though it is not clear what caused it. Gondwanan glaciation and meteorite impact are possible, though unproven, causes. Marine organisms were especially badly affected; brachiopods, graptolites, trilobites and fish were severely reduced and coral reef building ceased due to extinction of many coral species. The many groups of terrestrial organisms were relatively unscathed.

4.1.4.5 The Carboniferous period: 360×10^6 to 286×10^6 years ago

The Carboniferous period can be divided into the lower and upper Carboniferous which are equivalent to the Mississippian and Pennsylvanian periods which are recognized in the USA. In the lower Carboniferous, east Gondwana began to drift towards the south pole but overall the Earth experienced a warm climate. In the upper Carboniferous the South American / North African margin of Gondwana collided with Euramerica to produce Laurasia, the forerunner of Pangea. There were fluctuating sea-levels and a differentiating global climate as Gondwana continued its migration south.

Tropical and sub-tropical climates persisted around the equator but vast ice sheets developed in polar regions. The continental collisions caused some mountain building, e.g. the Palaeozoic Alps in Europe and the Altai and Tian Shan in Asia.

Life once again flourished in the oceans; brachiopods, ammonoids, echinoderms, bryozoa and corals dominated marine faunas. The trilobites of the earlier Palaeozoic became rare as did many types of fish which originated in the Devonian as they were replaced by a range of species of sharks. Life continued to colonize and proliferate on land. For example, forests predominated in the equatorial regions of Euroamerica. Marsh and swamp forests thrived in lowland and coastal areas with abundant water. The groups involved include the pteridosperms (including tree ferns), sphenopsids (horsetails), lycopods and cordaitales (ancestors of the gymnosperms). The *Lepidodendron*, a species of lycopod, is shown in Figure 4.6. These swamp forests gave rise to the coal deposits, as discussed in Section 4.3. Pteridosperms also dominated the terrestrial environment in the colder areas of Gondwana. This wealth of plant life drew down much carbon dioxide from the atmosphere and enriched it with oxygen; the concentration of oxygen was higher during the Carboniferous than in any other geological period. Animal life also diversified, especially the arthropods and tetrapods. Winged insects appear for the first time while tetrapods were particularly abundant in shallow aquatic environments giving ready access to the land. The rhizodont fish emerged and was a main predator. By the end of the Carboniferous period the first reptiles appeared, reflecting the adaptation of animal life to a terrestrial rather than aquatic environment.

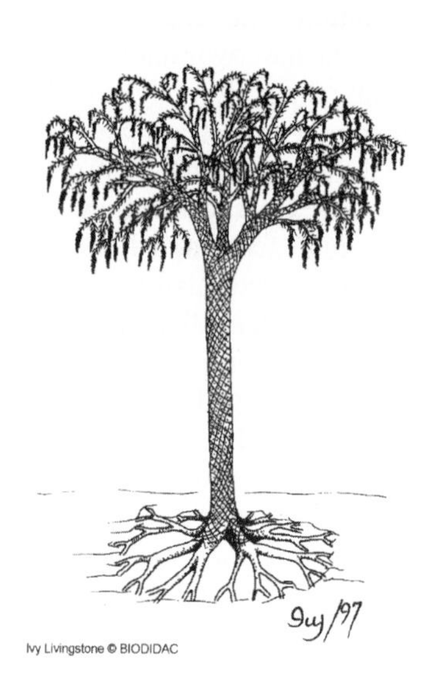

Figure 4.6. The *Lepidodendron*, reproduced with permission from BIODIDAC, 2005.

4.1.4.6. The Permian period: 286×10^6 to 245×10^6 years ago

Pangea, the super continent, was in existence as the Permian opened. Continental collisions produced mountain ranges and a diminution of shallow marine environments. Some shallow seas became isolated and dried up, leaving evaporates. Generally arid conditions, with marked wet and dry seasons, developed in the centre of the continent and in the area which eventually became Europe a shallow saline water body existed. Pangea was surrounded by an ocean known as the Panthalassic with the

Tethys Sea in the region that was to become south and central Europe. Climatically, global warming occurred overall, though there were oscillations of warm and cool episodes. The glaciers of the Upper Carboniferous retreated.

Life on land and in the ocean at the start of the Permian was similar to that of the Upper Carboniferous but substantive changes occurred as arid conditions developed. The lowland swamps and their lush vegetation disappeared leaving a terrestrial vegetation cover dominated by ferns and seed ferns. The lycopods etc. were replaced with conifers (gymnosperms) and ginkgos. The loss of such coastal environments and aridity caused a major decline in the tetrapods and only a few survived e.g. early crocodiles. In contrast, reptiles proliferated and became the dominant land animals. Three dynasties characterize the Permian: the pelycosaura of the early Permian, the dinocephalia of the middle Permian, which became extinct, and the therapsids which emerged towards the end of the Permian. These animals gave rise to the archosaurians of the Triassic period and were the ancestors of the mammals. Beetles and flies also emerged. There was another mass extinction event as the Permian, and the Palaeozoic, closed. Known as the Tatarian, this is considered to be the most severe of extinction events in Earth history; more than 90 percent of all marine life disappeared and c. 70 percent of terrestrial groups were lost with the greatest casualties in equatorial regions. It appears to have been a sudden and short-lived event of global significance. The cause of this turbulence is not known.

4.1.5 The Mesozoic Era: 245 x 10^6 to 65 x 10^6 years ago

The Mesozoic era lasted almost 200 x 10^6 years and is so called because its fossils attest to substantial changes in the Earth's fauna (see Figure 4.3). In contrast to the Proterozoic and Palaeozoic eras when most life forms were small, even microscopic, those of the Mesozoic period included many of the largest (and most popularized) animals, the dinosaurs as well as the emergence of birds. Some of these animals are shown in Figure 4.7.

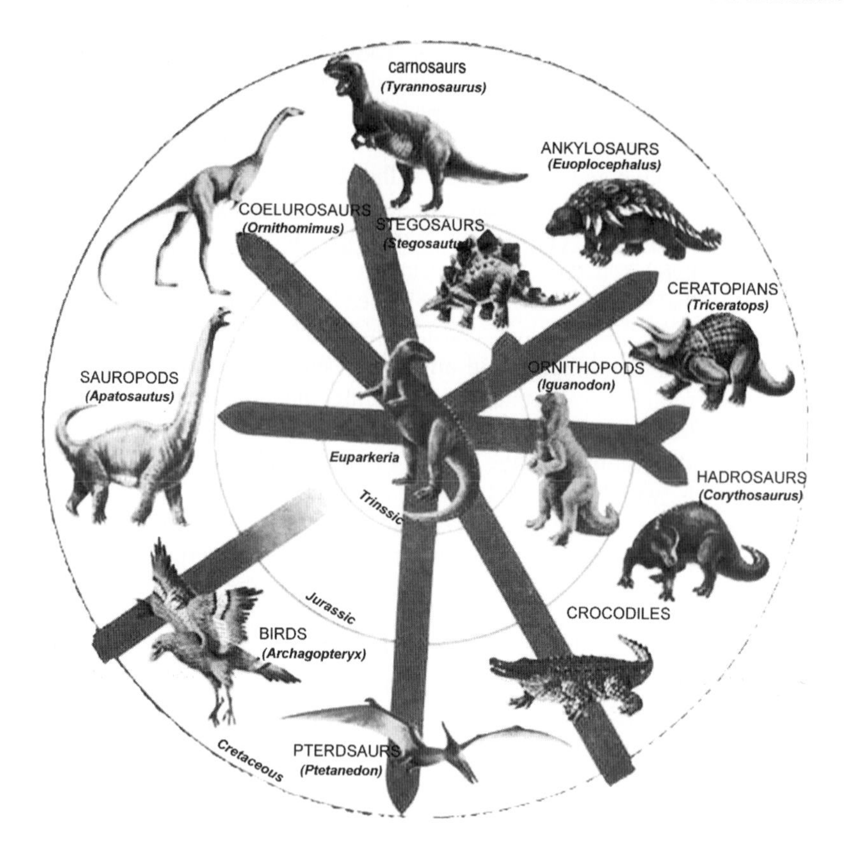

Figure 4.7. Animals of the Mesozoic era

The first mammals appeared in the later Mesozoic, evidence for which has been reviewed by Kemp (2005) who suggests that there is evidence for ten major groups by c. 230 x 10^6 years ago. Significant changes also occurred in the plant world as gymnosperms, which flourished in the early Mesozoic, were displaced to some extent by flowering plants, the angiosperms, which came into existence towards the end of the Mesozoic. Although life flourished overall, there were at least two extinction events. Structurally, the supercontinent of Pangaea, which formed during the Permian period, began to fracture and new continental units appeared. Marine organisms recovered from the Permian mass extinction episode and throughout the Mesozoic era accumulations of microscopic and other marine organisms were encapsulated within sedimentary sequences to generate some of the most important petroleum deposits. Moreover, there was a return of some animals to the sea; some reptiles became major predators in the marine environment. The following is based on Palaeos (2002) and the Museum of Palaeontology (2004).

4.1.5.1 The Triassic period: 245 x 10^6 to 208 x 10^6 years ago

During the 37 x 10^6 years of the Triassic period the Earth's geography, i.e. arrangement of the continents, climate and biota changed enormously as the Palaeozoic elements were superseded by those of the Mesozoic. Pangaea began to break up in the mid-Triassic period, altering the relationship between the continental landmasses and the oceans. While the old Pangaea had allowed plants and animals to migrate over vast areas, including from pole to pole, the new configuration, dominated by a northern landmass, Laurasia, and Gondwana in the south, restricted such interaction. Similarly, as North Africa began to separate from Europe, plants and animals became increasingly isolated. The major orogenic activity created a mountain range which extended from Alaska to Chile. Aridity associated with Pangaea (see Section 4.1.4.6) continued to persist but not everywhere, though overall climate was warm and dry and there is no evidence for glaciation. Coal deposits attest to warm but wet environments in some tropical regions and the biota of marine ecosystems, following the Permian extinction event, underwent a renewal, a process enhanced by the return of albeit limited shallow seas as Pangaea began to fracture.

Biotic changes during the Triassic were substantial. First, there was some recovery of organisms from the late Permian extinction. In the oceans ammonites recovered. On land forests were prolific and by the end of the Triassic gymnosperms, comprising the conifers, ginkgos and cycads, were dominant. Faunal changes were remarkable. Only a few reptile groups, previously so abundant, survived the Permian extinction event. Many that did survive were synapsids, mammal-like reptiles which were herbivores or carnivores. Two groups of synapsids developed: the dicynodonts and cynodonts. The former were approximately equivalent in size to cattle and were herbivores while the mammal-like cynodonts were varied in size and contained herbivores and carnivores. Eventually the diapsid reptiles came to prominence and thus the age of the dinosaurs opened. Pterosaurs which were flying vertebrates, ancestors of the birds, also evolved. An extinction event at the close of the Triassic caused the demise of many organisms, especially in the oceans from where all marine reptiles, except the ichthyosaurs, disappeared. The cause of the extinction event remains unknown but it was accompanied by volcanic activity.

4.1.5.2 The Jurassic period: 208 x 10^6 to 146 x 10^6 years ago

The Jurassic is the age of the dinosaurs, a megafauna which includes some of the largest animals ever known. Pangaea continued to break up though Laurasia and Gondwana remained linked allowing the dinosaurs and other animals to spread widely. In the late Jurassic Gondwana also began to fracture. There was some mountain building e.g. the Nevadan orogeny and the Rocky Mountain cordillera began to form. There is no evidence of glaciations and the climate was generally warm and wet which favoured the formation of lush forests of conifers, cycads and ferns. So diverse were the cycads that the Jurassic is also known as the 'Age of the cycads'. The bennettitales, shrub-like trees with a bulbous trunk were also abundant. The tree ferns (Cyatheaceae) grew to heights of c. 20 m. Numerous families of conifers had evolved, including many that are familiar today, e.g. the pines. Conditions were conducive to the evolution of many new plants and animals. The dinosaurs developed from the sauropods; some were carnivores, e.g. allosaurus, a theropod, while others were herbivores e.g. stegosaurus. Brachiosaurus and diplodocus were amongst the largest dinosaurs. Land reptiles had developed feathered limbs which allowed them to fly and thus birds were also characteristic of the Jurassic, e.g. archaeopteryx. The tetrapods diminished in number as they were replaced by familiar reptiles such as frogs and newts. Mammals existed as microinvertebrates with burrowing or climbing habits.

Life abounded in the spreading shallow seas in which limestones were being formed. The fossils in the rocks attest to the presence and abundance of the Ichthyosaurs which were among the largest marine animals along with sharks and rays. Cephalopods were widespread e.g. ammonites. Planktonic organisms, especially foraminifera and amoebae, were also abundant and their incarcerated remains created many oil deposits, including those of the North Sea. Gastropods, sponges, bryozoans, belemnites, brachiopods and corals were also abundant. The latter, especially the stone corals, were involved in extensive reef formation.

4.1.5.3 The Cretaceous period: 146×10^6 to 65×10^6 years ago

During the Cretaceous period, tectonic and biotic change culminated in a world that is familiar today. Pangaea continued to break up, isolating Laurasia from Gondwana; this influenced the course of evolution as regional differences in flora and fauna began to emerge, especially in land-based species and provided a greater extent of littoral habitats. Tectonic activity was pronounced in the mid-Cretaceous, leading to the formation of the Sierra Nevada (USA), the continued rise of the Rockies and the Alps. Sea-level was high at the start of the Cretaceous when climate was warm with little regional differentiation. No large ice sheets existed. Cooling occurred subsequently, marked regional climatic variation developed and sea-levels decreased.

Many of the dinosaur species of the Jurassic continued to flourish and new ones emerged. These included the first ceratopsian and pachycepalosarid dinosaurs; amongst these were the giant Tyrannosaurus and Gigantosaurus which were both carnivores. Recent work by Erickson *et al.* (2004) on the life history of tyrannosaurid dinosaurs suggests that *Tyrannosaurus rex* in maturity was one of the largest animals ever to evolve.

Mature adults weighed 5,000 kg or more, a weight reached with a maximal growth rate of 2.1 kg per day over c. 20 years; the life span of the animal was 28 years. Recent fossil finds from Liaoning in northeastern China have revealed the presence of carnivorous mammals and scotched the impression of a Cretaceous period in which only small rodent-sized mammals persisted. These discoveries, reported by Hu *et al.* (2005) comprise two almost complete skeletons of two mammals, *Repenomamus robustus* and *R.giganticus*. The former was c. 50 cm long and probably weighed 4-6 kg; the fossil contains the remains of a *Psittacosaurus* (a type of dinosaur) indicating that it was a carnivore. The latter was c. 1 m long and probably weighed c. 12-14 kg, rather like a medium-sized dog. The fossils are thought to be 128 x 10^6 years old. The first fossils of many insect groups, modern mammals, birds and flowering plants (angiosperms or anthophyta) are also found in Cretaceous rocks. Amongst the mammals were the first placental animals which characterize the early Cretaceous; mammals were flourishing overall. Angiosperms emerged and diversified, rapidly colonizing a wide range of habitats. They emerged in the early Cretaceous and radiated in the mid-Cretaceous. Forests proliferated and contained many family types that are familiar today, especially conifers. The cycads suffered competition from the angiosperms and many died out though it was the end of the Cretaceous before angiosperms colonized high latitudes in the southern hemisphere.

The end of the Cretaceous period is marked by a major extinction event. Though not as great as the extinction event at the end of the end of the Triassic period, as many as fifty percent of the Earth's species became extinct. Nevertheless it caused the demise of the dinosaurs and was the end of the age of reptiles. Why this event occurred is contentious but there may have been an asteroid impact which created the Chicxulub crater off the Yucatan peninsula, Mexico. The pterosaurs, many marine reptiles, some groups of foraminifera and ammonites also became extinct.

4.1.6 The Cenozoic (Cainozoic) era: 65 x 10^6 years ago to present

The Cenozoic era is sometimes known as the age of mammals because they spread throughout the Earth following the Cretaceous extinction event and many new species evolved. Kemp (2005) notes that after the Cretaceous/Tertiary boundary the number of mammal species in North America alone rose from c. 15 to more than 60. However, many groups of species e.g. the angiosperms, insects and birds, underwent evolution and radiation (see Figure 4.3). Not least of these developments is the evolution of the ancestors of modern humans. The latter themselves emerged c. 250,000 years ago. As this book illustrates, humans have harnessed carbon in all its forms and in so doing have transformed the life and environment of the Earth. Some consider that they are responsible for an extinction event as great as those of earlier geological periods e.g. Leakey and Lewin (1996).

By the time the Cenozoic opened, the configuration of the continents was as it is today. Consequently, regional flora and fauna were developing through isolation. The Cenozoic has two subdivisions: the much longer Tertiary period and the short Quaternary period. There was global cooling throughout which culminated in the substantial ice advances of the Quaternary period. These brought about major environmental and biotic changes worldwide at rapid rates in geological terms.

4.1.6.1 The Tertiary period: 65 x 10^6 to 1.8 x 10^6 years ago

According to the Museum of Palaeontology (2004) uplift continued to occur, especially in the Himalayas, Alps and southwest USA. For a while the Mediterranean Sea was sealed at the western end and much of it dried up. In east Africa several rift valleys formed and in the southern hemisphere Australia collided with Asia and New Guinea was formed. A land bridge, the Panama isthmus, was created between North and South America. There were periods of climatic variability though in general climate became drier and cooler than during the Cretaceous period. The world's great deserts were formed and toward the end of the Tertiary period ice was once again beginning to accumulate at the poles.

Many biotic changes were occurring. The mammals which emerged in the Mesozoic began to colonize the Earth, taking over from the dinosaurs and other reptiles (see Kemp, 2005 for a review). Four groups of mammals emerged, diversified and dispersed rapidly. These were the placental mammals (eutheria), marsupials, monotremes and the multituberculata. The placental mammals carry their young internally until birth, marsupials (e.g. kangaroo) carry their young in pouches until they can live an independent existence while monotremes (e.g. duck billed platypus) reproduce by laying eggs. A forth group has no living relatives; this is the multituberculata, so called because of cusps on their teeth known as tubercles. They were like small to medium sized rodents-like animals but are only known from the fossil record. All mammals are warm blooded, produce milk and have hair. Included in the placental mammals are the hominid ancestors of modern humans which diverged from the apes c. 5 x 10^6 years ago. Birds (Aves) also became increasingly diverse and abundant; descended from the dinosaurs, there are today c. 9,000 species worldwide. The dominant aquatic vertebrates were the ray-finned fish (actinopterigians) characterized by webs of skin over bony spines; they occupied all aquatic habitats, including the deep ocean.

There were major changes in the diversity, radiation and abundance of the angiosperms. All angiosperms reproduce via organs in the flower and they appeared suddenly c. 100 x 10^6 years ago. Their ancestry is unclear but they evolved and occupied a wide range of ecological niches rapidly in geological terms. It is not known if the first angiosperms were woody or herb species. They are classified into two groups depending on how many leaves are present around the embryo (the seed): the monocotyledons (this group is dominated by the grasses) have one leaf and the dicotyledons have two leaves. The latter are thought to have evolved first.

4.1.6.2 The Quaternary period: 1.8 x 10^6 years ago to present

Before the Tertiary period ended, substantial volumes of ice had accumulated around the poles and another period of ice-house Earth had begun by c. 3 x 10^6 years ago. The oscillation of warm and cold stages intensified during the Quaternary period (see Mannion, 1997a and Wilson *et al.,* 2000, for reviews). Evidence for this derives from many sources including sediments of the ocean beds, ice cores from ice caps and glaciers, terrestrial deposits from glaciers, cave deposits, peats, corals, and fossil soils. Age estimates using radiometric and relative techniques provide detailed time scales for Quaternary events. Each of the many cold stages lasted c. 100,000 years with an intervening warm stage (interglacial) of c. 20,000 years. During the former, ice sheets and glaciers extended far beyond their present borders to freeze high latitudes and creating extreme climatic conditions for life in non-glaciated areas. Tropical regions remained warm but the Earth overall experienced a drop in temperature of c. 10°C. All types of ecosystems were affected; reassembly of species occurred regularly as climate shifted from cold to warm and so the characteristics of the Earth's biomes were as dynamic as the climate. Sea levels also oscillated. High sea levels restrict biotic mixing while low sea levels provide land bridges for migrating flora and fauna. During the relatively short warm stages (interglacials; see Sections 1.4 and 2.3) ecosystem configurations similar to those of today were present but the faunal components were often quite different. For example the fauna of the last interglacial (the Eemian) in temperate Europe included many species which are now extinct e.g. straight-tusked elephant and woolly rhinoceros, as well as species that no longer occur in this region e.g. spotted hyaena and hippopotamus. Many extinctions of animal species also occurred against the backdrop of climatic change at the end of the last ice advance e.g. the Great Irish Elk and the mammoth. What role humans played in this has not been ascertained and there are polarized views focussing on climatic change and hominid hunting respectively.

Modern humans (*Homo sapiens)* emerged by at least c. 250,000 years ago, and possibly 400,000 years ago, in east-central Africa. Their ancestors, *Homo erectus,* had already expanded beyond their centre of origin to reach Europe and Asia and had begun the process of domesticating carbon by harnessing fire. Subsequently modern humans arose in Africa and proved to be particularly successful. Not only did they colonize all continents except Antarctica but they have also exerted immense influence over almost all other forms of life, including the extinction of many species. In harnessing the Earth's living resources for food, shelter, decoration, medicine etc., in the process of domesticating carbon, large scale environmental change has been brought about. Most of this has occurred in the last 10,000 years, in a geological epoch known as the Holocene (see Mackay *et al.,* 2003 for a review) during which climatic conditions similar to the present day developed. The Holocene can also be described as an 'anthropogenic' epoch, or 'anthropocene' because of the rise to dominance of humans and their increasing influence over the biosphere. Civilizations arose and declined, leaving behind an amazing archaeological record; cultural advances were made, ranging from the art of writing to the more sophisticated use of computers and information technology, and humans began to broaden their horizons through the exploration of space. This was a time, as this book reflects, when agriculture, industry and fossil-fuel use were initiated, urbanization began, plants and animals were introduced from one region to another, warfare left its mark on landscapes, a human imprint on global climate became manifest, the mechanism of heredity and genetic manipulation was ascertained and the environment became a significant political issue (see review in Mannion, 2002), mostly due to the domestication of carbon.

4.2 Sedimentary rocks

Most sedimentary rocks contain organic matter which is carbon that has been removed from the active carbon biogeochemical cycle and placed in the geological carbon cycle, as discussed by Berner (2003). The most common types of sedimentary rocks are listed in Table 4.1. The fossil fuels discussed in Sections 4.3 and 4.4 are also classified as sedimentary rocks but they comprise concentrated carbon derived from accumulations of plant or micro-organism biomass. Sedimentary rocks are formed through the action of water, wind or ice which provide the means of erosion and transport for fragments created by the weathering of rocks. Once deposited consolidation occurs, possibly with some addition at the place of deposition (e.g. the inclusion of the shells of marine organisms at the ocean bed or detrital organic matter settling out in a water body), and as pressure builds up the layers harden. This is lithification. Further pressure and/or heat may cause deformation of the horizontally-deposited layers. Rocks formed in this way are called clastic sediments. Sediments formed when minerals dissolved in water precipitate out are chemical sediments, as seen in Table 4.1.

Table 4.1. Classification of sedimentary rocks

CLASTIC SEDIMENTS	**CHARACTERISTICS**
Shales	Small particle size: <0.002 mm
Siltstones	Small particle size: 0.002 to 0.062 mm
Mudstones	Intermediate particle size: 0.062 to 0.05 mm
Sandstones	Large particle size: 0.05 to 2.00 mm
Conglomerates	Extra-large particle size: 2.00 to >80 mm
CHEMICAL SEDIMENTS	**CHARACTERISTICS**
Limestones	Calcium carbonate: organic and inorganic origin
Dolomites	Calcium/magnesium carbonates
Evaporites	Various salts e.g. gypsum, halite: inorganic
Hydrothermal deposits	Calcium carbonate; inorganic

Sedimentary rocks are widespread on the Earth's surface as shown in Figure 4.3. Apart from the fossil-fuel deposits, those most closely associated with the removal of carbon from the active biosphere are the carbonate rocks. As discussed in Section 4.1.2 the carbonate-silicon cycle has been important throughout geological time by removing carbon dioxide from the atmosphere and generating carbonate rocks such as limestone and dolomite. They may be formed inorganically, as in the carbonate-silicate cycle, or by both inorganic precipitation and the accumulation of organically produced calcium carbonate. Both forms of calcium carbonate are known as calcite. The organically-produced calcite is produced by marine invertebrates, such as ammonites (see Section 4.1.5.2) and foraminifera, which absorb calcium carbonate from the surrounding water to produce shells and other skeletal parts. Thus some limestones are fossiliferous whilst others are not. In both cases eroded rock fragments or fossil parts provide a matrix and then consolidation occurs as the interstitial spaces are filled up with a 'cement' such as a calcareous mud. Chalk is a form of limestone which formed from marine micro-organisms and which is fine grained and purer chemically than many limestones. Marble is a very hard form of calcite which has been altered by heat and pressure and is a metamorphic rock.

Most limestones formed between 500 x 10^6 and 65 x 10^6 years ago, between the Silurian and the Cretaceous periods (see Figure 4.1), especially during the Carboniferous and Jurassic periods when warm seas were abundant and marine life forms were especially diverse. Limestone deposition characterized the Lower Carboniferous while coal formation characterized the Upper Carboniferous (see Section 4.1.4.5). The Carboniferous limestones are dominated by crinoid remains. These are echinoderms (e.g. sea urchins and star fish) which flourished in colonies of fixed and mobile species that were very extensive in the shallow seas. It formed as shelves, reefs and in basins. In the shallow Jurassic seas marine organisms, e.g. brachiopods and corals, were abundant. The most famous fossil of all, that of archaeopteryx which is considered to represent the link between the dinosaurs and birds, comes from an area of Jurassic limestone in Solnhofen in southern Germany.

Limestones give rise to distinctive landscapes, as in Southern China, Northern Vietnam, parts of the Mediterranean Basin, parts of the UK, Mexico and the central USA. Sink holes (dolines), limestone pavements, cave complexes, hay stack hills and underground drainage systems are characteristic of such areas. Where these features are well developed, due to extensive erosion and solution of the bedrock, the landscape is known as karst.

4.3 Coal

Coal comprises the remains of land plants which once grew in conditions approximately equivalent to today's tropical rain forests where there is abundant surface water. The waterlogged conditions reduce rates of decomposition by inhibiting the activity of decomposers/detrivores, as in wetlands and especially peatlands, today (see Section 4.5). The plant species involved in coal formation vary with the geological period in which the deposits were formed. Following the evolution of higher plants in the Devonian period (see Section 4.1.4.4) there was an explosion of plant life in the Carboniferous period (see Section 4.1.4.5) and vast swamps developed under warm climatic conditions. The dominant plants included many different ferns including tree ferns, climbing ferns and epiphytic species (these grow attached on other plants) and lycopsid trees such as *Lepidodendron* (see Figure 4.6) and *Sigillaria* which could grow to c. 40 m and 30 m respectively (Kerp, 2001). Despite their height these plants contained little woody material. Other Carboniferous species, e.g. cycads and cordaites (see Section 4.1.4.5), are also represented in coal deposits of this age. Of all the geological periods, the Carboniferous was a time of the greatest carbon accumulation in the lithosphere (see Berner, 2003, for details). Later coal deposits from the Jurassic onward contain the remains of flowering plants and include species characterized by abundant woody tissue. In the USA, for example, coal deposits are of Carboniferous, Cretaceous and Tertiary age. Coal of the Carboniferous is present in the Appalachian region, Cretaceous coal occurs in the western Rockies and is the most abundant while coal of tertiary age is also founding the Rockies and along the Gulf coast.

The geology of coal has been examined by Thomas (2002) who points out that coal deposits are rarely uniform in composition. Most reflect changing water levels as well as changing sedimentation regimes and topography. Many are associated with so-called cyclothem deposits which involve the accumulation of organic matter as thick mats of peat, its burial by marine or fluvial (river derived) sediment deposition as water levels rose or subsidence occurred, subsequent retreat of water and recommencement of vegetation growth. Where marine incursions occurred the accumulated organic material became saturated with sea water which included sulphates. Such inundation may have occurred several times as sea levels oscillated or as subsidence of the land occurred. Hence, many coals are rich in sulphur; when burnt, the sulphur is released as sulphur dioxide and as this combines with water it produces rain with enhanced acidity. This 'acid rain' has an adverse effect on aquatic ecosystems, forests and urban fabrics (see Mannion, 1999 and Kemp, 2004 for reviews). The sediments separating the organic beds were themselves subject to the pressure of accumulation and developed into shales, sandstones and limestones depending on the environment of deposition. Peat formation itself involves physical and chemical alterations to the organic remains as some degree of homogenization occurs but the material remains relatively soft and actual plant parts (and spores) can still be recognized. Sometimes large trunks of trees are preserved. The final stage in coal formation is coalification which involves the transformation of peat into coal. The volume of organic material is reduced by at least a factor of three and possibly as high as ten and further physical and chemical changes take place within the buried organics as pressure and temperature increase to expel water and cause solidification. These processes have resulted in coals of different types, as shown in Table 4.2.

The classification is based on the degree of transformation which has taken place since the coal deposits were formed, for example lignite is a low rank coal which is described by the World Coal Institute (WCI, 2002) as 'coal with low organic maturity'. This means that it has less carbon and more moisture than higher ranking coals such as anthracite. The latter is the densest coal, has the highest carbon content of all coals and least volatile matter. It has the highest caloric value and because it produces no smoke when burnt it is thus favoured in areas classified as smokeless zones.

Table 4.2. The different types of coal (based on World Coal Institute, 2002)

LOW RANK	**CHARACTERISTICS**	**MAJOR PRODUCERS**
Lignite (brown coal)	Relatively low carbon content c. <70 percent	Germany, Russia, USA, China
Sub-bituminous	Harder than lignite, 10-20 percent moisture, 71-77 percent carbon	Widespread
HIGH RANK		
Bituminous	Coking and thermal coals, harder than sub-bituminous coals	Widespread
Anthracite	High carbon content c. 90 percent, low moisture content, hardest coal type with semi-metallic lustre	South Africa, China

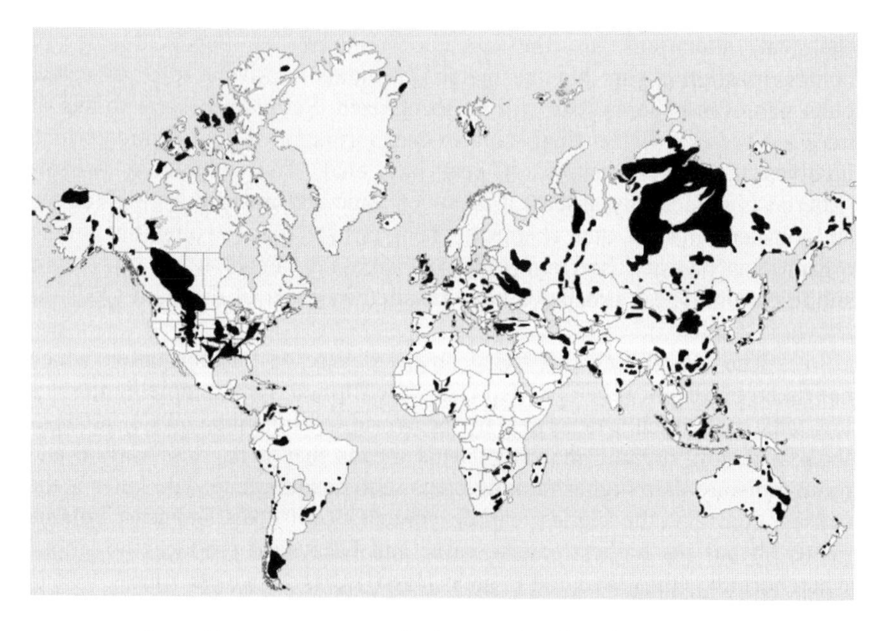

Figure 4.8. Map of coal deposits (based on Landis and Weaver, 1993)

As Figure 4.8 shows, the most abundant coal reserves occur in the USA, the Russian Federation and China. Germany and Poland have the greatest reserves in Europe. Reserves are much less extensive in the

Southern Hemisphere where South Africa and Australia are the major producers. The use of coal in industry underpinned the Industrial Revolution of the eighteenth century. Its widespread use from this period represents a major stage in the domestication of carbon and began an intense relationship between society, development and fossil-fuel energy. This relationship continues though oil and natural gas are now the favoured fuels (see Section 4.4). The mining and use of coal generate environmental problems such as the disposal of waste and 'acid rain', as well as disruption of the carbon cycle through the accumulation of carbon dioxide in the atmosphere. Society has thus reversed Nature's tendency to remove carbon from the atmosphere and in so doing has now to face the prospect of global climatic change.

4.4 Oil (petroleum) and natural gas

Oil and natural gas are the preferred fuels of the twenty-first century. They gained importance in the 1950s and have since overtaken coal on the grounds that they are cleaner and more efficient fuels and because oil has many uses other than as a fuel. These include lubricants and the production of organic chemicals such as solvents, and fibres such as nylon and plastics.

Both oil and natural gas are considered to be the products of micro-organisms which have been buried and compressed. Often, both are found together in the Earth's crust and natural gas may also occur with coal. Chemically, oil is more complex than natural gas and comprises a huge range of hydrocarbon compounds. It is a thick viscous liquid which varies in colour from black to dark yellow. When organic matter is incorporated into sediments its structure is altered through physical and chemical alterations. This material is known as kerogen (see Section 4.1.2). In abundance this can become oil and/or natural gas. The organisms identified as oil- and gas-forming are microscopic marine plankton (e.g. dinoflagellates) whose remains accumulate on the sea floor, especially along continental shelves. Subsequent burial by inorganic sediments begins the process of oil/gas formation. Head *et al.* (2003) state that most of the Earth's oil was produced by biodegradation under anaerobic conditions, possibly at temperatures up to 80°C. According to Craig *et al.* (2001) three major processes occur between the initial accumulation of organisms and oil/gas formation. These are diagenesis which occurs as the first 200 m of sediment accumulate and some oxidation of the organic matter takes place by anaerobic bacteria to produce gas, usually methane, which begins to percolate to the surface. This is a similar process to that which occurs today in swamps and marshes. Catagenesis is the next stage as the sediments become more deeply buried and are subject to increasing pressure and heat. Compaction takes place, drying occurs as water is squeezed out and the organic matter is increasingly converted to kerogen and to liquid petroleum. Biogenic gas production decreases because there are few, if any, organisms at such depths and pressures, but thermogenic gas production occurs as high temperatures cause some of the kerogen to break down. At even greater depths and pressures the process of metagenesis begins. This involves the production of a gas which is almost entirely methane. With further burial and heat graphite, an allotrope of carbon (see Section 1.1), is formed.

These processes have taken place at different geological times and under different environmental conditions. Oil/gas deposits derived from organisms in freshwater environments exist but are rare while oil/gas formed from marine organisms in a low-oxygen environment are most common and those formed from terrestrial plant matter give rise to coal (see Section 4.3) and gas. In addition, the content of natural gas may vary. According to the Natural Gas Organization (2004) natural gas deposits are of two types: wet gas and dry gas. The former is c. 70 to 90 percent methane with the remainder comprising hydrocarbons such as ethane, propane, and butane (see Section 2.2.1) plus small amounts of carbon dioxide, oxygen, nitrogen and hydrogen sulphide and traces of rare gases such as helium. Dry gas is almost entirely methane. Natural gas will tend to rise towards the Earth's surface but if it becomes trapped within a porous rock type, e.g. limestone or shale, over which a non-porous rock develops, or if faulting occurs to prevent the escape of gas/oil, a reservoir of gas will accumulate. Oil will accumulate in a similar way and both oil and gas are thus under considerable pressure when underground.

Oil deposits also vary in character. The Environmental Protection Agency of the USA (EPA, 2004) describes three types based on composition: paraffin oils, asphalt oils (dominated by naphthenes) and oil comprising a combination of the two. Paraffins are hydrocarbons and the molecules in petroleum (c. 10 different types) contain 10-16 carbon atoms. In contrast, naphthenes are aromatic compounds comprising carbon rings (see Section 2.2). All oil deposits contain elements other than carbon, including nitrogen and oxygen. Crude oil contains 84 to 87 percent carbon and between 0.1 to 5.0 percent sulphur; the more sulphur that is present the more 'acid rain' is produced on combustion. Overall, however, oil and natural gas contain much less sulphur than coal and are thus 'cleaner' fuels. The crude oil extracted from the Earth's crust must be refined before it is of use. This involves heating to c. 500°C; as heating progresses the components separate out as each has a different boiling point and can be retrieved using a refining tower; the least dense fraction floats to the top of the tower whilst the most dense fraction remains at the base (see Chevron, 2004 for a description). Table 4.3 gives the various fractions which reflect the hydrocarbons they contain, notably the number of carbon atoms which are present (see Section 2.2.1).

Table 4.3. The components of crude oil (based on Craig *et al.*, 2001)

COMPONENT	TYPE OF HYDROCARBON No. of carbon atoms	%
Gasolene	C_4 to C_{10}	27
Kerosene	C_{11} to C_{13}	13
Diesel fuel	C_{14} to C_{18}	12
Heavy gas oil	C_{19} to C_{25}	10
Lubricating oil	C_{26} to C_{40}	20
Remainder	$> C_{40}$	18

Geographically, oil and gas deposits are found in the world's sedimentary basins. The major oilfields/gasfields and associated pipelines are shown in Figure 4.9; there is a high concentration in the Middle East, in countries such as Saudi Arabia and Iraq, and in Central Asia, e.g. Iran. These countries also have the largest known reserves though they are not the main consumers. These issues are discussed in Chapters 5 and 6. In terms of the geological age of oil/gas deposits, most known deposits formed in the Devonian, the Carboniferous, late Triassic, Cretaceous and Tertiary periods (see Figure 4.1). Like coal, oil and gas are vital resources throughout the world. The extraction and use of oil and gas also represents a reversal of Nature's tendency to remove carbon from the atmosphere and while they provide a means of wealth generation and political bargaining power they cause environmental problems such as oil spills and contribute substantially to global warming.

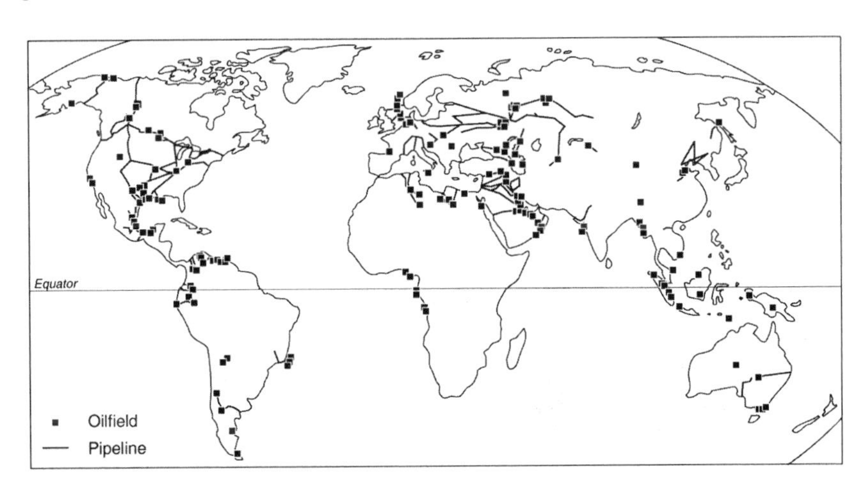

Figure 4.9. Map of oil deposits (based on The Times Consise Atlas of the World, 1995)

4.5 Wetlands

Wetlands are widely distributed on the Earth's surface today. While they are important because they help to regulate water availability, they are also important stores of carbon. In many ways they are the modern equivalents of coal as they reflect the accumulation of plant biomass and thus contribute to removing carbon from the atmosphere. There are many types of wetlands, the global distribution of which is given in Figure 4.10, and there are many ways of classifying them. The Ramsar Convention, an international agreement established in 1971 to oversee and promote wetland conservation, lists more than forty types (Ramsar, 2004). They include coastal and inland water bodies and permanently wet habitats in upland and lowland environments at all latitudes and human-made wetlands. Locally, wetlands are referred to as bogs, fens, marshes, peatlands, mires, muskegs and swamps. The factor in common is the presence of soil or substrate which is saturated with water for most of the year. In some cases this water derives from precipitation alone, in other cases it derives from surface drainage.

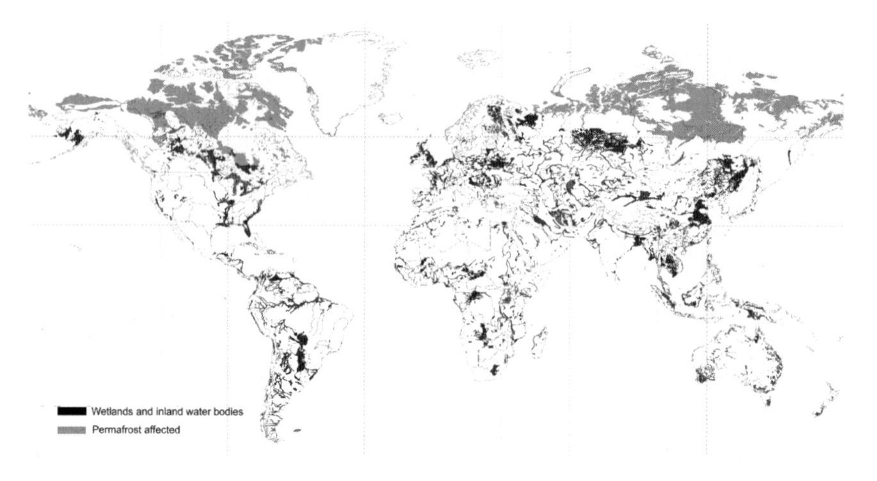

Figure 4.10. Map of wetlands (based on USDA, 2003)

The plants, animals and micro-organisms in wetlands are generally adapted to survival in wet habitats. One example is the species of the moss *Sphagnum* which thrive in such waterlogged, nutrient-poor and acidic environments because of various adaptations. The accumulation of biomass occurs because of this saturation which provides a medium that is unsuitable for decomposer organisms (detrivores, see Section 3.6.3). The medium is even more unfavourable if conditions are acidic. Thus the rate of breakdown of organic matter is lower than its rate of accumulation and so organic matter accumulates. Some wetlands contain more carbon than others; for example, bogs, peatlands and fens have more carbon than coastal lagoons.

Although wetlands occupy only 4 to 6 percent of the Earth's surface, they account for c. 20 to 25 percent of the carbon stored at the Earth's surface in soils (IPCC, 2001). A great deal of this is in peat. According the Irish Peatland Conservation Council (2003) "Peatlands may well contain 3 to 3.5 times the amount of carbon stored in tropical rain forests". Table 4.4 lists the extent of peatland by continent.

Table 4.4. The distribution of the world's peatlands (based on Irish Peatland Conservation Council, 2003)

CONTINENT/REGION	EXTENT km^2	COMMENTS
Africa	58,405	Upland and temperate areas
Asia Except Siberia	244,465	Indonesia, China, Malaysia
Siberia	7,600,000	The largest expanse of peatland (c. 60 percent of world total)
Australasia	23,000	Upland and temperate areas
Europe	956,950	Nordic countries; greatest losses due to human activity
North America	1,735,000	Mainly Canada and Alaska. Second largest expanse in world
South America	112,216	Brazil, Falkland Islands

Regionally, the largest expanses of wetlands are in the northern hemisphere in Canada and northern Eurasia, where between 70 and 90 percent of the carbon storage occurs; the remaining 10 to 30 percent is found in tropical wetlands. According to its Natural Resources Canada (2004) some 14 percent of Canada's land area comprises wetlands; this amounts to 1.27 x 10^6 km^2 which contains c. 154 Gt of carbon. These lands are particularly abundant at the southern part of Hudson Bay and along the river corridor linking the lakes of the Northwest Territory where low relief and abundant precipitation on acidic rocks create appropriate conditions for the accumulation of biomass. Peat accumulation of 5 to 6 m is commonplace in depressions; south of the tree line the vegetation cover of such areas comprises larch, black spruce and ericaceous shrubs; north of the tree line sedges and mosses dominate. At high latitudes, peat formation is associated with permafrost. This is permanently frozen ground. Annual temperature changes result in polygonal cracking and ice formation in the cracks creates hollows in which small ponds form in the summer months. The remains of mosses and sedges accumulate to produce peat.

Similar conditions exist in northern Eurasia. Here the greatest extent of wetland occurs as peatland in Siberia's boreal forest zone (Figure 4.10). As discussed by Tishkov (2002) this vast area contains an abundance of mires which can be classified into five groups on a latitudinal basis. Those of the tundra zone, the most northerly bordering the Arctic Ocean, are polygon mires, immediately to the south are palsa mires, then a discontinuous zone of aapa mires followed by a zone of raised string bogs with pine bogs/fens the most southerly. Polygon mires are associated with permafrost as in Canada (see above) while palsa mires take their name from the frozen mounds or ridges (palsas) which alternate with wet hollows and are prominent in the transitional zone between the tundra and the boreal forest. Aapa mires are treeless with a concave surface and alternating ridges and hollows and characterize the northern forest. Raised string bogs, the predominant mire type overall in Eurasia, and tree-covered bogs are more prevalent in the central and southern forest, including the watersheds of Western Siberia. Tishkov points out that the world's largest bog occurs in this area. This is Vasyuganskoe which is more than 5×10^6 ha (5,000 km^2) and comprises more than 14.3×10^9 tonnes of peat.

In tropical regions, carbon storage in wetlands is greatest in swamps associated with river floodplains rather than in upland peatlands. Swamp forest, for example, is associated with many large rivers such as the Amazon. Trees as well as grasses, sedges and reeds comprise the dominant vegetation and when waterlogged fail to decompose. Upland peats accumulate in the same way as their temperate or boreal counterparts while carbon storage also occurs in coastal communities such as mangrove swamps. As discussed in Chapter 5, all vegetation communities and soils store carbon but in most cases, except for wetlands, the carbon is recycled rather than stored. The world's wetlands are disappearing rapidly, mainly due to human activity. Drainage for agriculture and extraction for fuel are the major causes of decline. There is also concern that in a warmer world many wetlands/peatlands will begin to produce rather than store carbon as higher annual temperatures encourage drying out and oxidation. This would reinforce the impact of global warming through fossil-fuel use.

Chapter 5

5 THE HISTORY AND CONSEQUENCES OF CARBON DOMESTICATION

In geological terms the domestication of carbon is a recent event but it has had profound repercussions for the active biogeochemical cycle of carbon, the full extent of which has yet to be realized both for the environment and for society. Although the domestication of carbon has been occurring for the last 2.5×10^6 years, beginning with the ancestors of modern humans, there are several thresholds which can be identified and which are essentially technological innovations in human prehistory and history. These are: the harnessing of fire, the initiation of permanent agriculture, the expansion of Europe beginning in the sixteenth century, industrialization and fossil-fuel use in the mid-eighteenth century and the innovations of modern biotechnology. All are associated with levels of development, wealth generation, political power and environmental change in its broadest sense.

The domestication of carbon began with the harnessing of fire, though when this first occurred is controversial. It may have been as long as c. 2×10^6 years ago but such early evidence is disputed and less controversial evidence dates to c. 800,000 years ago. As stated in Section 1.5 the harnessing of fire was a major threshold in the domestication of carbon. Preceded by stone-tool making, the harnessing of fire was, nevertheless, one of the first technologies acquired by hominid ancestors of modern humans. It provided a means of manipulation and thus facilitated a degree of control over other organisms and the environment, a capability unique to hominids. The control of fire also provided a means of manipulating resources for food through cooking and may have encouraged the use of a wider resource base. Although the palaeoenvironmental record of early fire is controversial, there is increasing evidence for its use from c. 500,000 years ago. During the Holocene epoch (last c. 10,000 years) there is abundant evidence for widespread fires associated with temporary and permanent settlements and with the inception of agriculture. The latter was the next major threshold and although there was no hiatus between the hunter-gatherer precursors and the first agriculturalists, the shift in food procurement methods led to radical changes in society and environment. It has spread, in many forms, throughout the world to support billions of people through the provision of food, employment and income. Agriculture has been, and continues to be, a major agent of environmental change causing alterations in the physical environment and in the distribution and populations of other organisms.

Another significant development in the domestication of carbon was the expansion of Europe beginning in the late 1400s. The discovery and subsequent annexation of lands in the Americas, Asia, Africa and Australasia began the process which is today recognized as globalization. Instead of, or as well as, riches based on mineral wealth, the colonists accrued most benefit from biomass resources. These included the more exotic commodities such as silk and spices but exchanges of basic crops and domesticated animals also occurred which changed agricultural systems worldwide. Wealth and interdependence were generated, with an overall flow of carbon from the colonies to Europe.

The next major threshold is the Industrial Revolution which began in the mid-eighteenth century in Britain and which spread rapidly to Europe and North America. It was underpinned by the use of coal and began the large-scale release of carbon from geological repositories. Moreover, industrialization facilitated the more intensive use of other forms of carbon such as fibre crops, especially cotton; it encouraged food processing and the intensification of agriculture in European colonies as migration to the colonies began and as European-based agriculture became subservient to the burgeoning industrial base. The presence of coal provided employment in mining and transport and, along with iron ore, influenced the location of industry. Improved individual wealth brought by industrialization and growing populations increased demand for foodstuffs which in turn affected agriculture worldwide. Subsequently, attention shifted to oil and natural gas as energy sources so other repositories of carbon were tapped, especially following World War II. Today, these fossil fuels remain lynch pins in industrial countries, including those which have little or no oil/gas reserves. The politics of fossil-fuels are discussed in Chapter 7. Fossil fuels are now also widely used in agriculture which has become industrialized, especially in the developed world. As fossil-fuel reserves have a finite life there have been attempts to develop alternative energy sources such as solar power, wind power and biomass-based renewable fuels. None are yet major competitors for fossil fuels and the biomass fuels are themselves carbon-based.

The use of fire, development of agriculture, the expansion of Europe and the exploitation of fossil fuels all have had, and continue to have, significant repercussions in terms of the environment. All generate pollution in various forms and some of this pollution influences the carbon biogeochemical cycle, not least of which is the release of carbon, as carbon dioxide, to the atmosphere. This is the primary cause of global climatic change which will affect the entire biosphere and all the carbon in living systems. Other forms of pollution stem from fossil-fuel use, notably 'acid rain', heavy metal contamination, oil spills, coal-mining waste etc. Similarly agriculture generates pollution through its use of fossil fuel and the use of nitrate and phosphate fertilizers, and causes problems with hydrology/drainage systems through water use.

The final threshold in carbon manipulation is a recent development and involves the advent of modern biotechnology in the 1980s. This was when the first genetically-modified crops came to market. Since then many other transgenic crops have been developed though their acceptance around the world has been mixed. The essence of modern biotechnology is the ability to manipulate the genetic characteristics of plants and animals. Thus this manipulation of carbon is at the micro-scale in contrast to the macro-scale manipulation of fire, agriculture and fossil fuels. Modern biotechnology is an important tool in agriculture for crop and animal improvement but there are other applications as well. These include the genetic manipulation of humans, medical applications of genetic therapy and environmental applications such as pollution mitigation using suitable organisms, e.g. bacteria, which can scavenge heavy metals.

5.1 The harnessing of fire

Fire requires a fuel source which is carbon-based and it can be used to manipulate carbon-based aspects of the environment such as plants and animals which thus become resources. Indeed, it broadens the resource base by rendering palatable plant and animal tissues and can be used to manipulate particular plants and animals which respond to the burning of vegetation communities. There are also social advantages, notably the need for assistance to control fire, the advantages of sharing fire for warmth and defence, as well as the provision of a form of communication through beacons. Thus the ability to control fire was an important development in human and environmental history. Such control could have been achieved in several ways, including drilling wood with a bow, the use of iron pyrites to generate sparks, the conservation of fire from lightning strikes using wooden torches and the control of fires started by lightning. When this occurred is a matter for debate as the following account highlights.

The record of human evolution (see Lewin and Foley, 2004; Cameron and Groves, 2004, for reviews and discussions) is controversial because of disagreements about the classification of hominid fossils, notably the problems associated with the delineation of new species as opposed to the recognition of variation within existing species. Moreover, the discovery of new fossils prompts the reappraisal of the hominid lineage and sometimes generates further controversy, as illustrated by the recent finds in Ethiopia and Indonesia. The former involves finds of modern humans (*Homo sapiens*) from Herto which are dated at c. 160,000 years ago (White *et al.*, 2003 and Clark *et al.,* 2003); they provide hitherto-absent fossil evidence which correlates with the results of the genetic analysis of living people for an origin of modern humans at about this time. The Indonesian finds, from a limestone cave on the island of Flores, comprise the near-complete skeleton of a hominid including the cranium and the remains of several other individuals (see Brown *et al.,* 2004 and Morwood *et al.,* 2004, for details); they are particularly controversial because they are considered to represent a previously unknown hominid species which probably evolved from *H. erectus* but which also co-existed with *H. sapiens.* A general classification scheme for hominid evolution is given in Figure 5.1. This shows that the earliest ancestors of modern humans diverged from the apes c. 9 x 10^6 years ago and the australopithecines, including *Australopithecus afarensis* (commonly known as Lucy) reflect the advent of bipedalism and an upright human-like structure. However, the earliest evidence for planned tool making is dated at c. 2.5 x 10^6 years ago (Semaw *et al.,* 1997), possibly predating *Homo habilis* (a term which refers to dexterity), one of several hominid descendents of *A. afarensis,* who until recently was thought to have been the first tool maker. There is no evidence that *H. habilis* migrated out of Africa where fossil evidence indicates it evolved. *H. erectus* (a term referring to upright posture), evolved from *H. habilis* and is the hominid species to which the first use of fire is attributed. There are three sites in Africa, two in Kenya and one in South Africa, which have evidence for fire dated at c. 1.5 x 10^6 years. These are Koobi Fora and Chesowanya in Kenya and Swartkrans in South Africa. They are all controversial because the evidence can be explained as being due to naturally-occurring fire. In Koobi Fora, for example, the evidence comprises blackened sediment. However, recent work at Swartkrans, a cave site just north of Johannesburg in South Africa, on bones indicates that they were probably burnt at temperatures much higher than those usually associated with wildfires (see the BBC television report by Rincon in 2004 on ongoing work of Brain, Thackeray and Skinner). This suggests that deliberate burning within a hearth occurred, as originally postulated in Brain and Sillen (1988). Even if this is accepted as valid, early hominids may have been using fire much earlier. Moreover, it is not clear which hominid was responsible for the Swartkrans fire, though *H. erectus* is the most likely candidate.

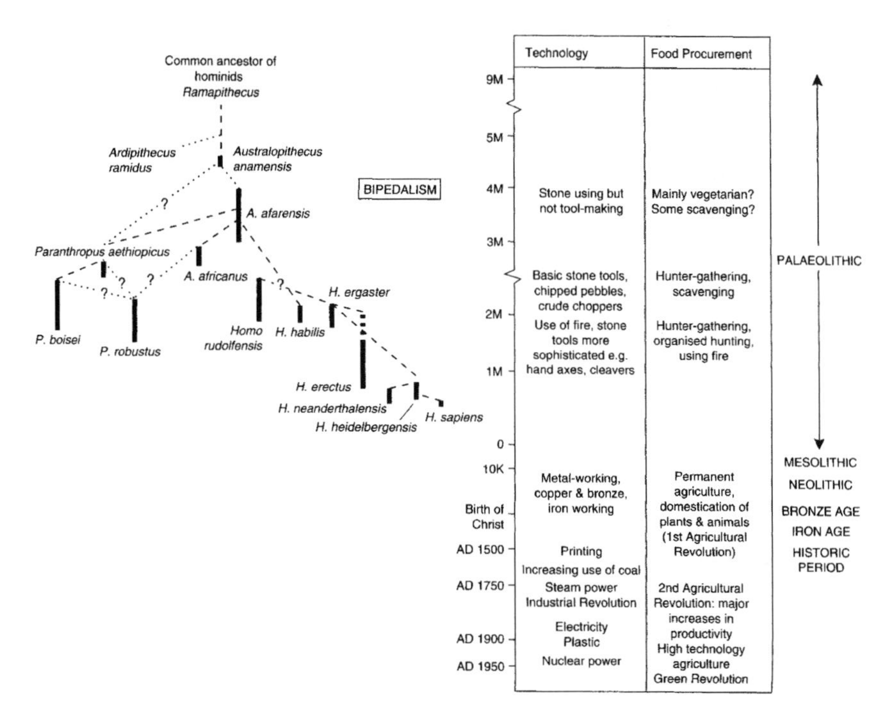

Figure 5.1. A possible record of human evolution.

Outside Africa, evidence for early fire is no less contentious. At Zhoukoudian, the famous site of early hominid occupation in China documented by Rukang and Lanpo (1994) contains a record of fire starting from c. 600,000 years ago. However, the blackened material discovered in the cave could have originated in other ways. More positive results have been documented recently by Goren-Inbar *et al.* (2004) from Gesher Benot Ya'aqov, an archaeological site on a bank of the River Jordan in what is now Israel. This gives dates of 790,000 years ago for wood and flint exposed to fire in ancient hearths. This is the earliest date for domesticated fire outside Africa and is attributed to the activities of *H. erectus.* Indeed, the ability to domesticate fire might have been a reason why *H. erectus* was able to migrate beyond its centre of origin. In Europe one of the earliest dates for human-controlled fire is from the open coal mine site of Schøningen, Germany. Here, three carved spears were discovered in association with the remains of ten butchered horses, stone tools and a likely hearth; the age of the site is thought to be c. 400,000 years old (Thieme, 1997). The assemblage attests to the hunting and carving skills of the site's occupying hominids, probably *H. erectus.* Recent work at a site south of Salisbury, Wiltshire in southern England, which was discovered as a result of road construction, revealed the vestiges of a river-side camp with the remains of animal bones and numerous flint axes as well as a likely hearth. The charcoal from this site has yielded dates of between 250,000 and 300,000 years (see details by Field, 2003) which is certainly the earliest date for fire in the UK. Some sites show evidence for repeated occupation and hearth construction, as exemplified by the Cave of Theoptera, Thessaly, in central Greece. Here Facorellis *et al.,* (2001) show that the earliest hearths are c. 46,000 years old while further hearths higher in the cave stratigraphy are c. 30,000 years old. These hearths reflect the presence of Middle and Upper Palaeolithic peoples respectively and were of the type *H. sapiens,* i.e. modern humans.

The history of human-induced fire in Australia is also contentious. Pyne (2001) describes Australia as the most fire-prone of all the continents because much of it has a climate and vegetation favouring fire, especially because of alternating wet and dry periods and the presence of abundant fuel in woodland, savanna and grassland communities. He believes that humans, whose timing of arrival is also contentious, seized the opportunities that natural fire afforded; it was, in effect, an important resource. One of the oldest records of fire considered to be associated with humans is that preserved in Lynch's Crater in Queensland (Kershaw, 1986). Dated at c. 39,000 years before the present, there is a peak in charcoal particle abundance which coincides with an increase in pollen grains from species indicative of disturbance. There is no direct evidence for human presence in the sediments of Lynch's Crater but human remains in the catchment also date to this time. Miller *et al.,* (1999) have suggested that human-induced fire may be even earlier at 50,000 years before the present and may be linked with the extinction of some of Australia's largest animals such as the emu-like bird *Genyornis newtoni.* Certainly, fire was used widely by Australia's Aboriginal people and early European settlers continued to use it for land clearance. Bush fires continue to be a major hazard today.

In the Americas the earliest date for human colonization is contentious and of course evidence for human-controlled fire is a vital component of such evidence. Until the late 1980s it was generally accepted that the so-called Clovis people were the first colonisers, having crossed from Asia to North America via the Bering Strait when sea levels were lower than at present and a land bridge between the continents existed. The Clovis people are so named after a site in New Mexico where distinctive fluted points and mammoth bones were found; the earliest Clovis sites, found mainly in the western USA, are dated to between 11,000 and 11,500 years ago. Some of these sites contain charcoal as evidence of fire manipulation. However, there has always been doubt as to whether the Clovis people, whose traditions derive from the hunter-gatherers of northeast Asia, were the first colonisers of such a vast continent (see discussion in Dillehay, 2003). Several sites in North and South America with evidence for human occupation, some including evidence for fire, are thought to predate the Clovis sites. One of these is Monte Verde, a montane site preserved by peat accumulation in the valley of Chinchihuapi Creek in southern Chile which was excavated between 1977 and 1985 by Dillehay of the University of Kentucky (Dillehay and Collins, 1988). According to Rose's (1999) review, the main site comprises the remains of a camp used by 20 to 30 people with stone and wooden tools and two large and many small hearths. The average of the radiocarbon dates for this material is 12,500 years before present. Although disputed by Fiedel (2000), the dating is generally accepted and thus places the occupation of Monte Verde, and the first use of fire in the Americas, a century before the advent of the Clovis people. Another aspect of Monte Verde is the presence of nearby deposits which include three possible hearths dated to c. 33,000 years ago. Should this be verified then the story of human colonization of the Americas and the advent of anthropogenic fire would change substantially.

Whilst current evidence supports a relatively recent history for the advent of anthropogenic fire in the Americas especially when compared with that of Africa etc., the capacity to control fire was a vital tool for advancing human control of the environment. Certainly, there is evidence throughout the world for the use of fire during the Holocene (last 10,000 years), especially in the context of hunting and gathering and for the clearance of land for permanent agriculture (see Section 5.2). This evidence derives from a variety of environmental archives, notably lake sediments, peat deposits, archaeological sites, caves and buried soils (see papers in Mackay *et al.*, 2003). It is, however, difficult to determine if the fire was wild fire or anthropogenic. Where other indicators of past environments, such as pollen grains, plant remains etc., are present the operation of anthropogenic fire may be conjectured. For example, Boyd (2002) has reconstructed the fire history of grasslands in south Manitoba, Canada, for the last 5,000 years. He attributes a peak in fire frequency, identified by an abundance of charcoal, c. 2,500 years ago to an intensification of human activity with fire possibly being used to control the movement of herds of bison. This would have allowed the indigenous hunter gatherer people to predict bison movement and thus to exert increased control over a vital resource. Similarly, but in a quite different environment, Maxwell (2004) has reconstructed a fire history from the sediments of three small lakes in northeastern Cambodia. Here fire was occurring more than 9,000 years ago but Maxwell notes changes in fire regimes c. 3,500 and 2,500 years ago which he believes may reflect a shift from natural to anthropogenic firing. In the UK there is evidence from a range of early Holocene sites for burning episodes which created small-scale forest clearances and which are attributed to human activity (see Simmons 2001). Fire remains a significant means of manipulating environmental resources today, notably in savanna, mediterranean-type, some forest and heathland/moorland ecosystems. This will be examined in Chapter 6.

5.2 The origins and spread of agriculture

Organized hunting and gathering involving a degree of control over certain preferred animal species was characteristic of the Upper Palaeolithic which came to an end as the last major ice sheets of the Quaternary (see Sections 1.4, 2.3 and 4.1.6.2) were retreating 12,000 to 10,000 years ago. By this time modern humans (*H. sapiens)* had colonized all the continents except Antarctica and had developed intricate food-procurement strategies. What happened next was to alter human and environmental history profoundly and can thus be highlighted as a threshold in carbon domestication, though it must be acknowledged that it was not a spontaneous development but part of a continuum underpinned by precursor hunting and gathering strategies. This was the domestication of selected animals and plants and the inception of permanent agriculture. There is considerable evidence from many parts of the world for independent agricultural beginnings, the earliest of which are in southern China (Far East) and the Middle East/Western Asia (Near East). There is also evidence for what was domesticated and when, but why it should have occurred remains enigmatic.

The precursor activities to permanent agriculture involved the manipulation of animal herds and wild plants. The many sites worldwide which contain animal bones more than 12,000 years old attest to the carnivory of early hominids and early modern humans. The cut marks on bones indicate selection and butchery and the archaeological record is a testament to the dominance of hunting over scavenging for much of hominid history, as advocated by Foley (see Lewin and Foley, 2004). Hominids rapidly developed a proactive as opposed to reactive strategy for obtaining meat, a process possibly related to the domestication of fire and to the social benefits of sharing information. However, the patterns of wear on fossil teeth indicate that plant material played a significant role in the diets of the descendants of the plant-eating australopithecines but little is known about the plant species utilized. This is because the remains of hard materials like bones, antlers etc. preserve better than plant parts which are soft and which degrade rapidly, especially under aerobic conditions. There are few sites where plant remains of hunter gatherers have been preserved. Thus little is known about the use of wild plant resources prior to domestication. This issue has been addressed by Hillman (1989) whose work at Wadi Kubbaniya in Egypt revealed the use of legumes (pea/pulse species), chenopods, sedges, and small-seeded grasses. Recently, Piperno *et al.* (2004) have reported their analyses of charred plant remains from Ohalo II, an extensive Upper Palaeolithic site on the southwestern shore of Lake Galilee, Israel, which comprises several huts, hearths and a human grave. The site is located in the area of the Middle East/Southwestern Asia where the domestication of many plants and animals occurred c. 12,000 to 10,000 years ago. It was occupied

c. 23,000 years ago and because of inundation by water plant and other remains are well preserved. The analyses reveal that wild cereals, dominated by barley and probably including wheat, were being selected; and, after grinding in stone receptacles, the resulting dough may have been cooked in the site's hearth.

Another conundrum in the history of domestication is that of the dog. The available evidence suggests that this is the earliest domesticated species of all plants and animals; but when it occurred and where is open to debate. Investigations of wolf and dog genetic characteristics (Vila *et al.*, 1997) have shown that the wolf is the closest wild relative and that the greatest genetic diversity occurs in eastern Asia, indicating that this was the area of domestication. Vila *et al.* also suggest, on the basis of molecular clocks, that the dog may have been domesticated as long ago as 135,000 years ago though this is disputed by Savolainen *et al.* (2002) who suggest that domestication occurred c. 15,000 years ago, a date which is in keeping with dog bones found in Eurasian archaeological sites. There is also the possibility that dog domestication occurred in more than one place. However, that dogs in the Americas were domesticated separately from those in Eurasia has been refuted by Leonard *et al.* (1997) whose genetic analyses show that dogs of pre-Columbian America are related to Eurasian wolves. This indicates that the dog migrated into America along with the early human colonizers c. 12,000 years ago (see Section 5.1 for details of the debate about early colonization of America). On current evidence the association between humans and dogs appears to be the oldest relationship between humans and animals. Dogs were useful for several reasons, especially hunting and guarding and thus helped humans to harvest protein. There is little evidence that they themselves were used as a source of food.

Diamond (2002) has also drawn attention to the fact that few plants and animals from the total pool of species have been domesticated. He states that of a total of 148 animal species weighing 45 kg or more, only 14 have been domesticated; of a total of 200,000 wild species of higher plants only c. 100 are valuable domesticates. It is also of note that those plant domesticates which have become major crops worldwide are all cereals and are members of the grass family. Questions arise as to the reasons for selection and the subsequent successful symbiosis between domesticate and domesticated. Although time, place, environmental and social conditions must have played roles in this process, the fact remains that the oldest domesticates remain the mainstays of modern agricultural systems. Wheat, rice and maize with cattle, sheep and pigs continue to be the main sources of carbohydrate and protein, and relatively few plant and animals species have been domesticated in recent times. Such issues, along with recent developments in molecular biology which have presented new ways to evaluate the relationships between wild and domesticated species, have been addressed in Zeder *et al.* (2004).

Figure 5.2 shows the main centres of crop origin where plant domestication occurred independently and at different times, as reflected on individual crop data in Table 5.1.

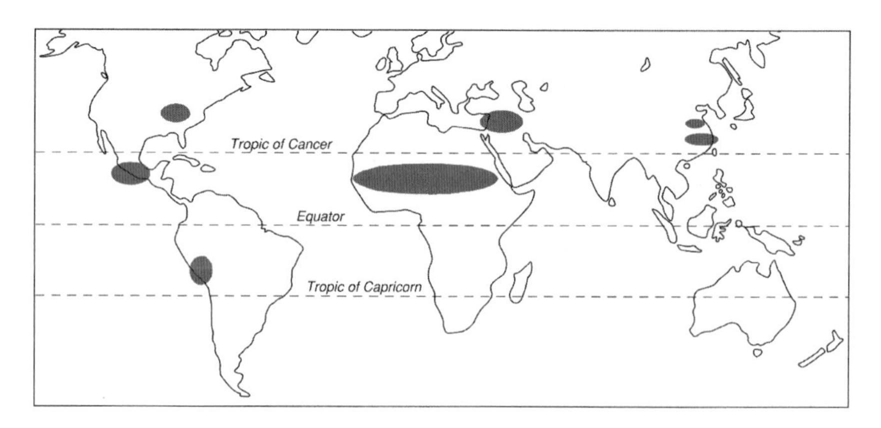

Figure 5.2. Major centres of crop origin (based on Smith, 1995)

Further discussion on these regional centres can be found in Bellwood (2004). Recent discoveries in South Korea indicate that Asian rice (*Oryza sativa*) was probably the earliest plant to be domesticated. The results are available in the Sorori Rice Cyber Museum (2003). Excavations in the village of Sorori, Chungbuk Province, have revealed the presence of peat beds in which 59 grains of burnt rice were found. Radiocarbon dates of the rice and peat lie in the range 12,500 to 14,600 years ago. Rival material for the earliest rice domestication has been presented by Zhao (1998) on the basis of phytolith remains preserved in the sediments in Diaotonghuan Cave, northern Jiangxi (Yangtze River valley) and in offshore sediments in the East China Sea (Lu *et al.,* 2002). Phytoliths are microscopic deposits of silica produced in the cell walls of certain plants and which preserve well in soils and thus in archaeological sites. Many plant families have distinctive phytoliths, including the grasses, and there is a growing literature on their use in palaeoenvironmental studies, including the distinction of wild from domesticated species. Consequently, the early finds in China are not easily comparable with those of Korea because the latter are actual remains of rice rather than phytoliths. Notwithstanding, both sets of data point to rice domestication between 15,000 and 13,900 years before the present.

Table 5.1. Some of the world's most important crop plants and approximate dates for domestication (from Mannion, 2002, with amendments).

Region / Crop	**Common Name**	**Earliest Date** (approx) Radiocarbon years BP
NEAR EAST		
Avena sativa	Oats	9.0
Hordeum vulgare	Barley	9.8
Secale cereale	Rye	9.0
Tritcum aestivum	Bread wheat	7.8
T.dicoccum	Emmer wheat	9.5
T.monococcum	Einkorn wheat	9.5
Lens esculenta	Lentil	9.5
Vicia faba	Broad bean	8.5
Olea europa	Olive	7.0
AFRICA		
Sorghum bicolor	Sorghum	8.0
Pennisetum glaucum	Pearl millet	4.0 ?
Oryza glaberrima	African rice	4.0 ?
Vigna linguiculata	Cowpea	3.4
Dioscorea cayensis	Yam	?
Coffea arabica	Coffee	?
Eragrostis tef	Tef	4.0
FAR EAST		
Oryza sativa	Asian rice	12
Glycine max	Soybean	3.0
Juglans regia	Walnut	?
Castanea henryi	Chinese chestnut	?
Southeast Asia and Pacific Islands		
Panicum miliare	Slender millet	?
Cajanus cajan	Pigeonpea	?
Colocasia esculenta	Taro	9.0
Cocos nucifera	Coconut	5.0
Mangifera indica	Mango	9.2
The Americas		
Zea mays	Maize (corn)	10?
Phaseolus lunatus	Lima bean	5.0?
Manihot esculenta	Cassava	4.5?
Ipomea batatus	Sweet potato	4.5
Solanum tuberosum	Potato	5.0
Capiscum annuum	Pepper	8.5
Cucurbita spp.	Various squashes	10.7?
Gossypium spp.	Cotton	5.5

A further point of interest in the East China Sea stratigraphy is the initial occurrence of phytoliths at 13,900 years ago and then their absence from the sediments dated between 13,000 and 10,000 years ago. This latter period was characterized by a return to relatively cold conditions following rapid warming as the last ice age ended. This climatic deterioration may have resulted in an abandonment of rice cultivation which was resumed after 10,000 years before the present.

As Table 5.1 shows, many crops originated in the Near East. Wheat, barley, oats, rye and various legumes were domesticated between 13,000 and 8,500 years ago. For example, Hillman *et al.* (2001) have identified domesticated rye and wheat from Abu Huyrera, in the Euphrates Valley (in present-day Syria), one of the earliest settlements in the Near East, which is dated to 13,000 years ago. The Near East was also the centre of domestication for the goat, pig, sheep and cattle, though new molecular evidence suggests that some animals, e.g cattle, sheep and pigs, may been domesticated at different times in the Near and Far East (Bruford *et al.*, 2003), and that cattle may have a third locus of origin in Africa. Other centres of plant and animal domestication are located in the Americas (Figure 5.2 and Table 5.1).

In Mesoamerica (Central America), for example, maize and tomato were domesticated. When this occurred and from which wild relatives have been contentious issues. The ancestor of maize is now thought to be a wild grass known as teosinte and the timing of the initial domestication has been put back to c. 10,000 years before present. According to Smith (2001) several other plants were domesticated at about the same time in Mesoamerica; these were the pepo squash and common bean. The former is probably the earliest domesticated plant in Mesoamerica. In South America several crops were domesticated in the region which is now Peru. As well as the potato and tobacco, a variety of cotton was domesticated and so were the alpaca, llama and the guinea pig. The potato was domesticated c. 8,000 years ago, possibly in the vicinity of Lake Titicaca. This region supports many varieties of potato and local farmers today continue to cultivate as many as 100 varieties on a single farm. Crop domestication also occurred in North America, notably in what is now the eastern USA. According to Smith (1995) the plant species domesticated in this region include goosefoot (*Chenopodium berlandieri*), which was probably domesticated c. 4,000 years ago, sunflower (*Helianthus annuus),* also domesticated c. 4,000 years ago, and marsh elder (*Iva annua*). The wild gourd (*Cucurbita pepo*) was also domesticated c. 4,500 years ago in this region though it had been domesticated much earlier in Mexico.

Another centre of crop domestication was sub-Saharan Africa but evidence is scarce. Diamond (2002) recognises three centres of plant domestication: tropical West Africa, the Sahel and Ethiopia. Tef, a cereal, derives from the latter while pearl millet and sorghum, also a cereal, originated in the Sahel, though it is not known precisely where or when the wild relatives of these crops were domesticated. Smith (1995) notes that the earliest evidence for sorghum domestication comprises an impression of a single grain on a fragment of pottery from Adrar Bous, central Africa, which is estimated to be c. 4,000 years old, and that pearl millet was probably cultivated from c. 3,000 years ago. An additional cereal, African rice (*Oryza glaberrima*), was domesticated in West Africa where it was a plant of savanna water holes. According to Linares (2002), its history is obscure but it was probably domesticated from its wild relative, *Oryza barthii,* between three and four thousand years ago, possibly at several centres in this area.

Agriculture spread relatively rapidly from these centres of domestication. Europe was especially influenced by domesticates from the Near East and few additional domestications have occurred. Africa also received domesticated species from the Near East, especially animals such as cattle, sheep and goats, and crop complexes evolved based on indigenous crops and those of Near Eastern origin. The latter domestication of yams also contributed to Africa's agricultural systems. Asia received animals and crops from the west and the east; wheat, barley, cattle etc. spread from the west while rice and chicken diffused from the east. In the Americas domesticated species, notably maize, spread from south through Mesoamerica to the north where indigenous species were also domesticated. In Australia, hunter-gatherer strategies were replaced on a large scale only with the advent of Europeans in the mid-eighteenth century (see Section 5.3).

The controlled acquisition of food through agriculture rapidly superseded hunting and gathering though the motives for this remain enigmatic. The two schools of thought focus on socio-economic (material) and environmental factors, as discussed in Mannion (2002). The former involves population increase which generated a need for improved food availability, a situation encouraged by and encouraging the development of sedentary and nuclear rather than nomadic and dispersed human groups. There is also the possibility of greed as the motive because a reliable surplus of food would, as it still does, promote security and power as well as providing a resource with which to barter for other goods such as tools. However, an environmental explanation is equally feasible. There was a period of rapid climatic change at the end of the last major ice advance, with initial rapid warming, a regression to lower temperatures and eventual Holocene warming. The archaeological and environmental evidence from the Near East, for example, has recently been reassessed by Wright and Thorpe (2003) who indicate that synchronicity between environmental change and agricultural innovation may reflect cause and effect. They show that warming as the cold stage began to experience amelioration led to population increase and that the subsequent regression to cooler and drier conditions prior to Holocene warming necessitated domestication to enhance food production. Could this be climatic determinism at work and could the motive for agricultural innovation involve both materialist and environmental factors? Expanding populations at a time of environmental change may have been the critical factors that turned humans from integral components to dominants and controllers within ecosystems. Whatever the reasons, the advent of permanent agriculture had profound repercussions for the carbon cycle. This became manifest in various ways, including the transformation of land cover into land use with the accompanying loss of carbon from the natural vegetation cover and the inauguration of agriculture as a carbon-processing system.

Today agricultural systems are varied and widespread, as discussed in Section 6.2.1.

5.3 Carbon exchanges during the Holocene

As discussed above, the technology of agriculture spread far beyond its centres of origin. Even during prehistoric times substantial parts of the Earth's surface were affected by its diverse practices which were disseminated by a combination of emigration from centres of origin and the diffusion of ideas. A major impact was that of the removal of the natural vegetation cover, especially the forest cover of temperate lands. As agriculture spread, carbon was released from the biosphere into the atmosphere. Ruddiman (2003) believes that this signal can be detected in the record of atmospheric carbon dioxide content of c. 8,000 years ago which is preserved in polar ice cores, and that an increase in atmospheric methane concentrations c. 5,000 years ago was due to the spread of rice cultivation (see Section 2.3). If this is the case then agriculture rapidly began to register an environmental impact. In addition, the production of abundant food contributed to division of labour and trade. The former fostered the development of new technologies such as metallurgy which also consumed carbon through the fuel required for smelting while trade involved the exchange of goods, many of which were the products of agriculture.

Carbon in various forms was thus shifted from place to place as seeds, animals, animal products and harvested crops. This probably began at a small local scale and gradually increased in frequency and magnitude. Such exchanges shifted plants and animals from indigenous to non-indigenous areas and began an exchange of crops, animals and associated plants (weeds), parasites and insects which constitute invaders and which may threaten indigenous species. Throughout prehistory and history there are myriad examples of such exchanges which have resulted in the geography of carbon within the biosphere today (Chapter 6). The spread of domesticated plants and animals itself displaced indigenous species but the most significant exchanges of carbon have been associated with empire building. The annexation of biomass resources from and the introduction of non-indigenous species to colonies represents an early form of globalization. The best documented examples of this are the Greek and Roman empires, the trade between Europe and Asia in the Middle Ages and the expansion of Europe which began in 1500s and which involved the subsequent colonization of the Americas and later Africa, and parts of Asia and Australasia.

The first ancient civilization of Greece was that centred on the city of Mycenae and which existed between 3,900 and 3,125 years ago in the central southern area of Greece. Archaeological finds attest to trade between other Mediterranean people in Italy and Egypt but little is known about the crops or animals produced or traded. The tablets with Linear B script from a palace in the city of Mycenae reflect the significance of sheep, goats and pigs and the use of cattle and horses. Reference is also made to perfumes, olives and olive oil but this may be biased in favour of palace activities rather than those of the overall economy of the empire. Between 3,200 and c. 2,900 years ago a so-called Dark Age ensued which was a period of little innovation. Thereafter the city states of Classical Greece emerged as trade once again became an important aspect of life in the eastern Mediterranean and by c. 2,500 years ago Athens had become the centre of an empire. Various texts from the time indicate that the staple crops were cereals and pulses with vine and olive crops being the most important economically. There is also evidence that some estates, many of which employed slave labour, were increasingly turning to the latter because of their trading value while staple crops were imported from other parts of the eastern Mediterranean and as far afield as southern Russia. Sheep and goats were the major livestock kept on higher areas and provided milk, wool and meat. According to Sallares (1991), Attica, the hinterland of Athens, between 20 percent and 50 percent of the land was cultivated and its thin soils were more suited to olive than to cereal production. The emphasis on sea power also meant that timber was in great demand while the development of coinage and an increasing reliance on metals for domestic and military purposes also increased pressure on woodlands. According to McNeil (1992) Plato referred to both deforestation and soil erosion in upland Attica, possibly on the mountain Hymettus, and there is evidence for some forest demise in mountains to the north (Gerasimidis and Athanasiadis, 1995) though much forest cover remained intact.

The creation and operation of the vast empire of the Romans between c. 2,260 and 1,600 years ago, as illustrated in Figure 5.3, involved carbon exchanges throughout the Mediterranean basin, central Europe and the Atlantic seaboard on a scale hitherto unknown. Not least of these was the acquisition of staple foods for Roman legions as the empire expanded, as well as the pressure on wood resources for smelting the large quantities of metal, especially iron, lead and copper, used for military and engineering purposes, and the introduction of exotic plants and animals by ex-patriots to the colonies (see Mannion, 1997a for a review).

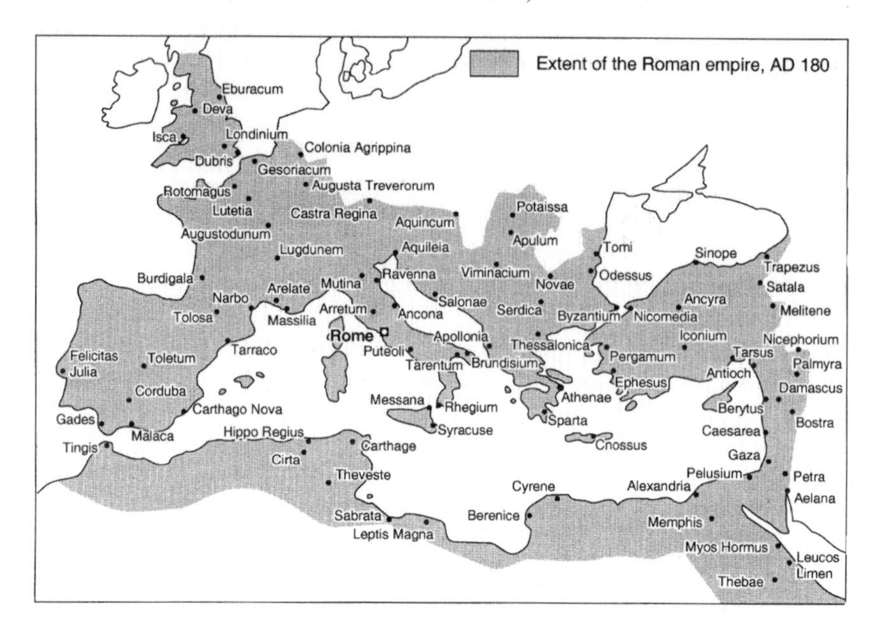

Figure 5.3 . The extent of the Roman empire in AD 180

In addition to appropriating the crop production of their colonies, the Romans undertook large-scale land improvements to increase agricultural productivity, i.e. the acquisition of edible carbon, at home and abroad. For example, drainage networks were constructed in the Po valley and in East Anglia, irrigation networks were established in areas such as southern Spain while other innovations included crop rotations with fallow periods, water conservation measures and the use of manures to improve nutrient availability. How much pressure the construction and metallurgical activities of the Romans brought to bear on woodlands throughout Europe cannot be determined, but it must have been substantial. Their metal-smelting enterprises alone were so extensive that there is a record of increased copper in the atmosphere c. 2,000 years ago as reflected in an ice core from Greenland (Hong *et al.,* 1996).

No part of the empire survived without introduced plant and animal species. Some of the species introduced to Britain, one of the more distant colonies of Rome, are listed in Table 5.2. Many species were and continue to be widely cultivated. In addition, archaeological sites of Roman age attest to the importation of exotic crops and commodities. For example, excavations near the Thames in London, which developed into a major trading centre in the Roman Empire, have revealed the presence of amphorae for olive oil, grape juice and wine as well as for figs, dates, various spices and salted fish. Exports from London included wool and wheat.

Table 5.2. Plant species introduced into Britain by the Romans (based on Cambell-Culver, 2001).

COMMON NAME	LATIN NAME
Vine	*Vitis vinifera*
Walnut	*Juglans regia*
Spanish/Sweet chestnut	*Castanea sativa*
Sour cherry	*Prunus cerasus*
Plum	*Prunus domestica*
Fig	*Ficus carica*
Medlar	*Mepilus germanica*
Leek	*Allium ampeloprasum*
Garlic	*Allium sativum*
Asparagus	*Asparagus officinalis*
Dill	*Anethum graveolus*
Opium poppy	*Papaver somniferum*
Stinking Gladwyn	*Iris foetidissima*
Common mallow	*Malva sylvestris*
Hemlock	*Conium maculatum*
Ground Elder	*Aegopodium pudagrana*

The Roman Empire was disintegrating c. 400 AD but trade continued, especially in goods from the East. According to Turner (2004a), the western appetite for spices, which were considered to be luxury goods, encouraged exploration and trade with the East. Arab and possibly Jewish merchants continued to bring West and East together as they shipped spices such as cinnamon and pepper. Such trade subsequently intensified with repercussions for society as a whole. Turner states that "Sown within the long-distance luxury trade were the seeds of the birth of Europe – its expansion around the turn of the millennium…No longer isolated and introspective, Europe was taking an expansionist and dynamic turn. Luxury, that most trivial of human pursuits, had revolutionary implications. Mediaeval Europe's first renaissance materialised out of the aromas for a nobleman's corpse, or the seasonings for his dinner, the tonics for his humours or fuel for his libido". Once again, trade in carbon brought about socio-economic transformations; this time the carbon was non staple, small volume and alluring in taste and smell (see Section 2.2.2. for the chemistry of flavourings such as ginger and vanilla). The following is a statement by the American Spice Trade Association (2004): "During the Middle Ages in

Europe, a pound of ginger was worth the price of a sheep; a pound of mace would buy three sheep or half a cow; cloves cost the equivalent of about $20 a pound. Pepper, always the greatest prize, was counted out peppercorn by peppercorn. The guards on London docks even down to Elizabethan times, had to have their pockets sewn up to make sure they didn't steal any spices. In the 11th Century, many towns kept their accounts in pepper; taxes and rents were assessed and paid in this spice and a sack of pepper was worth a man's life". This reflects the immense value of spices and the spice trade about 800 years ago.

The search for spices prompted exploration. For example, Marco Polo set out with his father in 1271 on a 27-year journey which took them to China via India and Java. Polo wrote about the production and abundance of spices, a tale that prompted European nations to engage in exploration and vie for supremacy as controllers of the lucrative spice trade during the next 400 years (see Turner, 2004b for a survey). The Portuguese followed by the Dutch proved to be masters of these ventures with Spain, Britain and France also making contributions. Indeed it was the lure of spices (and gold) that caused Columbus to seek a route to the 'Indies' of the East by sailing westward in 1492 and thus to discover the Americas. Columbus, reputed to hail from Genoa in Italy, acted under the patronage of King Phillip II and Queen Isabella of Spain. He did not find spices but opened up routes for exploration and made accessible a continent for annexation by Europe. Crosby (1986) has described this as the 'expansion of Europe'. It set in train new carbon flows as an exchange of crops, domesticated livestock and other biomass-based products took place. Following improvements in navigation techniques, which included the calculation of longitude in the 1730s (Sobel, 1995), the exchange of goods between the Old and New Worlds escalated, as did the emigration of people from Europe. The agricultural practices of Europe were exported to the Americas, notably the introduction of wheat, barley, cattle, sheep, horses and goats. The reverse flow involved maize, potato, tobacco, gourds and squashes. In addition, the paucity of wood reserves, especially in England, prompted the export of wood from North America, beginning first with white pine whose tall trunks provided alternatives to English oak for ship masts. These developments reflect another phase of globalization. Not only did carbon-based resources pour into Europe from its colonies in the Americas and elsewhere at a time of rapid population growth but the colonies also provided opportunities for the millions of immigrants who settled therein to seek wealth and well-being. These processes were continued in the mid-eighteenth and mid-nineteenth centuries with the annexation of Australia and New Zealand by the UK and the establishment of extensive European-style agriculture. In these cases carbon started to flow from the South to the North; this was another piece in place in the process of globalization that is today a term used to describe the homogenization of consumption (not only carbon as food but all types of consumer goods manufactured mainly through fossil fuel use) through the operation of multi-national companies.

5.4 History of fossil-fuel use and industrialization

Throughout prehistory and history the fossil fuels of peat and coal have been used on a local basis, often in regions where wood is in short supply, e.g. arctic and sub-arctic regions and islands. In the Outer Hebrides of Scotland and in western Ireland, for example, the widespread peat deposits have been exploited for millennia as a ready source of fuel, as is also likely to be the case in northern Eurasia. Similarly, there are reports that coal was used locally in Europe as in Britain during the Roman period (Freese, 2003). Small mines were in operation in the 1200s in Europe, providing coal for forges, smithies etc. Water wind and animal-derived energy were also produced and used locally but until the eighteenth century most people worldwide were dependent on straw and wood as well as charcoal, which is derived from wood, for fuel. The widespread use of fossil fuels, and thus the start of carbon-fuel dependence and another threshold in the domestication of carbon, began in earnest with the beginning of the Industrial Revolution in Britain/Europe c. 1750s (see King, 2001 for a review). This was a time of considerable innovation though there is much debate as to why Britain was such a centre of innovation, as discussed in various papers in Prados de la Escosura (2004).

Whatever the underpinning socio-economic conditions, the invention of labour-saving devices and horse and water power were significant in the development of small-scale manufacturing based primarily on cotton, wool and metal. This occurred mainly in the counties of Lancashire and Yorkshire and the English Midlands where farming communities undertook weaving to supplement their meagre farm earnings. These rural workers were supplied with raw materials by town merchants; they wove the cloth and then the merchants sold it on. With the invention of labour-saving devices, beginning in 1733 with the flying shuttle of John Kay and thirty years later the spinning jenny which is attributed to James Hargreaves, the textile industry gradually became mechanized which increased the volume of cloth produced substantially. Eventually textile production became a major activity and wealth generator through exports. Metal smelting, in contrast, had been a feature of the countryside since the Bronze Age (beginning c. 3,500 years ago) and it gradually became increasingly widespread as iron ores were discovered and as iron tools became important in everyday lives. The process of importing raw materials, manufacturing them and then exporting the finished goods thus began on a large scale and increasingly involved many different types of raw materials, including metal ores.

A ready supply of abundant fuel was an important component of innovation and industrial expansion. According to Fouquet and Pearson (2003), it took Britain a considerable time, notably 150 to 200 years, to shift from a 60 percent reliance on biomass fuels (mainly wood) to less than 10 percent, i.e a major shift to fossil fuels. This occurred between c. 1500 and 1700 but what is not clear is whether the advantageous geology of Britain with its abundant coal measures (see Sections 4.1.4.5 and 4.3 for the geology of coal) facilitated rapid industrialization or that industrialization forced the exploitation of coal. From Hatcher's (1993) history of coal use both would appear to be true. He states: "Between the mid-sixteenth century and later seventeenth century coal was transformed from an occasional or specialized source of heat…into the habitual fuel of much of the nation...the expansion of coal consumption was truly spectacular…British coalmining, with its output soaring well in excess of tenfold in the century and a half after 1550...rose from virtual insignificance to become one of a select group of staple industries". Hatcher's reasons for this include a rapidly growing population which increased the pressure on existing resources of wood and straw, the shift of people and industry from the countryside to the towns, the concentration of which facilitated the dissemination of fuel collected from distant sources, as well as rapid growth in industries requiring fuel. Two other developments were also important stimuli of coal use: the commercial steam engine and the production of coke from coal to replace charcoal in the casting of iron. Both occurred in the early 1700s. Metal production, textiles and coal thus stimulated industrial development in the heartlands of England and Scotland in the eighteenth century. By 1800 another innovation provided further impetus. This was the gas industry which was also based on coal. The first gas street light in London began operations in Pall Mall in 1807 (Victorian London, 2004). By 1812 the London Gas Light and Coke Company had been formed and by the 1820s most towns in the UK with populations of 10,000 or more enjoyed street lighting by gas. Britain led the world in coal production and, as Table 5.3 shows, extraction increased by almost a factor of 100 by 1913; by this time the largest consumer was manufacturing followed by iron and steel production and domestic use (Fouquet and Pearson, 1998). Coal fuelled coal! The invention of machinery such as the steam engine also facilitated the recovery of coal from mines; steam pumps removed water from underground and thus improved productivity while steam trains provided a means of transport for large volumes of coal from mine to town and city.

Table 5.3. Coal production in the UK between 1700 and 1913 (based on Fouquet and Pearson, 1998)).

YEAR	Amount (10^6 Tonnes)
1700	2.55
1750	4.20
1800	12.30
1855	57.47
1903	176.17
1913	209.6

Table 5.4 gives coal supply data for 1999 to 2003. It shows that total supply has increased over that period but that the total supply of coal is much less now than in 1900 and more like the production of 1850. This is because other fuels, notably oil and natural gas, have replaced coal in many sectors of UK industry as they have in many other developed nations. In addition more coal comes from deep rather than opencast (near surface) mines and for a nation geologically endowed with coal deposits the UK now imports coal because it can be obtained more cheaply this way.

Table 5.4. Coal supply in the UK 1999-2003 (based on data from the Department of Trade and Industry, UK, 2004).

Supply (10^6 tonnes)	**1999**	**2000**	**2001**	**2002**	**2003**
Production	36.16	30.60	31.51	29.53	27.75
Deep-mined	20.88	17.18	17.34	16.39	15.63
Opencast	15.27	13.41	14.16	13.14	12.12
Other sources (3)	0.914	0.598	0.417	0.450	0.500
Imports	20.29	23.44	35.54	28.68	31.89
Exports	-0.76	-0.66	-0.55	-0.05	-0.54
Total supply	**55.44**	**58.66**	**64.04**	**58.49**	**62.22**

Some of the coal produced in the UK between 1700 and 1850 was exported to its colonies, including parts of what is now the USA. Here wood was abundant but it took just 30 years, from c. 1880 to 1910, for a major shift to coal to occur (Fouquet and Pearson, 2003). By this time the Industrial Revolution was underway in the USA as was general economic development. This was aided by the construction of the railroads which opened up this vast country, linked the established eastern cities and those of the newly colonized west and provided the wherewithal for transporting grain (mainly maize and wheat), meat and minerals, including coal, to the markets and manufacturing centres of the east. The latter required coal for domestic and industrial purposes and the railroads needed it for the production of steam. Moreover, the USA has abundant coal deposits as shown in Figure 4.8. Once again carbon, as coal, encouraged the production of carbon as grain and meat.

According to the Energy Information Administration, USA (2004a) the first commercial coal was produced in Richmond, Virginia, in 1748 and the first coal-fired electricity generating plant was constructed to supply the city of New York in 1882. By this time coal was being produced on a large scale, especially from coalfields east of the Mississippi. Since then coal production has generally increased annually as is reflected in Table 5.5, which gives production data for the period 1950 to 2000.

Table 5.5. Coal production in the USA, 1950-2000 (data from Energy Information Administration, USA, 2004a).

YEAR	Supply (10^6 tonnes)
1950	560.4
1960	434.3
1970	612.7
1980	829.7
1990	1029.1
2000	1073.6

The USA remains a major world producer of coal, occupying second position to China. More than 90 percent of the coal production is consumed within the USA, especially for the generation of electricity, while the main customer nations for coal exports are Canada, Brazil, Japan and Italy.

Globally, c. 5,000 million tonnes of coal were produced in 2002. Apart from the USA (discussed above), China and India are major producers. China produces a massive 1,520 million tonnes and India 390 million tonnes; both countries are exploiting their abundant coal resources (see Figure 4.8). Both countries are also undergoing 'industrial revolutions' as they develop from largely agrarian to industrial economies. Data from the EIA (2004a) given in Table 5.6 show that in both cases coal production has increased markedly since 1980.

Table 5.6. Coal production from China and India, 1980-2002 (data are from EIA, 2004a).

YEAR	CHINA	INDIA
	Supply (10^6 tonnes)	
1980	684	126
1990	1190	248
2000	1314	369
2002	1520	393

Although oil and gas are important, coal is the major source of energy for both nations. In China, for example, coal supplies c. 70 percent of the nation's energy needs. Coal production is expected to increase in the next decade as industrialization continues in both nations. Globally, coal production is also set to increase as other nations industrialize, notably in southeast Asia and South America. Other major producers of coal are Russia, Australia, South Africa and in a number of European nations, e.g. Poland and Germany, coal mining remains an important industry with a long history.

Since c. 1950 the world's fossil-fuel energy sector has undergone considerable change. Apart from an overall increase in energy demand and production, some nations have shifted away from coal to oil (petroleum) and natural gas (see Section 4.4 for descriptions of geology and composition). This is because new reserves of these commodities have been discovered and because both oil and natural gas are 'cleaner' than coal. As discussed in Section 5.5 (below), all fossil fuels create pollution but coals, especially lignite and sub-bituminous coals (Table 4.2), cause additional problems such as acid rain and leaves residues, e.g. slag derived from mining residues and fly ash which remains after combustion, which must be accommodated. The major oil and gas fields of the world are shown in Figure 4.9 and Figure 8.7 gives an analysis of world energy production by energy source.

The data show that energy from oil and gas together has substantially exceeded that from coal since 1970, a result of increased geological exploration after World War II. Oil, also known as 'black gold', and natural gas are not only sources of energy and economic commodities but are also political instruments; any turmoil in producing nations, not least the Middle East, leads to oil price increases which in turn influence the prices of goods and travel and thus impinge on the daily lives of billions of people. These commodities reflect the interdependence of nations, albeit nations with different ideologies, and are thus another component of globalization. The global circumstances of oil/gas production are different from those of coal production a century ago insofar as the proximal availability of coal underpinned the Industrial Revolution in the UK, Europe and the USA, the oil/gas resources, especially those of the Middle East, while necessary for advancement of western economies (or until the West breaks its oil/gas bonds with the Middle East through alternative energy sources) do not necessarily contribute to indigenous development. Many major oil-producing countries remain relatively undeveloped. Changes in oil production since 1960 are shown in Table 5.7.

Table 5.7. Oil production data, 1960-2003 (based on EIA, 2004a).

YEAR	Saudi Arabia	USA	WORLD
	Supply (10^6 barrels per day)		
1960	1.31	7.40	20.00
1970	3.80	9.64	45.80
1980	9.90	8.60	59.60
1990	6.41	7.36	60.75
2000	8.40	5.82	68.34
2003	8.85	5.74	69.50

Overall production has more than tripled, with an especially large increase occurring during the 1960s. The largest producers of oil are now Saudi Arabia and the USA, though the latter's output has declined since 1970 while the former's output has increased substantially. The USA is also the largest consumer of oil but Saudi Arabia's oil is exported to the USA and to Europe and Japan, in common with oil from several other Middle East nations, e.g. United Arab Emirates and Iraq. European nations also obtain oil from North Sea reservoirs while much of the production in Russia is consumed internally. The largest reserves of oil are in the Middle East where it is estimated that c. 700 x 10^9 barrels exist; the second largest reserves are in the Americas and comprise between 120 x 10^9 and 314 x 10^9 barrels. It is thus likely that carbon, as oil, will continue to flow from the Middle East to Europe, Japan and North America, though increased demand is also arising from industrializing countries such as China and India.

The picture is similar for natural gas. As Table 5.8 shows, world production has increased by a factor of 2.5 since 1975 with a huge increase in production in Saudi Arabia, which is now the second largest producer in the world. Much of this, like Saudi Arabia's oil, is exported. In terms of reserves, estimates given in EIA (2004a) indicate that the Middle East has the greatest reserves at 2,000 x 10^{12} to 2,500 x 10^{12} barrels while the next largest reserves are in Eastern Europe in combination with the former USSR where reserves comprise c. 2,000 x 10^{12} barrels.

Table 5.8. Natural gas production data, 1975-2002 (based on EIA, 2004a).

YEAR	Saudi Arabia	USA	Canada	WORLD
	Thousand barrels per day			
1975	140	1633	309	2791
1985	375	1609	337	3938
1995	701	1762	581	5402
2000	750	1911	699	6333
2002	1000	1880	698	7042

Coal, oil and natural gas are not only used as sources of energy but are also the sources of starting chemicals for many synthetic substances which have become part of daily life. Examples are plastics, synthetic rubber, various fibres and pharmaceuticals. These are the end products of petrochemicals which are themselves produced from oil and natural gas. Coal is also used to produce chemicals, such as benzene, naphthalene, anthracene, creosote and coal tar pitch which are used in a diverse range of products such as dyes, pesticides, inks, surfactants and cosmetics. Coal chemicals are produced from coal tar, a result of the distillation of coal. The petrochemical industry began to develop in the 1940s when war-time conditions fuelled the need for synthetic materials to replace natural products. Since then industrial growth has been considerable; the products of petrochemicals are now commonplace worldwide and the industry is of substantial economic value. Two of the fractions produced by the refining of crude oil are fuel oil and naphtha . The latter is an important feedstock (i.e. source) for petrochemicals along with butane from natural gas as well as ethylene and propylene from crude oil. Further chemical preparation involves 'cracking', so called because it breaks down the large molecules of these hydrocarbons into smaller molecules such as those listed in Table 5.9. This table also shows that the world's largest producers are North America and Europe. These chemicals are then used to generate plastics, paints, soaps, artificial fibres etc. The chemical structures of some of these compounds are shown in Chapter 2.

Table 5.9. Global petrochemical production (data from Petrochemistry Net, 2004).

World Petrochemical Production Data for 2003				
Product Tonnes x 10^3	Asia	Western Europe	North America	South America
Ethylene	17,947	20,686	28,698	3,516
Propylene	12,453	14,769	15,228	N/A
Benzene	9,279	7,892	7,950	1,066

Overall, the use of fossil fuels is a cause and consequence of an industrializing world and consumption will undoubtedly increase in the future. The likelihood of exhaustion of these non-renewable resources is already giving cause for concern. This, plus technological advances in alternative fuels, such as biomass fuels, wind, nuclear, solar and tidal power may eventually lead to a decline in fossil-fuel consumption but this is some time into the future. Moreover, sources of chemicals for the synthetic materials on which the world relies so heavily must also be found.

Meanwhile, there is abundant evidence for the environmental problems caused by the intensive use of fossil fuels, as discussed below.

5.5 Pollution history

Since its inception, more than seven thousand years ago when copper smelting began, metal smelting has left an environmental record. It still does so today, along with other types of industry. The major pollution types resulting from fossil-fuel use and industrialization are acid rain, heavy metal release and deposition, and the accumulation in the atmosphere of heat-trapping gases. All are inter-related insofar as fossil fuel is essential for metal smelting and both the fuel, notably coal, and metal ores contain heavy metals which are released while some of the carbon dioxide and sulphur dioxide (this latter is generated when coal rich in sulphur is burnt) released combines with water to produce acid rain. Each pollution type is widespread, though acid rain and heavy metal deposition are most important in the industrialized countries of the Northern hemisphere. Heat-trapping gas accumulation in the atmosphere is generating a global problem through climatic change. Other pollution types which have a bearing on carbon and its domestication include those by artificial substances such as dichlorodiphenyltrichloroethane (DDT) and polychlorinated biphenyls (PCBs) which are the products of the post World War II chemical industry. Lake, estuarine and fluvial sediments, peats and ice sheets/glaciers, soils, museum specimens of plants and animals and trees contain records of past human activity.

Records for the pre-Industrial Revolution (pre-1750) period are limited and are confined mainly to peats, sediments and ice cores from the Arctic and Antarctic; for example, reference has been made in Section 5.3 to elevated copper concentrations in a Greenland ice core c. 2,000 and 1,200 years ago which are attributed to the bronze smelting of the Roman empire and metal smelting of the Mediaeval period respectively. Such intense smelting would have required abundant wood, as did iron smelting throughout the historic period until coal use intensified. This and similar records of elevated metal concentrations in ice cores (see Hong *et al.,* 1998) are a proxy but unquantifiable reflection of the intense use of wood/charcoal (and possibly some coal use) during the periods of Greek and Roman supremacy. A mid- to late- Holocene record of heavy metals in peat has' been presented by West *et al.,* (1987) from two mires on Dartmoor, southwest England. The record from Tor Royal shows that heavy metals began to increase c. 4,500 years ago when Neolithic groups became active in the region and that a notable peak in lead occurred during the period associated with Roman occupation (c. AD 0-400). Lead concentrations also increase substantially with the Industrial Revolution and then decline in layers associated with the last 50 years as lead mining declined in the UK and then as the practice of using lead as an additive in petrol was discontinued. The record in the second mire, Crift Down, contains a record of increased tin deposition during the tenth to the fourteenth centuries, a period which corresponds with local tin mining.

The industrialization of the post-1750 period is well recorded in the archives of lake sediments, peats and ice. For example, a considerable body of research exists on lake sediments in Europe and North America which attests to increasing acidification associated with fossil-fuel consumption. The evidence comprises heavy metal profiles, carbonaceous (soot) particle occurrence (a direct record of coal use) and changes in the algal communities, notably diatoms, which are sensitive to changing hydrogen ion concentration, i.e. pH which determines acidity. Some of this material has been reviewed in Mannion (1997a and 1999) and relates to two major projects in North America and the UK undertaken in the 1980s to determine the extent and timing of lake acidification. These were the US-based Paleoecological Investigation of Recent Lake Acidification (PIRLA) and the UK-based Surface Waters Acidification Programme (SWAP). Both projects involved the investigation of numerous cores from a variety of lakes and both revealed that widespread acidification had occurred in lakes in upland areas of acidic bedrock, e.g. in the Adirondack Mountains and in the uplands of southwest Scotland. pH declines of between 0.4 to 1.6 were found to have occurred, beginning in Scotland in the 1840s and in northeast USA in the 1880s. All such changes, along with similar trends found in Scandinavia and upland Europe, are attributed to industrialization. Curbs on emissions, the shift to oil and natural gas, which contain less sulphur, and increased energy efficiency have since resulted in a partial reversal of these acidification trends in many lakes and reflects the fact that management of carbon and its biogeochemical cycle can be environmentally beneficial.

The post-1750 record of pollution is equally evident in ice cores. First, the heavy metals released from fuel use and metal smelting are registered as air contaminants on a regional basis; secondly, increases in acidity occur because acids generated from coal use are deposited in polar lands remote from the production sources; thirdly, both Arctic and Antarctic ice cores provide a record of human interference with the carbon and related biogeochemical cycles. McConnell *et al.*, (2002) have described trends in lead concentrations from a Greenland ice core covering the last 250 years. Prior to 1870 there was a steady increase in lead concentrations and by 1890 they were 300 percent greater than in pre-industrial times. There was a notable reduction in the ice deposited during the worldwide depression of 1929 to 1934 when industrial lead use declined, and then a substantial increase in the ice deposited after World War II due to the addition of lead to petrol. In contrast, the ice of the 1970s and later has relatively low concentrations due to the fact that lead was no longer being used in this way.

Indeed, such trends are evident in many lake sediments, peats and ice cores which have prompted Renberg *et al.*,(2001) to suggest that lead profiles may be used to identify the sediment/ice deposited during the Roman and Medieval periods and in the post-1970 period independently of age estimation. Work by Yalcin and Wake (2001) also reflects how lands distant from the source can be polluted. Their analysis of a core from the Eclipse ice field near Mt. Logan in the Yukon shows an increase in the concentrations of sulphate and nitrate, i.e. acid rain, began in the 1940s and that the source of this pollution is industry in Eurasia, possibly in the former Soviet Union.

Reference has already been made to the temporal record of atmospheric concentrations of carbon dioxide and methane in polar ice cores in Section 2.3. As shown in Table 2.11, there has been a c. 35 percent increase in the former since c. 1750 and more than 150 percent increase in methane. Both are components of the carbon cycle and both are heat-trapping gases. Their records reflect increasing fossil-fuel use and thus of industrialization. As Petit *et al.*, (1999) have pointed out, the atmospheric concentrations of these gases are now considerably higher than at any time in the ice core record of the last 420,000 years. Few now doubt that this pollution of the atmosphere is a major cause of global warming as shown in Figure 5.4. The last few decades have been the warmest on record, with 1998 being the warmest at 0.58°C above the 1961-1990 mean.

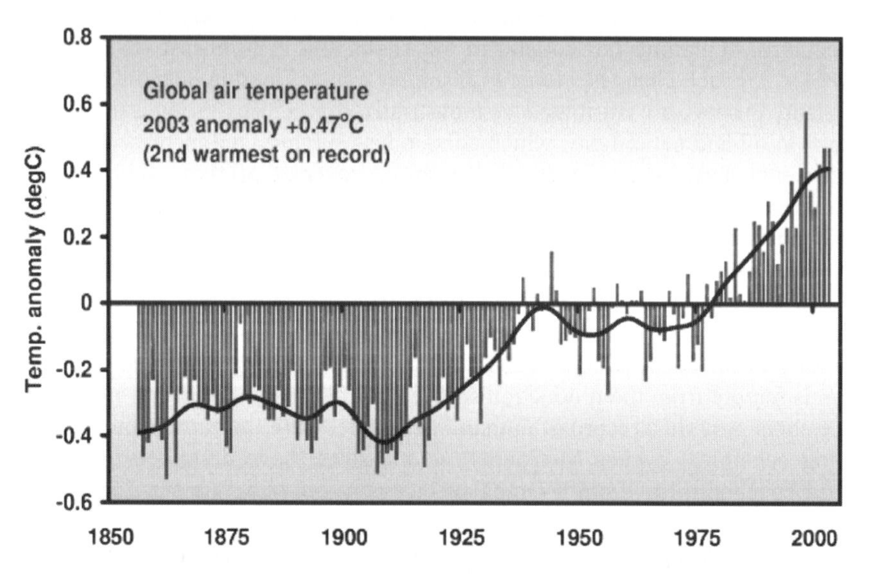

Figure 5.4. The record of temperature change since 1850 (based on data in Jones *et al.*, 1999 and Jones and Moburg, 2003. Reproduced from an information sheet published by the Climatic Research Unit of East Anglia University, available at www.uea.ac.uk/cru/info/warming/).

Carbon dioxide and methane reduce the amount of heat radiated back into space from the Earth's surface and thus act like a greenhouse. Anthropogenically produced, they amplify the existing greenhouse effect of naturally occurring concentrations of these gases. Other gases, such as chlorofluorocarbons, water vapour and nitrous oxides, also contribute to global warming. Given that the world's consumption of fossil fuels shows no abatement, as discussed in Section 5.4, this warming trend can be expected to intensify in the future. Global climatic change will affect the physical geography and ecology of carbon distribution, i.e. the distribution, composition and functioning of the world's ecosystems and agricultural systems (see Chapter 6), as well as human settlement patterns, health and economic activities.

5.6 Biotechnology

Biotechnology is the manipulation of living organisms for specific tasks; it has been in existence ever since the ancestors of modern humans began to manipulate biotic resources through selection and the use of fire. Agriculture is itself a biotechnology and modern agricultural systems (Chapter 6) are the result of millennia of plant and animal breeding, crop protection and production techniques. Biotechnology has also been used for millennia in brewing and food processing, e.g. the use of yeasts and fungi. Today biotechnology still involves the manipulation of organisms, or components of organisms. Not only have established organisms such as bacteria been harnessed (as entire organisms) to undertake tasks, such as the removal of insect pests in certain crops, but it has become increasingly precise to focus on the manipulation of cell components, notably the DNA, of organisms in order to genetically modify them to improve their usefulness for specific purposes. Such applications were made possible owing to the discovery by Watson and Crick in the early 1950s (Watson and Crick, 1953) of the double helix structure of DNA. Thus began the crossing of another threshold in the domestication of carbon and one which extends to the manipulation of human DNA. There are many applications of modern biotechnology: agriculture, medicine, some industrial processes, and environmental remediation.

In agriculture, the applications of modern biotechnology are many and varied. Both crops and animals can be manipulated in various ways, micro-organisms such as beneficial bacteria can be manipulated and various pests and diseases can be controlled. The objective is to improve the harvest through the enhancement of flows of energy and nutrients to the crop by reducing or eliminating competitors which include plant (weeds), fungi and insect pests (see Chrispeels and Sadava, 2003 for an overview). Many techniques can be used to achieve improved crop varieties, including conventional breeding and tissue culture. The former involves cross-pollination between species; it requires the reproduction of plants *in vivo*, i.e. conventionally cultivated as whole plants, and is the method whereby most crop plants were improved until recently. The latter involves the production of whole plants from tissue cells (sometimes involving single cells only) which are grown *in vitro*, i.e. in a growth medium under laboratory conditions. First, this method facilitates the production of many identical plants, or clones, by capitalizing on the totipotency of many species. This is the capacity of individual cells to produce all the different cell types required by a mature plant for survival and so plants with desired characteristics can be reproduced rapidly and in large volumes. Examples of crops which have been so produced include a virus-free cassava and a salt-tolerant millet. Tissue culture is especially useful for the production of tree crops with advantageous characteristics because it bypasses the need for the plants to reach maturity. For example, Viljoen *et al.,* (2004) examine some of the improvements achieved in banana and plantain breeding, including the production of disease resistant species, and the potential that tissue culture holds for the dissemination of such plants to small-scale farmers in Africa who cannot afford crop protection chemicals. Altman (2004) also points out that forestry would benefit from the similar exploitation of trees produced for commercial purposes.

Tissue culture also facilitates genetic engineering which involves the modification of DNA, usually to introduce foreign gene components into plant or animal cells so that the individuals produced will express the desired trait. Genetic engineering is now a fact of modern life though many people object to its use on a variety of grounds, including the possible contamination of the wider environment with altered genes and cross fertilization with related wild species and manipulation by agribusiness companies (see Smith, 2004 for a discussion). As with all technologies, biotechnology has its real and potential disadvantages. Nevertheless, many genetically-modified crops are already being cultivated. They include cotton, soybean, maize and canola (rape). Cotton has been genetically modified to exhibit resistance to the boll weevil, one of its major pests and a cause of significant crop loss in the field. The genetic modification involves the insertion of gene components from the bacterium *Bacillus thuringiensis* (Bt) (see Metz, 2003). This organism has the ability to kill through the production of chemicals lethal to the bollworm and commercial preparations of the bacterium have been used for several decades as a pesticide. Thus the genetic engineering takes the biocontrol a stage further by engineering cotton to produce its own Bt-derived deterrence. Much controversy has been generated since the first Bt-cotton was marketed in the mid 1990s by the U.S. agrochemical company Monsanto. First there are concerns that bollworms will rapidly develop resistance and thus reduce productivity again and second there are concerns that the company has a monopoly over the supply of seedlings and may thus exploit farmers. Such concerns, along with potential escape to the wild, apply to all transgenic crops. Transgenic cotton is grown in the USA, Mexico, India, China, South Africa and Argentina and all report success, notably increased yields and reduced costs for small and large-scale farmers as fewer conventional insecticides need to be used. Other transgenic crops include herbicide-resistant cereals such as maize and wheat and insect-resistant soybean. According to James of the International Service for the Acquisition of Agri-biotech Applications (James, 2004) there are 14 nations primarily involved with growing transgenic crops and that the worldwide extent of transgenic crops has increased by c. 20 percent between 2003 and 2004 to a total of 1.5×10^9 hectares. Table 5.10 provides details of hectarage and crop type. It shows that the USA and Argentina are the leaders, though James notes that developing countries are beginning to adopt transgenic crops (see Section 8.2.3.)

Table 5.10. Extent, type and location of transgenic crops in 2004 (based on James, 2004).

COUNTRY	ha x 10^6	MAJOR CROPS
USA	47.6	Soybean, cotton, maize, canola
Argentina	16.2	Soybean, cotton, maize
Canada	5.4	Soybean, maize, canola
Brazil	5.0	Soybean
China	3.7	Cotton
Paraguay	1.2	Soybean
India	0.5	Cotton
South Africa	0.5	Soybean, cotton, maize
Uruguay	0.3	Soybean, maize
Australia	0.2	Cotton
Romania	0.1	Soybean
Mexico	0.1	Soybean, cotton
Spain	0.1	Maize
Philippines	0.1	Maize

Other aspects of biotechnology in agriculture include the engineering of crops to produce their own nitrogen supplies by encouraging liaisons with nitrogen-fixing bacteria or through direct engineering with nitrogen-fixing genes. There are also possibilities for engineering crops to produce speciality chemicals, substances beneficial to human health and pharmaceuticals. One aim of plant biotechnology must be the ultimate goal of enhancing plant productivity through the genetic improvement of photosynthesis. Additional possibilities for genetic engineering in agriculture include the modification of domesticated animals to enhance protein, milk or wool production and to improve disease resistance. The production of these commodities would also benefit from improved pasture with a high nutrient and water content. These and other applications of agricultural biotechnology are discussed in Persley and MacIntyre (2002) and Parekh (2004).

Further applications of biotechnology include bioremediation (see Singh and Ward, 2004a and 2004b for further details), which is the use of organisms to improve environmental quality through their ability to combat contamination with noxious substances. Water and soil quality can be improved and resources may be recovered. The ability of certain bacteria to break down organic matter is an important component of sewage treatment, for example. There is a growing interest in phytoremediation, i.e. the use of green plants to remove environmental contaminants from water or land. This capitalizes on the ability of specific plants to concentrate harmful substances such as heavy metals, e.g. zinc and nickel. It could also be possible to recover the concentrated metals from the plants and to use genetic engineering to improve their scavenging capacity or to develop the ability in other plants. Contamination with polychlorinated biphenyls (PCBs) and other organic substances such as diesel oil and phenols may also be remediated by certain bacteria as could the desulphurizaion of fossil fuels. Some fungi have similar abilities to degrade complex molecules; the white rot fungus, for example, can break down lignin. Some bacteria have the capacity to scavenge metals such as copper and uranium and could, in theory be used in a process known as biomining; this exploits the ability of some bacteria to use sulphide-rich substrates as an energy source. These are the chemoautotrophs, also known as chemolithotrophs (see Section 3.3). Many Archeans (Section 3.2) have the capacity to produce methane as they break down organic matter; their presence in landfill sites can generate sufficient methane that it can be collected for domestic fuel. Organisms have also been harnessed to produce food, an example of which is Quorn®. This is the mycelium of the fungus *Fusarium graminearum* which is marketed as a substitute for meat. Similarly, organisms can be manipulated to produce fuels such as ethanol. In Brazil ethanol is produced by the action of bacteria on sugar cane residues and there is potential for ethanol production from a variety of substrates such as wood pulp and sorghum. These and other possibilities have been reviewed by Brown (2003 and IEA, 2004). Biomass fuels are generally cleaner than and could reduce dependence on fossil fuels; this issue is revisited in Chapter 8.

Chapter 6

6 THE GEOGRAPHY OF CARBON

The Earth's surface is characterized by abundant carbon in various repositories. These include the dynamic vegetation and animal communities that comprise global terrestrial biomes and their soils as well as aquatic ecosystems. However, these environments have been substantially modified by anthropogenic factors such as agriculture, forestry, fossil-fuel use, industrialization and urbanization. The natural and anthropogenic, or alternatively the ecological and economic, facets of carbon-cycle dynamics of the global environment involve carbon fluxes/flows all of which contribute to the ever-changing geography of carbon and its biogeochemical cycle (Section 2.3). In the last 50 years there has been increasing recognition that human impact on the Earth's surface is a threat to the very existence of organisms and even to humanity itself. Consequently several methods to express this impact have been devised, notably to illustrate the degree of national/regional/local sustainability. One such measure is the 'ecological footprint', the major inputs of which reflect carbon manipulation and the use of other natural resources such as water. The quantity and quality of the latter are indeed a concern worldwide but because carbon lies at the heart of anthropogenic impact a carbon index would be equally appropriate.

Natural biomes/ecosystems are dynamic entities that have emerged from the environmental changes of the 10,000 years of the present interglacial, the Holocene. They can be classified in various ways with an emphasis on the dominant vegetation type which is a function of climate and soils. Biomes also comprise fauna and micro-organisms which are also stores of carbon in living matter while soils, as well as providing a habitat, contain a large volume of carbon as dead organic matter. There are fluxes of carbon between the biome/ecosystem components and the atmosphere, via the processes of photosynthesis, respiration and decay (Section 3.6), as components of the carbon biogeochemical cycle which is described in Section 2. 3. Whilst natural environmental change has been ongoing, its pace has been relatively slow when compared with the rate of modification by human action, especially in the last 5,000 years. This anthropogenic factor continues to exert a significant and increasingly severe impact on the Earth's carbon geography. One aspect of this impact is agriculture which has implications for the distribution and function of natural ecosystems as these are either modified or replaced. Such alterations generally result in a reduction in carbon storage. As discussed in Section 5.2, agriculture has been a major cause of environmental change since its inception c. 10,000 years ago and one which has escalated in the last 350 years. Agricultural systems are not uniformly distributed on the Earth's surface because they are constrained by the same factors as natural ecosystems, i.e. climate and soils. Thus their distribution is another manifestation of the geography of carbon and its move from place of production to place of consumption.

Fossil-fuel production and consumption also influence the global geography of carbon. Fossil-fuel producers are not always the major consumers and so carbon flows occur between the two as is the case with many products of agriculture. The magnitude of flow, which is economic rather than ecological, depends largely on the degree of industrialization so this too is a consideration in the geography of carbon. The demand for wood and wood products worldwide also means that forestry is another source of economic carbon flows. The type of forestry defines how sustainable it is; poorly managed removal of primary forest will result in huge carbon losses, especially if the land is then used for agriculture, whereas managed plantation forestry not only provides a product but also acts as a carbon sink. Forestry is thus one means of managing the carbon cycle and is an important component of proposed carbon trading between nations. This is a component of the Kyoto Protocol (see Chapter 8) to tackle climatic change. It requires developed nations to reduce their greenhouse gas emissions to at least 5 percent below 1990 levels by 2008 and for those nations who do not reach this goal may offset a proportion of their emissions through forest planting. Concentrations of people in urban areas are the main sink for these carbon flows. Urban areas produce no carbon directly but they rely on carbon imports, especially wood, food and fossil fuels. They constitute what the US ecologist, Eugene Odum (1975) described as 'fuel-powered urban-industrial systems' which exist on the basis of imported organic and inorganic resources and produce organic and inorganic waste. Carbon dioxide is a major output, resulting in a gradual increase in its concentration in the atmosphere which is the chief cause of global warming. As entities, cities generate wealth but they are inherently unsustainable; they only become sustainable through reliance on their hinterlands for food, fuel, waste disposal etc. This hinterland may include other nations. The relationship between the producing and consuming components of a region or nation can be quantified to give a measure of sustainability. This provides a basis for quantifying human impact at various scales or by specific activity.

6.1 World ecosystems

An ecosystem comprises biotic and abiotic components between which there is interdependence and an exchange of materials. Biotic components are organisms, including micro-organisms while abiotic components include soil, rock and atmosphere. The relationships and exchange of materials include photosynthesis, respiration and food chains/webs (Sections 3.6 and 3.7). The major groupings of organisms on land are known as biomes many of which are named after the dominant vegetation type e.g. boreal forest, temperate grasslands. An exception is the world's wetlands as these can occur at any latitude, as discussed in Section 4.5. The world's biomes also reflect climatic characteristics as annual temperature and precipitation regimes play an important role in plant growth and thus in ecosystem function. According to Groombridge and Jenkins (2002) terrestrial ecosystems occupy 147 x 10^6 km^2 which is c. 29 percent of the Earth's surface; the oceans occupy 71 percent or 360 x10^6 km^2. Moreover, about 67 percent of the landmass is in the northern hemisphere with an extensive area in the arctic and sub-arctic while in the southern hemisphere there is an equal division of land above and below the Tropic of Capricorn and there is little tundra. In the marine environment the groupings of organisms are not so discrete with more gradual transitions between warm and cold waters. Marine organisms and the oceans are nevertheless major components of the carbon cycle (Section 2.3; Figure 2.17). Both terrestrial and marine environments contain a vast store of carbon, as discussed below.

6.1.1 Terrestrial biomes

Classifying the Earth's biomes is not an easy task and many variations exist in published works, especially since the seminal work of Holdridge *et al.* (1971) who formalized the relationship between mean annual temperature, precipitation and potential evaporation to define 30 'plant formations'. These ranged from dry tundra to rain forest with grasslands/scrub in between. A much simpler classification would be a division into forest and non forest. Most publications adopt a scheme in between these two approaches. Accounts of global biomes can be found in Aber and Melillo (2001), Groombridge and Jenkins (2002), Christopherson (2005) and Stiling (2002) amongst others, whilst the world's forests have been reviewed by the World Commission on Forests and Sustainable Development (WCFSD, 1999); for interactive forest maps see Global Forest Watch (2004). Since this book is about carbon the eight-category scheme adopted by the IPCC (2001) is used here (as in Table 6.1) in order to examine the role of each biome in the geography of carbon. For comparative purposes Table 6.1 gives data on carbon stocks and net primary productivity in the major biomes as presented by the IPCC (2001).

Table 6.1. Data on carbon storage and net primary productivity in the world's major biomes (based on IPCC, 2001a).

Biome	Area (10^9 ha)		Global Carbon Stocks (PgC)[a]						NPP (PgC/yr)	
	A) WBGU	B) MRS	A) WBG			B) MRS	C) IGBP		Atjay	MRS
			Plants	Soil	Total	Plants	Soil	Total		
Tropical forests	1.76	1.75	212	216	428	340	213	553	13.7	21.9
Temperate forests	1.04	1.04	59	100	159	139[e]	153	292	6.5	8.1
Boreal forests	1.37	1.37	88[d]	471	559	57	338	395	3.2	2.6
Tropical savannas + grasslands	2.25	2.76	66	264	330	79	247	326	17.7	14.9
Temperate grasslands + shrublands	1.25	1.78	9	295	304	23	176	199	5.3	7.0
Deserts + semi deserts	4.55[h]	2.77	8	191	199	10	159	169	1.4	3.5
Tundra	0.95	0.56	6	121	127	2	115	117	1.0	0.5
Croplands	1.60	1.35	3	128	131	4	165	169	6.8	4.1
Wetlands[b]	0.35	-	15	225	240	-	-	-	4.3	-
Total	15.12	14.93[c]	466	2011	2477	654	1567	2221	59.9	62.6

A) Wissenschaftlicher Beirat der Bundesregierung Globale Umweltveränerungen (WBGU) (1988): forest data from Dixon et al. (1994); other data from Atjay *et al.* (1979).

B) MRS: Mooney, Roy and Saugier (MRS) (2001). Temperate grassland and Mediterranean shrubland categories combined.

C) IGBP-DIS (International Geosphere-Biosphere Programme – Data Information Service)

a. Soil carbon values are for the top 1 m, although stores are also high below this depth in peatlands and tropical forests.

b. Variations in classification of ecosystems can lead to inconsistencies. In particular, wetlands are not recognised in the MRS classification.

c. Total land area of 14.93 x 10^9 ha in MRS includes 1.55 x 10^9 ha ice cover not listed in this table. In WBGU, ice is included in deserts and semi-deserts category.

Two sets of data are given to illustrate how different researchers have arrived at different figures; this is because definitions of biome types vary. Note that wetlands were examined in Section 4.5 and croplands are discussed in Section 6.2. It must also be noted that the living components of the Earth's biomes contribute to carbon flows other than those related to the large scale biosphere – atmosphere exchanges etc. These include the provision of medicines, fibres, perfumes, foods and biomass fuels.

6.1.1.1 Tropical forests

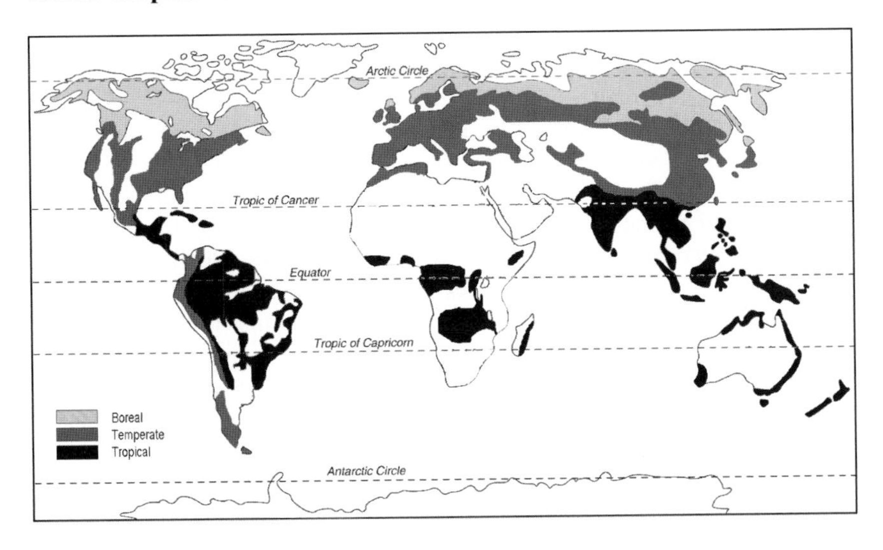

Figure 6.1. Forests of the world

Forests occupy a large proportion of the landmasses between the Tropics of Cancer and Capricorn, as shown in Figure 6.1. These equatorial forests comprise many different forest types and are amongst the most biodiverse ecosystems on Earth. Many areas of tropical forest have been identified as biodiversity hotspots (Conservation International, 2004); not only do they have high biodiversity but high endemism and acute threats through human activity as in Madagascar and Sundaland (the Indo-Malay peninsula with Indonesia). They can be divided into two main groups: tropical moist (rain) forests and tropical dry forests. The former occur where there is abundant rainfall all year round while the latter occur where there is at least one pronounced dry season. Some 50 percent of tropical forests are in Central and South America with most extensive formations in Amazonia. Extensive areas also lie in central Africa and southeast Asia with smaller, but distinct formations in terms of species present, in the Hawaiian Islands and Australia. Tropical moist forests are five times more extensive than tropical dry forests. Most forest types within these broad categories are rich in species, especially trees, though the moist forests are more diverse than the dry forests; the exception is the mangrove forests of tropical coastal regions where tree diversity is low because of water logging and/or poor nutrient supplies. Nevertheless, productivity (see below) remains high. Although tropical forests occupy c. 11 percent of the Earth's land area they contain 40 to 50 percent of the organisms.

Why tropical forests are so species-rich is a matter for debate; possibly, a combination of factors is responsible, such as abundant water and solar energy (no cold season), the lack of major soil disturbance during the last 2 million years of oscillating cold and warm (ice ages and interglacials) as these regions did not experience direct glaciation, and the need to diversify to avoid major outbreaks of disease in warm and often humid environments. Tropical forests are layered, a characteristic which generates many ecological niches and fosters adaptations such as those of the many epiphytes (these plants live attached to other plants) which allow them to grow attached to the trunks and leaves of tropical species. In descending order the layers are: the emergents are individual trees which tower above the canopy, the canopy layer of trees which is continuous, a discontinuous understorey of trees beneath the canopy, shrubs, tall palms, ferns etc. beneath the trees and a discontinuous ground layer of herbaceous species. The epiphytes and climbers, e.g. lianes or vines, which attach themselves to trees via thorns or tendrils, may extend through several layers. The fauna is also diverse, and as with the plants there is a high degree of endemism. Mammals include many primates; Groombridge and Jenkins state: "In Africa, the Guineo-Congolean forest block contains more than 80 percent of African primate species, and nearly 70 percent of African passerine birds and butterflies". The true range of insects remains unknown though diversity is high as is the case for reptiles and fish. Even less is known about the diversity of micro-organisms. Generally, biodiversity and numbers of endemic species are higher in moist rather than dry tropical forests.

Soils are often shallow and many trees have roots within the litter layer which comprises the decomposing organic matter of leaves, wood etc. at the soil surface. Many tropical soils are also nutrient deficient because of a long history of leaching. Consequently, many plants obtain most of their nutrients from the litter which undergoes rapid decomposition by micro-organisms in the favourable humid and warm environment. This is known as tight intra-system biogeochemical cycling and contrasts with temperate and boreal forests in which the soil plays a greater role in nutrient provision and litter a lesser role. Tropical forests play a major role in the carbon cycle because of the carbon they store and exchange with the atmosphere. As a carbon store, the living plants, especially the trees and litter, are important; only in tropical wetland forests is the store below 1 m significant. As shown in Table 6.1, the IPCC (2001a) quotes a range of between 428×10^9 t and 550×10^9 t for the total carbon storage in tropical forests. No other terrestrial ecosystem houses as much carbon as tropical forests, the destruction of which results in a shift of carbon from the biosphere to the atmosphere and reduces the absorption of carbon through photosynthesis, as is discussed in Section 6.2. Today, tropical forests are being removed at an alarming rate, mainly for agriculture and lumber. Tropical moist forests alone account for c. 33 percent of the global net primary productivity, i.e. the net storage of organic matter which is the total produced by photosynthesis minus the amount used for respiration etc (see Section 3.7). It follows that primary productivity is higher in tropical forests than elsewhere with an estimated range of between 13.7 to 21.9 x 10^{12} gC yr^{-1} (Table 6.1). See Whitmore (1999) for details on tropical forests.

6.1.1.2 Temperate forests

Figure 6.1 illustrates the global distribution of temperate forests and shows that the biome is most extensive in the northern hemisphere, especially where humid conditions occur in mid-latitudes. Deciduous broadleaf forests dominate these regions where seasonality is pronounced though needleleaved species may also be present (see Allaby, 1999 for details). In the southern hemisphere broadleaved evergreen species predominate. The considerable heterogeneity of temperate forests reflects both recent geological history and location as maritime or continental. Many areas now covered with temperate forest were glaciated or subject to periglacial activities during the last ice advance and only became forested as the present interglacial conditions developed. Thus they are relatively recently formed ecosystems when compared with tropical ecosystems.

Temperate forests occupy c. 1.1 x 10^9 ha (Table 6.1). The four major groups are characterized by different species and associations. In North America oaks, hickory, chestnut, beech, maple and various pines are dominant with a decrease in numbers of species and size of individuals towards the drier continental interior. In the East Asian formation oak, maple, lime, walnut, bay and ash are the dominants with some conifers such as spruces, firs and pines. In Europe oaks predominate with elms, lime, field maple, beech, birch and conifers such as pines, spruce and fir. The floristics of the temperate forest in the southern hemisphere are quite different as here the southern beeches are dominant and in parts of South America, New Zealand and Tasmania they are temperate rain forests. In all formations there is layering, possibly with a sub-canopy layer, an understorey of shrubs and a ground layer of mainly herbaceous species with mosses and lichens. Seasonality is marked by loss of leaves in the winter months and is recorded in the trunks of trees as the tree rings reflect different growth patterns. Overall, these forests are not as diverse as tropical forests but are more diverse than boreal forests (see below). According to Groombridge and Jenkins (2002) the temperate forests of East Asia are the most varied in terms of numbers of tree species. There are 66 different species of oak, for example, compared with 37 species in the North American formation and only 18 in the European formation. Many more species of magnolia, willow and cherry also characterize East Asia. The major animals include various deer, bears, many small animals such as badger, beaver and voles. Invertebrates are abundant, especially insects.

Soils in these regions vary enormously with rock type and climate. Many have developed on glacial deposits, which, when weathered release nutrients. The addition of organic material from plants/trees gives rise to high water and nutrient retaining capacities and a generally abundant soil flora and fauna ensure good mixing. Some leaching may occur, resulting in pale upper soil horizons. In Europe brown earths dominate temperate regions; these are rich in iron oxides which gives the deep brown colour. There is a high organic content in these neutral to slightly alkaline soils with a loamy, friable texture. Where bedrock is acidic podzols may form in which the leaching of iron results in bleached upper horizons and red/yellow lower horizons where iron is deposited. In East Asia, fertile soils have developed over loess, a fine sediment deposited during the cold stages of the Quaternary period and in some areas red earth soils have formed. The soil reservoir of nutrients is much more important in temperate than in tropical forests and this reservoir is replenished through weathering of bedrock and incorporation of the litter fall into the soil where it is decomposed by micro-organisms. Most nutrient uptake occurs in spring and summer when growth is greatest while litterfall, as leaves and twigs, is most abundant in autumn.

Table 6.1 shows that temperate forests are a significant store of carbon, a store that would be considerably larger if it were not for the high degree of alteration of this biome. These forests contain c. 6 percent of the total biome carbon store which is much less than other forest types. There is approximately twice as much carbon in the soil as there is in the above ground biomass. Productivity is seasonal as photosynthesis takes place when temperatures are in the range 5 to 25 °C . Thus little productivity occurs in winter months with a surge as temperatures rise in the spring. Net primary productivity is in the range 6.5 to 8.1 x 10^{12} gC yr^{-1}. This is about twice the rate of boreal forests but half that of tropical forests and is a broad reflection of climatic influences on productivity.

These ecosystems have been substantially modified by human activity as forests have been replaced by agriculture over the millennia. The temperate forests of Europe are the most altered ecosystems of all; in some regions almost no forest remains. Not only have forests been removed for agriculture and fuel but industrialization and urbanization have also exacted a toll. This means that many animals have become extinct, e.g. wolf, and others, such as birds of prey, are rare and/or threatened.

6.1.1.3 Boreal forests

The coniferous forests, which are also known as the taiga, occupy almost as much land area (c. 1.37 x 10^9 ha) as the tropical forests (Table 6.1); most lie in the northern hemisphere in a broad circumpolar zone which extends across North America and Eurasia as shown in Figure 6.1. Climates vary from sub polar maritime to cool temperate to continental and the tree formations vary accordingly with densest growth in the regions with a milder climate, i.e. maritime, and a longer growing season than in continental interiors where the growing season may be as little as 50 days. The dominant species are needle-leaved evergreen conifers. The North American formations, comprising c. 418 x 10^6 ha in Canada alone, extend from Alaska to Newfoundland and are dominated by white and black spruce and lamarack (a larch). According to the ten-fold classification scheme of Global Forest Watch: Canada (2004), the most extensive forest types are first, southerly spruce and lamarack forests with balsam fir and jack pine and a few broadleaved species such as aspen and poplar and, second, northerly open spruce/lamarack forests with extensive wetlands and bare rock. The remaining formations occupy upland, coastal or Great Lake locations in which species such as Douglas fir/lodgepole pine, western hemlock/sitka spruce and red and white pine/eastern hemlock occur respectively. In Eurasian boreal forests the less extreme climatic areas of Western Europe are dominated by pine, especially Scots pine, silver fir and Norway spruce with birch and aspen. Larch becomes increasingly dominant further east and Dahurian larch forms a monoculture in the high regions of Eastern Siberia. Other species in the Asian boreal forests of Russia and China include Siberian fir and Siberian larch; at lower altitudes and towards the east coasts diversity increases to include various species of fir, pine and spruce plus a few broadleaved species such as aspen, maples and birches. Layering with shrubs and a ground flora occurs throughout, though is sparse in areas of harsh climate where mosses and lichens are the main components of the ground flora. Overall, plant biodiversity is lower in boreal than in temperate or tropical forests as is the faunal biodiversity. Small mammals and ungulates predominate while sudden surges in populations of some insects can cause defoliation.

Given the area covered by boreal forests, soils are very varied. They occur on some of the oldest rock types on Earth, as on the Canadian Shield, and on some of the youngest which include sediments deposited by the last ice sheets. Most are acidic with pHs of 3.2 to 5.5 and weathering rates depend, like annual growth rates, on climatic conditions. The relatively high acidity contributes to the movement of iron and aluminium through the soil profile and their deposition in discrete layers that may consolidate to produce impervious horizons and so cause drainage problems. In poorly drained areas organic matter may accumulate to produce acidic peat, a process which is favoured by the acidic litter produced by conifers and the persistence of low temperatures for long periods. Permafrost, i.e. permanently frozen ground is present in some areas, especially where a ground flora of moss is well developed. Trees persist only if shallow rooted. The organic content varies from low to high; decomposition rates are low because of climatic conditions and because of the content of coniferous needles which are waxy and lignin-rich and which do not break down easily.

In terms of biogeochemical exchanges, the pools in the soil, vegetation and litter are all significant. The soil/litter compartment comprises a major nutrient store, as is reflected in the data of Table 6.1. This shows that the soil/litter stores more than 80 percent of the total carbon in the biome. Whilst it is a measure of the store of organic matter it also contains other essential nutrients such as calcium, sodium, potassium etc. Thus its slow decomposition means slow nutrient release. Moreover, a major limitation on growth everywhere is nitrogen availability but it is especially limited in he boreal (and tundra) zones because the relatively harsh climatic conditions restrict the nitrogen-fixing capacities and decomposition capacities of soil bacteria. The conifers etc. survive because they are evergreen and thus their nutrient requirements are low in comparison with deciduous trees. Needless to say nutrient uptake is greatest during the early growing season and least during the harsh winters to which most species have adaptations. These include low metabolic rates and varied rates of photosynthesis which may cease altogether in winter months. The latter influences primary productivity and as Table 6.1 shows, this is c. 50 percent of that of temperate forests and c. 25 percent of that of tropical forests, a situation which reflects the limitations of annual temperature regimes. Nevertheless, boreal forests house between c. 400 and 600 x 10^{12} gC, which makes them a very important global carbon store. This may increase in the future as global warming causes treelines of the boreal forest to move north into tundra regions (see Körner and Paulsen, 2004 for a discussion of the importance of temperature on treeline location worldwide).

Boreal forest ecosystems provide the raw materials for lumber/pulp industries in Canada, Scandinavia and Russia. Unlike tropical forests, they receive relatively little publicity but are subject to as many threats; deforestation is rife in the Russian Far East and Siberia and there are concerns about the sustainability of North American and Scandinavian boreal forests because of lumbering industries, as discussed in Section 6.2.2 below.

6.1.1.4 Savannas (including tropical grasslands)

Savanna communities are present in tropical regions where there is a pronounced dry season of between two and ten months, and where fire is an important component of the ecosystem dynamics. They are varied in structure and composition, the common feature comprising a continuous grass cover. Trees are present in many savanna types, ranging from scattered individuals to closed canopy woodland. The height of grasses and trees along with tree density are features that have been used to devise many different types of classification schemes. Many savanna communities occur as inclusions in other ecosystem types, especially in dry tropical forests.

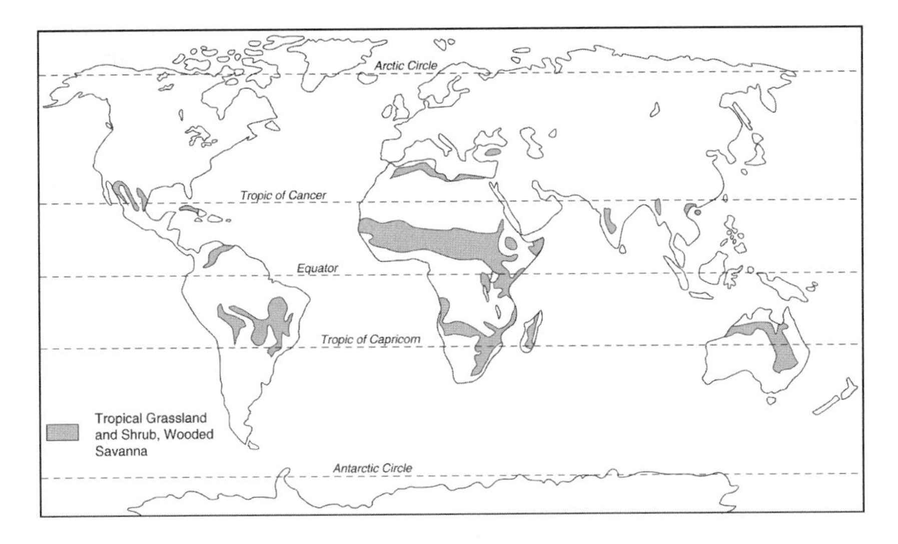

Figure 6.2. Savannas

The global distribution of savannas is illustrated in Figure 6.2 which shows that the greatest extent is in Africa where woodland savanna predominates, followed by Central and South America. Most savanna species have adaptations to cope with drought. For example, savanna grasses photosynthesize using the C_4 pathway (Section 3.6.1) to conserve water. Many tree species lose their leaves during the dry season while others have sclerophyllous (tough, leathery) leaves and so reduce water loss through transpiration; many have an insulating corky bark which withstands fire. According to Mistry (2000), woody species are the most diverse, notably the closed woodland savannas of Brazil with c. 10,000 vascular species while the Australian savannas are least diverse. The plant formations of South America are the most varied, sal is prevalent in Asian savannas, acacias are widespread in Africa and Australia where eucalyptus is also important (see descriptions in Mistry, 2000). The co-existence of trees and grasses appears to be a paradox though it is likely that disturbance from drought, fire and/or grazing, including the impact of colonial insects such as termites and ants, and the seasonality of rainfall facilitates this co-existence. In terms of the animal communities of savannas those of Africa are quite distinct. Here herds of large herbivores prevail. These include three of the so-called big five: buffalo, rhinoceros, elephant as well as others such as zebra, giraffe, warthog and c. 70 species of antelopes and other bovids as well as lion and leopard. These animals provide the basis for a considerable tourism industry in Southern Africa. This is not the case in American or Australasian savannas though some Asian savannas have elephant and tigers.

Savannas occur in regions which have not experienced major natural disturbance in recent geological time. They are prevalent on plateaux separated by escarpments and drained by rivers. Rock type is variable but most soils have been well weathered and leached. Nutrient stores in soils are thus low and some soils are saline due to evaporation causing water to rise through the soil profile and deposits salt when it evaporates. In terms of biogeochemical cycling, the soil compartment is smallest while that of the litter is largest; plants thus obtain most of their nutrients from the litter as in tropical forests. Fire also plays a significant role in nutrient cycling as do insects such as termites. Although some nutrients, and carbon dioxide, are lost to the atmosphere, fire mineralizes litter and generates ash which contains soluble salts; once the rains begin the nutrients are washed into the soil to support growth. Termites influence nutrient cycling in savannas and are thought to be responsible for as much as 20 percent of litter decomposition, the components of which are redistributed in the soil or in termite mounds which can exceed 5 m in height and which become distinctive landscape features. Some termites also have a symbiotic relationship with nitrogen-fixing bacteria. Ants also redistribute organic matter and aerate the soil; both ants and termites distribute seeds.

As Table 6.1 shows, savannas (including tropical grasslands) occupy more than 2×10^9 ha, approximately 17 percent of the Earth's land surface, excluding polar regions. Thus they are as important spatially as boreal and temperate forests together. They do not, however, comprise such a large carbon store even if the larger of the two estimates given in Table 6.1 is accepted. This is because the tree cover is much reduced in comparison with the forest biomes, a factor reflected in the comparative data for the plant and soil components. Also in contrast with forest biomes, between three and four times more carbon is stored in savanna soils than in the vegetation. This is mainly due to grasses and their tillering habit which means that they develop extensive and intertwining root (tiller) systems so that more organic storage occurs as below ground biomass. Overall, primary productivity is high in comparison with other biomes, which reflects the large size of the savanna biome. Productivity is, however, seasonal with peaks occurring during the rainy season and is low during drought as moisture availability is a major limitation on growth. Secondary productivity is high in savannas, especially in Africa, due to high populations of large herbivores. Many of these are migratory so their biomass is the result of grazing in several biomes.

Disturbance in savanna ecosystems has a long history; indeed it was in such ecosystems in eastern Africa that humans evolved. Today, savannas are home to c. 20 percent of the world's population many of whom are reliant on subsistent agriculture. Grazing and shifting cultivation are the major land uses. Some areas of tropical dry forests have been converted to savannas because of human impact.

6.1.1.5 Temperate grasslands and shrublands (including Mediterranean-type ecosystems)

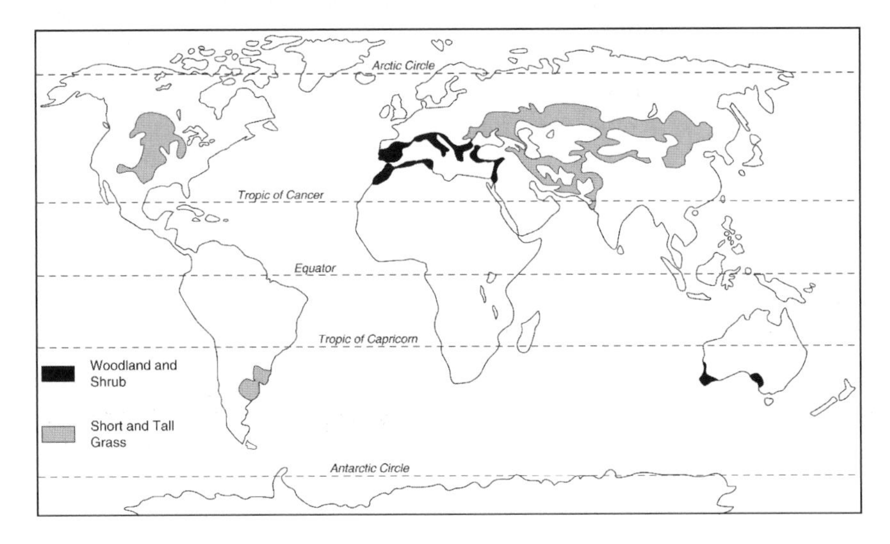

Figure 6.3. Temperate grasslands and shrublands

The middle latitudes are not only characterized by temperate forests but also by temperate grasslands (see Archibold, 1995, for a description) and formations similar to those of the Mediterranean basin which are mainly shrublands such as chaparral, garrigue, matorral and maquis. The distributions of these grasslands and shrublands are given in Figure 6.3. In general, the former occupy continental interiors where annual precipitation is too low to support trees while the latter comprise tree, grass and shrub communities which are typical of regions with hot dry summers but warm winters with abundant rainfall. The temperate grasslands have been intensively modified to become the main grain-producing regions of the world, e.g. the prairies of North America, the pampas of South America and the steppes and plains of central Eurasia. The mediterranean-type ecosystems have become the world's wine-producing regions. Few shrubs or trees are present in the grasslands where gradients of plant communities occur in response to increasing continentality, decreasing rainfall and length of the dry season. In North America the east west gradient involves a continuum from tall grass through mixed grass to short grass prairie. In Eurasia there is a forest-steppe zone in the north as forest grades into grassland which in turn grades into the tuft-grass zone dominated by tussock grasses while the most southerly steppe are characterized by sagebrush and grass. In the pampas, grassland grades into semi-desert towards the west. Smaller but distinct temperate grasslands occur in South Africa and New Zealand. Biodiversity is low in comparison with forest biomes. In both grassland and Mediterranean-type ecosystems fire is an important component of ecosystem dynamics, especially for nutrient cycling, as occurs in savannas (see Section 6.1.1.4).

This is not the case in the Mediterranean-type ecosystems where 'hot spots' of biodiversity occur, possibly because of topography and heterogeneous soils which vary from coast to mountain, a history of natural disturbance and sufficient moisture availability and in the Mediterranean basin itself, a meeting of continents. For surveys see Allen (2001) and Rundel *et al.* (1998). The vegetation communities are mixed and layered with evergreen shrubs and trees with adaptations to drought and fire e.g. sclerophillous leaves and corky bark. Table 6.2 gives data on biodiversity in the five Mediterranean/Mediterranean-type ecosystems; it shows that variation occurs within this grouping but that the relatively small area occupied, i.e. one to two percent of the Earth's landmass, is home to a rich array of organisms with many endemic species.

Table 6.2. Floristic data for Mediterranean-type ecosystems (from Vogiatzakis *et al.*, 2005)

	Area 10^6 km^2	Native flora (plants)	% end-emic	No. under threat	% under threat	% area conserved
California	0.32	4300	35	718	16.70	1.6
Chile	0.14	2100	23	?	?	2.6
Mediterranean Basin	2.30	23300	50	4251	18.24	3.1
South Africa	0.09	8550	68	1300	15.20	14.4
South-west Australia	0.31	8000	75	1451	18.10	6.5
Total area	2.95					

In the grasslands, soils generally have a rich litter layer, are well drained, have a good structure and an abundant soil flora and fauna; this is why they make good agricultural land. In Mediterranean-type ecosystems the heterogeneous topography and geology give rise to soils which vary from deep and fertile with a high concentration of nutrients to those which are skeletal, acidic and nutrient poor. In the Mediterranean basin the abundance of limestone gives rise to typical terra rosa soils; these are rich in bases and the high iron content provides the deep red colouration. In terms of biogeochemical cycling, the soils of grasslands play a major role in ecosystem function because they contain considerable organic matter and nutrients. This is partly because the soils themselves are a good source of nutrients but also because they contain an abundance of underground roots/tillers of grasses which provide nutrients when decomposed. In contrast, in Mediterranean-type ecosystems more nutrients are held above ground in the living biomass, than in litter and soils, a situation that reflects the presence of shrubs and trees.

As global stores of carbon these biomes approximate that of temperate forests, as indicated by the data of Table 6.1. All three biomes have been substantially altered by human activity so these data do not reflect the true carbon-storage capacity of these ecosystems. Indeed, the human impact in these biomes has reduced their carbon storage and resulted in a net output of carbon dioxide to the atmosphere. Moreover, there is a major difference in the carbon storage above and below ground with the former being considerably greater in Mediterranean-type ecosystems than in grasslands where carbon storage is predominantly below ground.

6.1.1.6 Deserts/semi-deserts

Figure 6.4 shows the distribution of the world's deserts. They are located in continental tropical and sub-tropical regions where daily temperature changes can be extreme and the annual precipitation is low, usually less than 250 mm, and erratic; water availability is thus the major constraint on life form presence. Some water is provided as dew if moist air is cooled sufficiently at night.

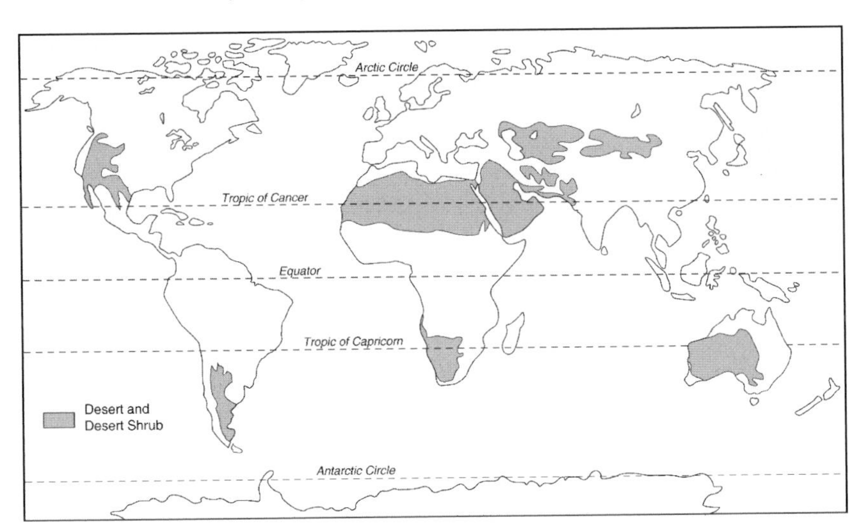

Figure 6.4. Deserts

Deserts occupy c. 30 percent of the Earth's land surface and are sometimes described as hot or cold depending on altitude and the predominance of rainfall or snowfall. Rates of evaporation are high during daytime which further limits water availability. The major desert areas are listed in Table 6.3.

Table 6.3. The world's largest deserts (based on Missouri Botanical Garden, 2002).

Desert	**Location**	**Area:Km^2**
Sahara	North Africa	9,065,000
Gobi	Mongolia-China	1,295,000
Kalahari	Southern Africa	582,000
Great Victoria	Australia	338,500
Great Sandy	Australia	338,500

In deserts, vegetation is sparse or absent and both plants and animals have adaptations to cope with the extreme conditions. Trees are absent and vegetation comprises shrubs and herbs, especially annual species. Shrubs and perennials may be succulents or non-succulents. The former have thick leathery leaves to curtail water loss and include the cacti of the Americas and the euphorbias of Africa. The non-succulents are more prolific and include a range of grasses and shrubs. These and the annuals have survival strategies which include the capacity to react quickly to rain so that seeds germinate when there is sufficient water; life cycles can be completed in short periods so that seeds are produced to be stored in the soil until the next rain which may be several years later. Seeds of the annual herbs can remain viable for long periods but their rapid germination after rain causes the 'desert to bloom' for short periods giving rise to an ephemeral vegetation community. The vegetation cover is low and discontinuous and some species of annuals grow in association with shrubs which provide shade and a nutrient store in the litter. The adaptations of desert animals, which comprise mainly small rodents, lizards and insects, include the production of a near crystalline urine, seed predation and nocturnal feeding habits to avoid the heat of the day. The dry conditions also constrain decomposition by micro-organisms

Desert soils are variable but in general they are nutrient poor, well drained with little water-retaining capacity, and a low organic content. High daytime temperatures result in high rates of evaporation which, through capillary action, brings sodium and magnesium salts to the surface. High salinity also deters plant growth and in extreme circumstances salt flats or pans, such as those of northern Botswana, form. During periods of high winds clouds of salt and clay are produced. Rock, sand or gravel substrates may occur as a result of weathering which is exacerbated by diurnal temperature extremes; alternating cold and hot temperatures cause rock cracking and eventual disintegration. The movement of sediment by wind is also commonplace, giving rise to dunes, and channels may be created during the short-lived wet period (see Goudie, 2002 for details). The largest store of nutrients occurs in the soil; that of the biomass is second in importance while the smallest though sometimes absent store is in the litter.

Deserts and semi-deserts store between 169 and 199 x 10^{12} gC (Table 6.1) with most of it occurring in the soils. This is low in comparison with forests, especially considering the large area occupied by deserts, but it is greater than the carbon store in tundra environments (Section 6.1.1.7). Net primary productivity is also low, at c. 10 percent that of other tropical biomes such as tropical forests or savannas which reflects the lack of water. Short bursts of high productivity follow rain. Some alteration of desert ecosystems has occurred, especially through the grazing of livestock, the encroachment of crop growing in years of good rain and the establishment of irrigation systems for agriculture, as in Egypt.

6.1.1.7 Tundra

Tundra is the term given to high latitude and sometimes high altitude (also known as alpine) ecosystems where temperature is the main constraint on growth. Details of tundra environments can be found in Archibold (1995) and Wielgolaski (1997) who give a wide range of examples. As Figure 6.5 shows, the tundra biome is especially extensive in the northern hemisphere and is restricted to the Antarctic Peninsula in the southern hemisphere.

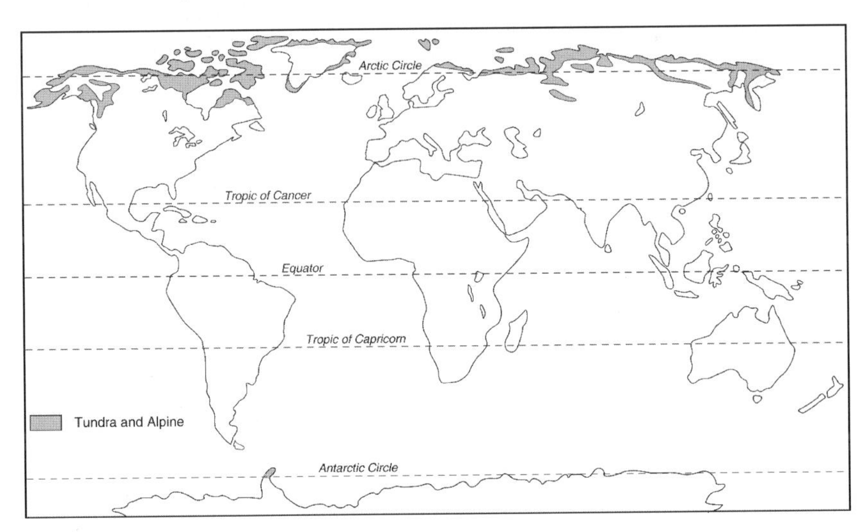

Figure 6.5. Tundra

Like the boreal biome, the tundra is young and occupies polar and sub-polar lands that were subject to glaciation or periglacial processes during the last ice advance. Trees are absent and shrubs, herbs, mosses, liverworts and lichens predominate. There is a relatively short growing season of between two and six months when temperatures rise above freezing point. The most extreme conditions occur in northerly mid-continent areas as in northern Siberia where winter temperatures as low as -40°C occur; milder conditions occur in tundra near coasts, as in southern Greenland and Iceland. Permafrost is widely present; much ground remains frozen throughout the year though surface layers melt during the summer months creating stagnant pools. Annual precipitation is below c. 250 mm, more than half of which falls as snow. There is so little precipitation in some areas that they are known as polar deserts. The more extreme the climatic and soil conditions, the less diverse is the flora and fauna with a low vegetation cover and much bare ground. In less extreme situations, shrubs, which include dwarf birch, alder and willow in northern Europe, are prolific; the proportions of grasses, sedges, herbs, mosses etc increase as climate becomes increasingly harsh. Below ground plant parts are important; they store nutrients essential for rapid growth when conditions allow. Plant biodiversity is low in comparison with other biomes though there is considerable variation. According to Archibold (1995), the greatest tundra biodiversity is in Alaska where 600 plant species are found.

As in deserts, plants and animals have evolved adaptations to cope with the climatic extremes. The plants species grow close to the ground so as not to expose flowers, fruits etc to the wind. Most grasses and sedges grow as tussocks which develop their own microclimate and afford protection for new shoots. Many herbs grow in cushion form for the same reasons. Life cycles are generally completed during the months when temperatures are above freezing point and some species have the capacity to alter their pigmentation, swapping green for purple, which encourages heat conservation. Some species can begin to grow when under snow or even benefit from micro-habitats beneath ice. For many, reproduction occurs vegetatively while sexual reproduction occurs through insect pollination with wind pollination becoming increasingly important as conditions deteriorate. Faunal biodiversity is low, though large populations of migratory wading and water birds are common. Small mammals include the lemming, hares, and grazing animals such as caribou and reindeer. The latter are an important component of the economy of nomadic peoples in the Eurasian Arctic; they graze in lichen-rich tundra pastures in the summer and retreat to the boreal zone during the winter. Insect species are also well represented e.g. aphids, mites, mosquitoes, collembola and beetles but diversity decreases with extremity.

An extreme climate and permafrost mean that nutrient cycling is limited. The dominant pathway is that between the plant cover and litter; the latter has the largest store of nutrients while the soil is the least important of the three compartments. Most cycling takes place during the growing season. Low temperatures and acidic soils formed on acid bedrock restrict decomposition, which diminishes as climate becomes harsher. This is also the case for net primary productivity which, as Table 6.1 shows, is lower in this biome than in any other. Tundra accounts for only c. 2 percent of the global annual total. It also stores less carbon than any other biome, with most carbon present in the litter layer and below ground plant parts in near-surface soil.

The harsh conditions have protected the region from human impact to a great extent, despite the Soviet aim of taming its northern lands. However, the construction of pipelines to carry oil and gas (see Section 6.2.3) in such a fragile environment have been detrimental; as new sources are developed, especially in Alaska and Siberia, the impact will intensify. Such regions are also likely to be altered through global warming which will result in some melting of permafrost and vegetation change as woody species migrate north. There is also the likelihood of carbon losses as surface drying occurs due to global warming (see Mack *et al.,* 2004, and Section 4.5)

6.2 Marine ecosystems

Some 71 percent of the Earth is covered in water, which is equivalent to c. two and a half times the terrestrial environment. As Miller (2004) has discussed, the oceans' physical, chemical and biological components are important in the functioning of the Earth system. The interconnected oceans play an important role in climate regulation through currents that distribute heat and through the exchange of carbon dioxide with the atmosphere. The latter has two components: the chemical diffusion of carbon dioxide from air to water and vice versa, plus the absorption of carbon dioxide by photosynthesizing marine organisms and the release of carbon dioxide through respiration (see Section 2.3). The store of carbon in the oceans is thus an important component of the carbon cycle (Figure 2.17). Unlike the biomes of the terrestrial environment those of the oceans are far less distinct. This is because the oceanic environment provides a more homogenous milieu than land while water, currents, tides, salinity etc. limit the abundance, movement and productivity of marine organisms. Nevertheless there are broad geographical distinctions on a latitudinal basis, zonations related to water depth and different types of habitats related to oceanic structure and geology.

Geographically, the chief distinction is the dominance of diatoms in the colder waters of the Southern and Arctic Oceans, especially where upwelling occurs to bring nutrients from the ocean deep, and the abundance of coccolithophores in warmer tropical and subtropical waters. Both are microscopic and free floating; the diatoms have a frustule (skeleton) made of silica while the coccolithophores comprise plate-like structures and are made of calcium carbonate. Both absorb carbon through photosynthesis and are thus primary producers known as phytoplankton. Dinoflagellates are another type of plankton; some species are autotrophs and others are heterotrophs. Throughout the oceans very tiny organisms, known as nanoplankton, are responsible for c. 70 percent of oceanic primary productivity. The most important of these is *Prochloron,* a cyanobacterium (see Section 3.3). The relationship between organisms and water depth concerns the penetration of light which in turn influences photosynthesis. Only those autotrophs in illuminated water, the so-called photic zone, can photosynthesize. The depth of this zone varies considerably; calm clear waters have the deepest photic zone whilst it is shallow in turbulent waters with sediment or pollutants. However, many such organisms do not have the chemical or structural mechanisms to maintain a presence in this zone as they are free-floating organisms without independent propulsion and are thus subject to the influence of winds, waves and currents to carry them into or out of the photic zone. Free-swimming organisms that live below the photic zone or on ocean beds/sediments are heterotrophic; the food chains/webs being fuelled entirely by the solar energy trapped in the photic zone by autotrophs.

The structure of the oceans, as shown in Figure 6.6, gives rise to a range of habitats which are characterized by distinct groups of organisms. The factors influencing such communities include depth of photic zone, water quality, sediment characteristics and nutrient availability.

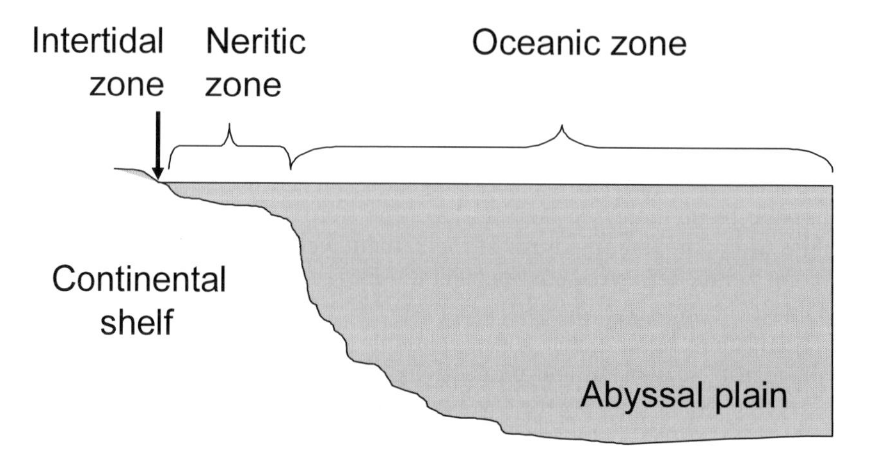

Figure 6.6. Marine zonation

The phytoplankton are the basis of all marine food chains. Their primary productivity is constrained mainly by nutrient availability and is thus not uniform throughout the oceans. Moreover, the importance of the grazing pathway of energy transfer varies spatially along with detrital energy transfer (see Section 3.7); estimates of the latter vary from c. less than 10 percent to c. 90 percent.

In the oceanic zone, the greatest biodiversity corresponds with intermediate levels of phytoplankton biomass (Irigoien *et al.,* 2004) which contrasts with that of terrestrial ecosystems. Consumers include free-swimming zooplankton and fish as well as organisms which live on or in the sediments of the continental slope and abyssal plain. The latter is the deepest part of the ocean and characteristic organisms have adaptations to cope with the high water pressure and lack of light. Examples include copepod crustaceans, molluscs, worms and comb jellies, most of which feed by collecting or filtering detrital particles which sink through the water from the photic zone. The deeper the water, the fewer and more specialized are the species. The neritic zone over the continental shelf is much more biodiverse because it is characterized by relatively shallow water much of which is illuminated. This plus the input of nutrients from the land means that primary productivity is higher per unit area than in the oceanic zone. There are free-swimming organisms and those that live on the bottom sediments. The latter are benthic species and include worms, crabs and clams; some are grazers while others are detrivores, consuming organic matter which may be partly decomposed by bacteria. The productivity of the intertidal or littoral zone is by far the highest of the three; neither nutrients nor light are constraints on photosynthesis. Phytoplankton are not the only primary producers in this zone which is home to salt marshes, mangrove communities and coral reefs. Indeed the latter are amongst the most productive ecosystems in the world. The distribution of these ecosystems is shown in Figure 6.7. Mangroves and coral reefs enjoy a tropical and subtropical distribution while salt marshes are in temperate and boreal latitudes.

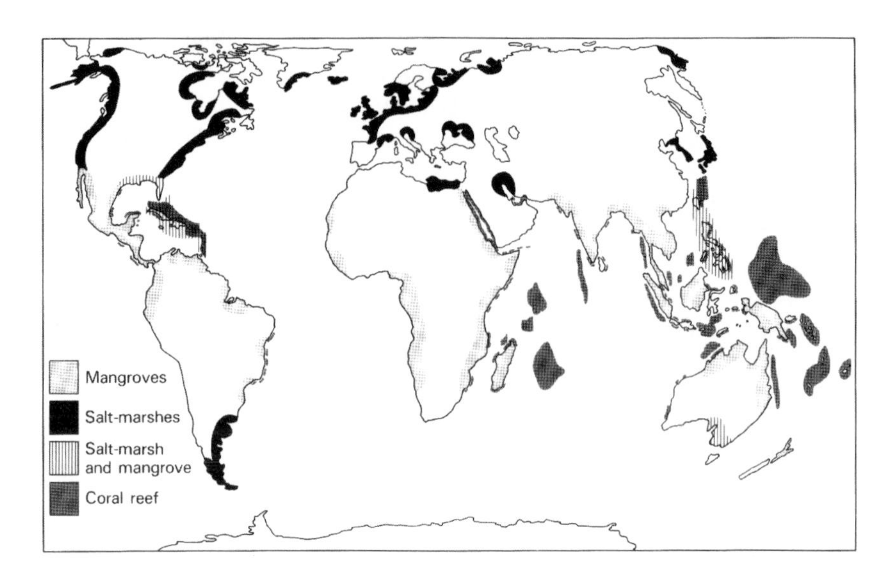

Figure 6.7. Distribution of mangroves, salt-marshes and coral reefs

The data of Table 6.4 show that gross (total) primary productivity (GPP) for the biosphere as a whole is 104.9 x 10^9 tC yr^{-1}, of which terrestrial productivity is slightly higher than marine productivity. However, in view of the much larger area occupied by the oceans, gross productivity is considerably less than in the terrestrial component when calculated by unit area. Many factors contribute to this difference but it is particularly significant that only c. 7 percent of the total solar energy suitable for photosynthesis (i.e. photosynthetically active radiation) is used in the marine environment as compared with 31 percent in the terrestrial environment. In addition, marine primary producers comprise only 0.2 percent of the total biosphere producer biomass and have a rapid turnover time. This occurs in a matter of days as compared with years in the terrestrial biosphere. Net primary productivity mirrors that of gross primary productivity. However, a so-called biological pump operates in the oceans. This involves the cascade of organic matter produced through photosynthesis within the marine ecosystem. Photosynthetic carbon is either broken down by the respiration of marine organisms which returns carbon to the water as dissolved inorganic carbon (DIC), or it is used by the producers, herbivores, carnivores or detrivores. Organic detritus, comprising faecal pellets and dead organisms, sinks through the water to be used by deep-water organisms (see above), a process of carbon transfer known as 'export production'. These organisms respire and so convert the carbon to DIC at depth. This process results in atmospheric carbon dioxide concentrations of c. 200 ppm lower than would otherwise be the case (IPCC, 2001).

Table 6.4. Global gross primary (GPP) productivity data (from Field *et al.* 1998).

	GPP x 10^9 tC yr^{-1}	GPP as % of total
Terrestrial	56.4	53.8
Marine	48.5	46.2
Total	104.9	100

Overall, the oceans contain c. 38 x 10^{12} tC (see Figure 2.17). The amount of carbon in the oceans has increased since the Industrial Revolution due to absorption from the atmosphere via the inorganic and organic processes referred to above. The IPCC (2001) indicates that c. 2 x10^9 t of carbon produced by human activity is absorbed annually and that about one third of the total additional carbon produced through fossil-fuel consumption since c. 1750 has been absorbed in this way. Consequently, the oceans constitute a considerable buffer against global warming though how long this can continue is a matter for conjecture. Some scientists believe that the oceans are already close to saturation and that increased global temperatures will, amongst other factors, reduce their capacity to absorb carbon.

6.3 The alteration of the world's biomes

A substantial proportion of the world's biomes have been altered by human activity. The success of *Homo sapiens* has been achieved at the expense of other organisms as carbon has been domesticated through technological innovation. Human impact is evident even in Antarctica and that impact will become more widespread as global climatic change intensifies. The major activities causing change are agriculture, manipulation of forests, and the activities linked with industrialization such as fossil-fuel use and urbanization. All of these activities have resulted in the emission of carbon to the atmosphere and have impaired the capacity of the biosphere to absorb carbon.

6.3.1 Agriculture

Since its inception c. 12,000 years ago (Section 5.2), agriculture has been introduced to all but the most inhospitable regions and has been the most important aspect of human activity in the transformation of the Earth's biomes. Some of this transformation has been subtle, involving the manipulation of components of natural ecosystems such as stock rearing in semi deserts, while other types of transformation have involved complete replacement of vegetation cover with pasture or cultivated crops. Transformations of both kinds have occurred over many millennia, as discussed by Mannion (1997a and 2002) but the pattern of modern agricultural systems began to emerge after 1700 with the expansion of Europe (see Section 5.3). Using various historical sources, statistics on changing forest cover (see Williams, 2003 and references therein, and Section 6.2.2) which is one measure of land-cover/land-use change, have been derived while a more direct reconstruction of cropland increase has been presented by Ramankutty and Foley (1999). The latter's data are shown in Figure 6.8.

These data show that immediate post-1700 cropland expansion in Europe and China probably continues an already established growth trend, while the Former Soviet Union, tropical Africa and South Asia experienced a significant but less extensive pulse of cropland creation between 1700 and 1850. This was a 'take off' time for many parts of the world, especially in the eastern part of North America. By 1900 croplands were expanding in the west of North America, Central America, and South America and Australia. Overall, croplands expanded from c. 4.05×10^6 km^2 in 1700 to 17.92×10^6 km^2 in 1990, an increase of c. 342 percent.

During the 290-year period of their study, Ramankutty and Foley estimate that forests/woodlands decreased by c. 8.80×10^6 km^2 or 16.7 percent of their extent in 1700. A similar percentage (17.4 percent) decline in grasslands, savannas and steppes also occurred during this period, i.e. a decline of c. 5.61×10^6 km^2. To this must be added the alteration of land from its natural state for pasture/grazing, though no quantifiable data are available to show the extent or timing. Moreover, further notable increases in agricultural systems have also occurred since 1990 as recorded by the Food and Agriculture Organization (2004a).

Data on this are given in Table 6.5, which shows that the most significant increases in arable and permanent pasture have occurred in the developing world with increases of c. 7 percent and 3 percent respectively since 1990. A trend in the opposite direction has occurred in the developed world where the amount of land under agriculture has declined since 1990 following declines throughout the 1980s, with a noted decline in arable land. This is mainly due to the production of a food surplus and the initiation of set-aside policies to encourage alternative land uses such as recreation, tourism and a return to nature. In contrast, rapidly increasing populations together with the need to generate income from agricultural produce have stimulated agricultural expansion in the developing world.

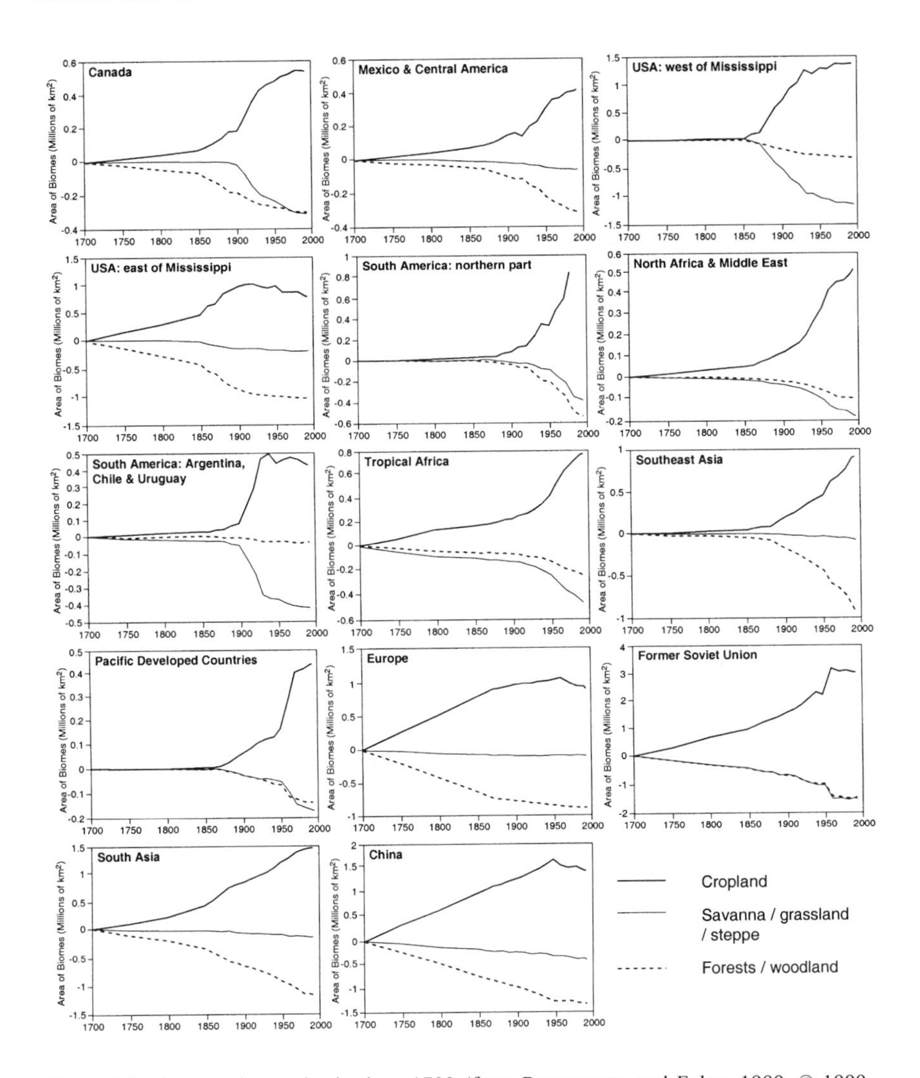

Figure 6.8. Increase in croplands since 1700 (from Ramancutty and Foley, 1999, © 1999 American Geographical Union)

Table 6.5. Changes in the area of agricultural land 1990-2002 (from FAO, 2004a)

Year	Developed world		Developing world	
	Arable	Permanent pasture	Arable	Permanent pasture
	thousands of ha			
1990	649,485	1,191,949	742,080	2,221,469
1995	632,734	1,212,610	758,766	2,263,703
2000	616,402	1,205,389	779,925	2,283,546
2002	na	na	792,537	2,286,880

Agricultural systems are many and varied and can be classified in various ways; based on crop type, production type, crop variety and energy inputs: one such scheme is given in Figure 6.9. An agricultural system can be determined by any combination of these factors coupled with the constraints of the physical environment and the influence of a range of socio-economic factors. The major types of global agriculture are given in Figure 6.10 to provide a snap shot of just how much of the Earth's land surface has been affected by agriculture.

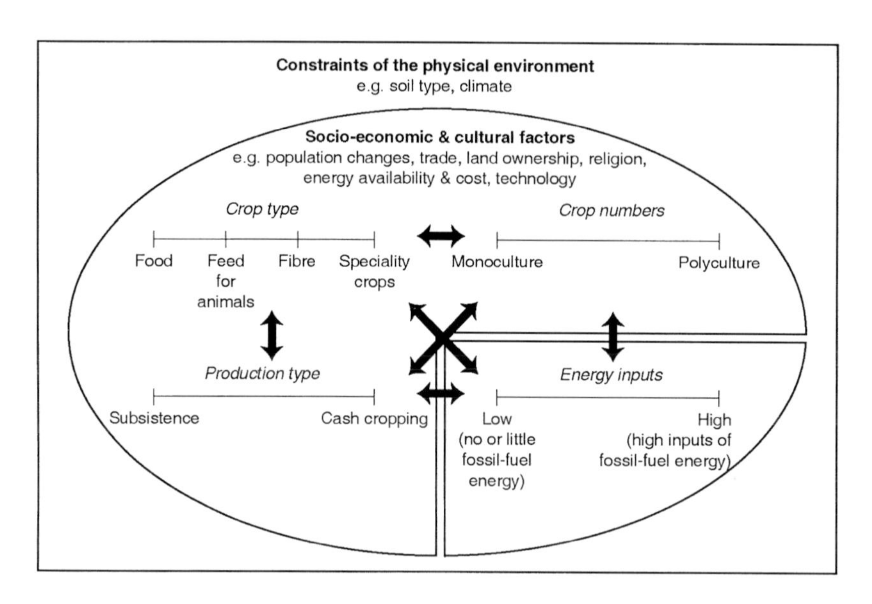

Figure 6.9. A scheme for classifying agricultural systems (from Mannion, 2002)

Measures of the alteration of the natural environment are examined in Section 6.4. As a carbon processing system, agriculture is, self evidently, beneficial for the sustenance of society but the inevitable loss of natural vegetation and soil alteration has resulted in an output of carbon to the atmosphere as well as a reduction in the capacity of agricultural land to store carbon when compared with undisturbed ecosystems. How much carbon has been shifted from the biosphere to the atmosphere because of land-use change? Houghton (2003) has summarized data on carbon emissions due to land-use change since 1850 and while much of the total 156 GtC, (with c. 60 percent coming from the tropics) is due to forest losses (see Section 6.2.2), he notes that croplands produce c. 1 to 1.5 GtC annually. Agricultural lands also store carbon, mainly in the soil; data quoted by IPCC (2001) for croplands vary from 131 to 169 GtC.

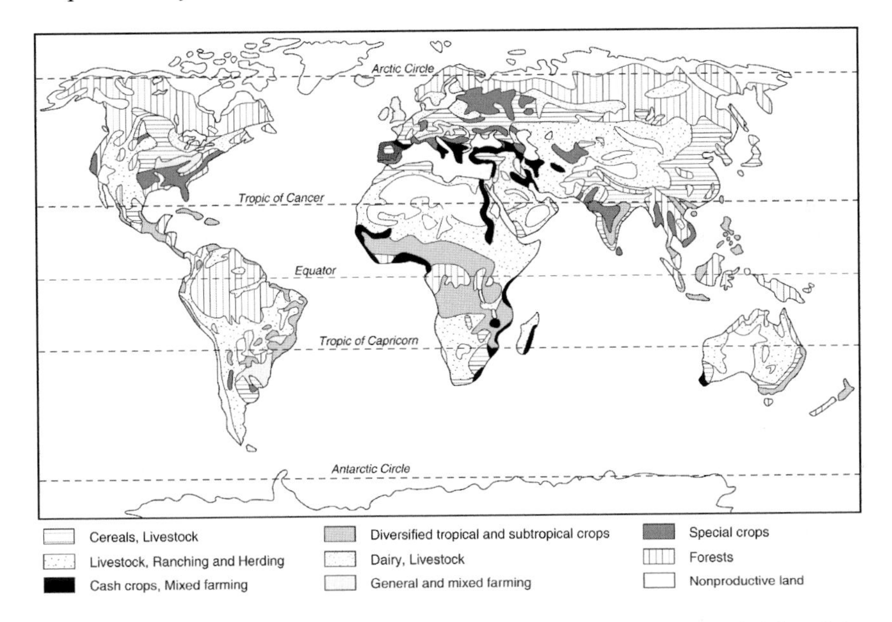

Figure 6.10. Major types of global agriculture (based on Oxford Hammond Atlas of the World, 1993)

Many modern agricultural systems, especially those of the developed world, use fossil fuels to enhance the carbon processing of agricultural systems. Thus many agricultural systems are consumers of fossil-fuel energy which is used for mechanization, crop and animal protection, such as pesticides and animal health products, and the production of artificial fertilizer. Table 6.6 gives details of energy expenditure on a high-technology farm in the USA. The table shows that fertilizer use accounts for a large share of energy inputs; in some agricultural systems this may rise to c. 50 percent. The energy is used to produce ammonia from hydrogen and atmospheric nitrogen using the Haber process which was developed in the late 1800s. (see Leigh, 2004 for a history). Thereafter nitrates are produced. When added to agricultural soils they enhance the supply of nitrogen, which is often a limiting factor to plant growth, and thus increase productivity.

Table 6.6. A typical energy balance for a farm in the USA (based on Pfeiffer, 2004).

ACTIVITY	ENERGY %
Inorganic fertilizer	31
Field machinery	19
Transport	16
Irrigation	13
Livestock raising (not feed)	0.8
Crop drying	0.5
Pesticides	0.5
Other	0.8

The production and consumption of artificial fertilizers have increased enormously since 1945 as agriculture industrialized; a further sevenfold increase in nitrogenous fertilizer consumption has occurred in the last 40 years. According to Jenssen and Kongshaug (2003), current production consumes c. 1.2 percent of the world's energy and releases c. 1.2 percent of the total volume of greenhouse gases, notably carbon dioxide and nitrous oxides.

Table 6.7. Changes in fertilizer consumption, 1961-2001 (based on FAO, 2004a).

	Item (1000 t)			
World Consumption	**Nitrogenous Fertilizers**	**Phosphate Fertilizers**	**Potash Fertilizers**	**Total Fertilizers**
1961/62	11,588	10,931	8,664	31,182
1971/72	33,536	22,435	17,340	73,310
1981/82	60,452	30,946	23,749	115,147
1991/92	75,633	35,241	23,732	134,606
2001/02	81,970	33,050	22,711	137,730

The use of phosphate and potash fertilizers, both derived from specific sedimentary rocks through mining rocks, have also increased substantially (Table 6.7). Both are produced by the mechanical processing of rock which requires fossil-fuel energy and both generate dust as a pollutant. There is unlikely to be any abatement in fertilizer production/consumption in the next few decades and possibly an increase as agricultural systems in the developing world seek to increase productivity.

The use of fertilizers to capture carbon can have disadvantages. Apart from the fossil-fuels their production involves, they may also cause the cultural eutrophication of ground water, wetlands, aquatic systems and the nearshore marine environment. This occurs because not all of the fertilizer is taken up by the crops for which it is intended; once in the soil a proportion is removed below the root zone by percolation and may enter ground water. High nitrate levels can thus develop in ground water with implications for domestic water supplies as high nitrate concentrations may constitute a health risk. In addition, some nutrients from the fertilizer enter drainage systems where they have an enriching effect. Nitrogen and phosphate, for example, are limiting factors to growth but once enrichment occurs in aquatic systems algal reproduction is stimulated. Algal blooms may form and inhibit the exchange of gases, notably oxygen, between the water and the atmosphere. Anaerobic conditions ensue and other organisms such as fish and crustaceans will die. This process of enrichment is known as cultural eutrophication. It is a problem in regions with industrialized agriculture and in coastal areas, especially in enclosed basins, which receive enriched drainage waters such as the Mediterranean Sea and Chesapeake Bay off the east coast of the USA. Mitigation is possible through close management of the volumes and timing of fertilizer treatments.

Table 6.8. The extent of irrigated land 1970-2002 (data from FAO, 2004a)

x 10^3 ha	**1970**	**1980**	**1990**	**2000**
AFRICA	8483	9491	11235	12711
ASIA	106666	132377	155009	192692
EUROPE	10583	14479	17414	25341
LATIN AMERICA	10191	13811	16794	18591
NORTH AMERICA	16421	21178	21618	23170
WORLD	168034	210220	244988	275188

Irrigation is another important element of agriculture worldwide and, along with fertilizer use and crop breeding, it has contributed to substantial gains in crop productivity in the last 30 years. As Table 6.8 shows, the extent of irrigation has increased globally by c. 60 percent since 1970, with the largest increases occurring in Asia and Europe. However, it too has its drawbacks. Injudicious management has resulted in problems of salinization of soil and water in irrigated regions, often with implications for surrounding regions. Salinization occurs because irrigation channels expose an increased volume of water to the atmosphere and evaporation causes an increase in salinity in the channel water and in the soil water. Too little water also encourages salt formation in soils while too much water causes waterlogging which results in a rise in the water table. As water evaporates an upward movement of water is established which leaves behind salts in the root zone as the water evaporates. This will inhibit growth and increasing salinity tends to reduce productivity due to the low tolerance of many crops to salinity. In extreme cases a salt crust may develop over soils and inhibit percolation and growth. Table 6.9 gives data on the distribution of salinized land and shows that the greatest problems are in Australia and Asia.

Table 6.9. The distribution of salinized soils (based on data quote in FAO, 2000).

Regions	Total area x 10^6 ha	Saline soils	%	Sodic soils	%
Africa	1899.1	38.7	2.0	33.5	1.8
Asia and the Pacific and Australia	3107.2	195.1	6.3	248.6	8.0
Europe	2010.8	6.7	0.3	72.7	3.6
Latin America	2038.6	60.5	3.0	50.9	2.5
Near East	1801.9	91.5	5.1	14.1	0.8
North America	1923.7	4.6	0.2	14.5	0.8
Total	12781.3	397.1	3.1%	434.3	3.4%

Whilst acknowledging problems associated with data on soil salinity, the WRI (2003, quoting a PAGE Agroecosystems Report of 2000) suggests that about 1.5 x 10^6 ha of irrigated land are lost to salinization annually and cost c. \$11 x 10^9 per year in reduced productivity which amounts to just under 1 percent of both the global irrigated area and annual value of production. Thus irrigation can be counterproductive; in extreme cases land cannot support agricultural crops, nor can it revert to its natural state. Both its carbon-producing and carbon-storing capacities are impaired. Good management, involving the provision of an adequate amount of water and a replacement of open, often unlined, drainage channels with covered channels or sprinklers, is the solution. One of the most acute examples of poorly-managed irrigation is that of the Aral Sea basin in Central Asia. Here, water extraction for irrigation has been so great that little water now reaches the sea from the Amu Darya and Syr Darya Rivers. The level of the Aral Sea has dropped as it has lost c. 70 percent of its volume (see Middleton, 2004 for an account); fishing industries have been lost, health problems have ensued and salinization is a major problem; prospects for mitigation are slim.

6.3.2 Forest loss and forestry

Human impact has affected all biomes but especially the forests, as documented by Williams (2003). The spread of agriculture between 10,000 and 2,000 years ago was achieved at the expense of the Mediterranean forests and shrublands and the temperate forests of Europe. The last two hundred years have witnessed a continued but slower demise of these communities, and in recent times a shift to conservation and renewal, while the tropical and boreal forests are now the foci of attention and are being lost at an alarming rate. The forest history of the world is noteworthy for a variety of environmental and cultural reasons but not least because of their importance as a store of carbon. As stated in Section 6.2.1, clearance for agriculture has been, and continues to be, the primary cause of forest loss, though the acquisition of wood for fuel, shelter and other wood products as well as mining are also important. In many respects trees provided the resources which are now obtained from oil. Forest utilization has resulted in a reversal of Holocene trends to absorb carbon in the biosphere. However, some activities are redressing forest losses, though only on a small scale. These include afforestation and plantation agriculture as well as incentives to preserve forests such as tourism and bioprospecting for organisms that might be future crops, pharmaceuticals etc. In addition, conservation measures in some regions are contributing to forest preservation and there are new possibilities opening up with likely arrangements under the Kyoto Protocol to counteract carbon emissions with carbon sinks such as forests. This is discussed in more detail in Chapter 7.

There are major discrepancies between estimates for biosphere carbon storage pre and post significant human activity. Thus any consideration of overall change is spurious but some consideration of these changes remains important in order to better understand carbon-cycle dynamics with and without human intervention. Otto *et al.* (2002) have estimated that since the last glacial maximum (c. 21,000 years ago) there has been an increase of 827.8 to 1106.1 Gt of carbon in the biosphere and as shown in Table 2.11 there was a shift of atmospheric carbon dioxide concentrations from c. 180 ppmv to c. 280 ppmv as the interglacial opened. The increased storage of carbon in the biosphere occurred in forests and peatlands, though partitioning this storage is uncertain. However, Behling (2002) has determined that storage increased substantially in South America's forests, as shown in Table 6.10.

The greatest increase occurred in the Amazon rain forest, which Behling suggests expanded in area by c. 39 percent. This is also indicated by Mayle and Beerling (2004), who calculate that LGM carbon storage in Amazonia was c. 135 Gt, and although this represents only c. 50 percent of its carbon storage today, it amounted to twice as much as a proportion of the total terrestrial carbon store. What is not clear, however, is where all this carbon was sequestered during the cold stages, though one possibility is the oceans. Moreover, deforestation through human action over the last c. 5,000 years and especially in the last three hundred years has impaired the world's forests as carbon stores and, as in Section 6.2.1, there has been a release of carbon from forest biomass and associated soils as forested land has been appropriated for agriculture. Indeed, Ruddiman (2003) has suggested that forest clearance for agriculture c. 8,000 years ago marked the start of global warming (see Section 2.3). However, land-use change accelerated after 1800 (see Figure 6.8) and since 1850 some 156 GtC (Houghton, 2003, see above) have been released to the atmosphere due to global deforestation.

Table 6.10. Changes in carbon storage in tropical forests of South America (data from Behling, 2002).

	Increase in carbon storage (10^9 tonnes)	**% change**
Amazon rain forest	28.3	20
Atlantic rain forest	4.9	55
Araucaria forest	3.4	108
Semi-deciduous forest	6.3	46

The causes of deforestation are many. Mather *et al.* (1998) believe that c. 50 percent of the deforestation throughout human history is due to population growth, a major stimulant of agriculture. Related factors include land tenure, poverty, fuel needs and trade. Many countries have experienced what Mather (1992) describes as a forest transition stage where forest areas, having once been reduced in extent are being increased through afforestation and conservation policies. This is particularly the case in many developed countries whereas many developing countries continue to deplete their forest resource, often on a large scale and because there are ready markets in the developed world. Thus, deforestation is greatest in the tropics and subtropics due to clearance for agriculture and the provision of wood and wood products. This produces a net flow of carbon to the atmosphere and to developed countries. However, many developed countries have long-established forest industries e.g. the USA, Canada, Scandinavia and Russia. These also give rise to carbon shifts but generally to other developed nations while management and reforestation encourage carbon sinks to compensate for losses.

The forest industries of North America, for example, are based on boreal and temperate forests (see Figure 6.1). The forest industries of Canada and the Russian Far East illustrate the size and importance of forestry in the boreal/temperate zone. Canada is the world's leading producer/exporter of timber products.

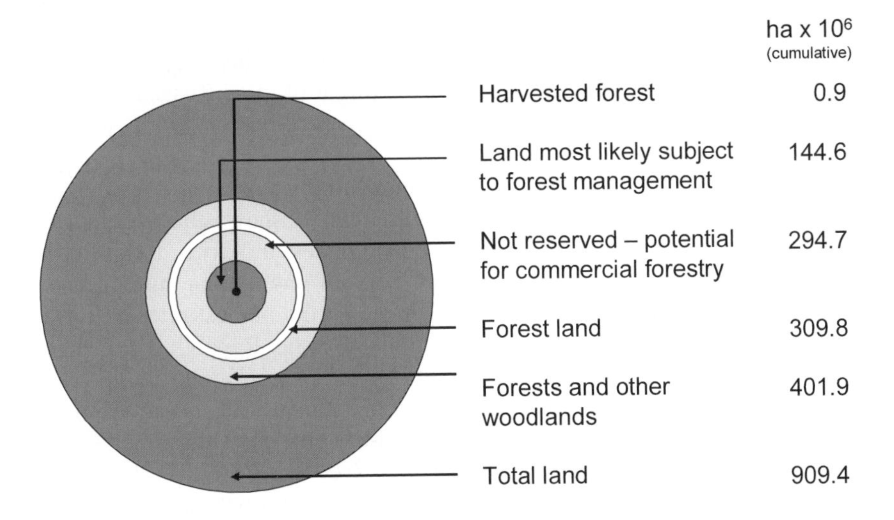

Figure 6.11. Canadian forest industry (based on Natural Resources Canada, 2004)

Data on forest lands are given in Figure 6.11 which shows that forests occupy 402 x 10^6 ha of which 94 percent are publicly owned. Timber industries are based on the granting of licences by provinces for 25 years; management includes, extraction and replanting. Approximately 1 x 10^6 ha are harvested annually mainly through clear cutting which involves the removal of large tracts at a given time, exposing soils to oxidation. Replanting often involves only one tree species and so the multiculture of nature is being replaced with the monoculture of human use. The industry is worth c. $80 billion of which c. 50 percent constitutes export earnings (data are for 2001). According to Environment Canada (2002) forestry, and land use, are a carbon dioxide sink but one that has declined by c. 30 percent since 1990. Another area that receives relatively little attention is the boreal zone of the Russian Federation. Forest industries are a traditional way of life but in the last two decades the forests of Siberia and the Russian Far East have become a focus of exploitation. According to Newell (2004) logging has intensified in the Russian Far East since the end of the Soviet era. This is because of reduced demand from former Soviet states and increased transport costs as well as increasing demand from Pacific-rim countries such as South Korea, Japan and China. In 2000 Newell states that 12 to 15 x 10^6 m^3 of timber was produced compared with 10.5 x 10^6 m^3 in 1995. Moreover, extraction using clear felling has increased and the demise of local wood processing, with associated job losses, has resulted in increased waste; small logs, branches etc are left behind as only the best and largest logs are required for export. Even during the Soviet era c. 40-60 percent of cut wood was wasted, a figure far in excess of that in Canada or Scandinavia but this has increased further as woodchips production has decreased and as transport costs have increased. Illegal logging is also a serious and apparently widespread problem, including in areas where logging is not allowed by law. Despite these problems the plight of these Russian Far East forests receives little media attention when compared with tropical deforestation.

The deforestation of the tropics and subtropics is better documented. Table 6.11 gives broad-scale data by continent. It shows that the rate of forest loss between 1990 and 2000 has been greatest in Africa, followed by South America with an annual overall loss of more than nine million hectares worldwide. According to the FAO data the highest rates of forest loss have occurred in the Ivory Coast, democratic Republic of Congo, Indonesia, Malaysia, Nicaragua, Haiti and Brazil. According to Malhi *et al.* (2002) tropical deforestation releases 1.7 GtC annually to the atmosphere, the single largest source of carbon from land-use change. This amounts to c. 27 percent of the emissions from fossil fuels (see Section 6.2.3). However, sequestration of carbon occurs due to forest regrowth and enhanced productivity caused by the fertilization effect of high atmospheric carbon dioxide concentrations.

Table 6.11. Changes in forest cover 1990 to 2000 (data from FAO, 2004a).

		Forest area, 2000				Forest cover change, 1990-2000	
Country / area	Land area (10^6 ha)	Total forest (10^6 ha)	% of area	Area per capita (ha)	Forest plantations ('000 ha)	Annual change ('000 ha)	Annual rate of change (%)
Africa	2978	650	21.8	0.8	8036	-5262	-0.8
Asia	3084	548	17.8	0.2	115847	-364	-0.1
Europe	2259	1039	46.0	1.4	32015	881	0.0
North and Central America	2136	549	25.7	1.1	17533	-570	-0.1
Oceania	849	198	23.3	6.6	2848	-365	-0.2
South America	1755	886	50.5	2.6	10455	-3711	-0.4
World	13064	3869	29.6	0.6	186733	-9391	-0.2

Given increasing demands for wood and wood products plus impacts from mining deforestation is set to continue for some time to come. Moreover, logging and mining tend to open up hitherto inaccessible forests for agriculture so in many regions secondary growth is restricted and carbon storage is not restored at least in the short term.

Afforestation/reforestation in various forms is also occurring in many countries but at a much lower rate than that of forest removal. This plus forest regrowth and increased sequestration of carbon due to the fertilizing effects of increased atmospheric carbon dioxide concentrations are counteracting carbon emissions from fossil fuels to some, albeit small, extent. This trend is likely to increase in the future as governments formulate policies of carbon emission mitigation under the umbrella of the Kyoto agreement on Climatic Change (see Chapter 7).

6.3.3 Industry, fossil-fuel use and urbanization

Industrialisation, fossil-fuel use and urbanisation are interlinked. As discussed in Section 5.4, industrialization on a large scale began in the UK in the mid-eighteenth century and then spread to Europe and North America. Based mainly on the processing of local and imported raw materials, the industrial character of these nations has altered substantially. The major trend has involved a shift from heavy industries such as iron and steel production and textiles to lighter industries such as pharmaceuticals, finance and information technology. Much of this shift has been necessitated by competition; other nations began to industrialize and because of cheap labour they began to produce manufactured goods at lower cost. Today, industrialization is occurring in Asia, notably in China, India following similar developments in Taiwan, South Korea and Hong Kong. Industrialization requires energy. Consequently, patterns of world energy consumption mirror patterns of industrialization. One point of contrast, however, with the eighteenth and nineteenth centuries is the fact that the main energy consumers are not necessarily the main producers. The result is a flow of carbon from producers to consumers. Industrialization also requires labour. When it began in the UK and Europe in the mid-1700s there was a parallel migration of people from the countryside to the industrial centres. A similar process is occurring today where industrialization is underway. The shift of people from the countryside to the cities is also occurring in many nations where industrialization is non-existent or at a low level, as in many African nations, because of the apparent lure of possible work and because of population increase. Thus urbanization is occurring worldwide.

From c. 1950 industrialization began to occur in regions beyond those with a traditional industrial base. Several Asian countries began to establish manufacturing and became known as the Asian Tiger economies.

A combination of cheap labour, improving educational standards and investment in infrastructure and industrial plant provided the conditions for successful economic growth in Taiwan and South Korea. These economies continue to grow and have been joined by other nations which include Malaysia, Thailand, the Philippines and Indonesia.

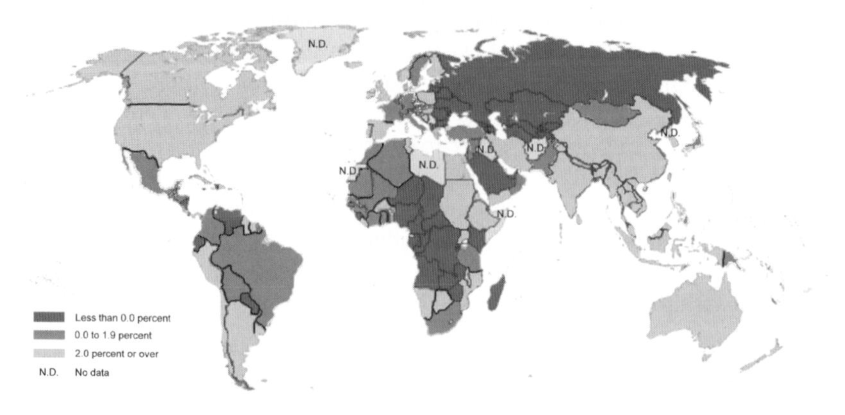

Figure 6.12. GDP growth per capita for 1990-2001 (based on World Bank, 2004)

However, the last two decades have witnessed rapid growth in China and India which are two of the most populous countries in the world. As Figure 6.12 shows, all of these nations have enjoyed considerable increases in the per capita Gross Domestic Product (GDP) which is the value of all final goods and services produced within a country's borders on the basis of head of population. China and India achieved the highest growth rates of c. 3 percent or above. Indeed China has achieved growth rates of c. 9 percent annually since 1978 when economic reforms began. Many nations experienced much lower growth but started from a considerably higher base, e.g. Western Europe, North America. However, most Sub-Saharan nations and many Middle Eastern nations achieved no or little growth, thus accentuating the gap between rich and poor.

Industrial development has been underpinned by fossil-fuel use, and especially increasing dependence on oil. Hall *et al.* (2003) state that "The global use of hydrocarbons for fuel.... has increased nearly 800-fold since 1750 and about 12-fold in the twentieth century". Data on historical fossil-fuel production and consumption are given in Section 5.4 (see Tables 5.3, 5.4, 5.5, 5.6, 5.7 and 5.8; Figure 8.7). Fuel production and consumption have risen substantially in the last 30 years. According to the International Energy Agency (IEA, 2004b) energy supply amounted to 6,034 Mt oil equivalent (includes c. 14 percent non fossil fuels) in 1973 and rose to 10,230 Mt oil equivalent (includes 21 percent non fossil-fuels) by 2002.

Table 6.12a. The major exporters and importers of coal in 2002 (based on IEA, 2004b)

Coal exporters	**Amount (Mt)**		**Coal importers**	**Amount (Mt)**
Australia	208		Japan	162
China	93		Korea	72
Indonesia	90		Taipei	54
S.Africa	71		Germany	35
Russia	60		UK	32
Colombia	46		Russia	24
USA	39		India	24
Canada	26		USA	23
Kazakhstan	25		Netherlds	22
Poland	20		Spain	22
Rest World	40		RestWorld	239
Total World	718		**Total World**	709

Table 6.12b. The major exporters and importers of oil in 2002 (based on IEA, 2004b)

Oil exporters	**Amount (Mt)**		**Oil importers**	**Amount (Mt)**
S.Arabia	289		USA	515
Russia	188		Japan	206
Norway	140		Korea	108
Venezuela	110		Germany	105
Mexico	95		Italy	90
Iran	95		India	82
Nigeria	92		France	80
UK	87		China	69
Canada	80		Spain	58
USA	79		UK	57
RestWorld	663		RestWorld	667
Total World	1918		**Total World**	2037

Table 6.12c. The major exporters and importers of natural gas in 2002 (based on IEA, 2004b)

Gas exporters	**Amount 10^3 Mm^3**		**Gas importers**	**Amount 10^3 Mm^3**
Russia	186		USA	111
Canada	102		Germany	84
Norway	71		Japan	81
Algeria	64		Ukraine	66
Netherlds	48		Italy	62
Turkmenistan	43		France	43
India	41		Austria	34
Austria	27		Netherlds	26
Malaysia	25		Korea	25
USA	20		Spain	23
RestWorld	157		RestWorld	225
Total World	784		**Total World**	782

The flows of carbon worldwide are reflected in the data given in Table 6.12 a-c. Of particular note are the major flows of carbon occurring from Russia, Saudi Arabia and Central Asian nations to the USA, Europe and industrial Asia. The rapidly industrializing nations of India and China are also significant importers of oil. According to IEA (2004b) oil consumption in China has increased from 2,000 barrels per day in 1980 to c. 5,500 barrels per day in 2003; coal still provides c. 70 percent of China's energy requirements of which some 66 percent is used in the industrial sector. India's growth has not been quite as great as that of China but has nevertheless been substantial. IEA's country analysis shows a growth rate of 4 percent in 2002, rising to 8.3 percent in 2003 and a projected growth of c. 6 percent for 2004 and 2005. Oil use has increased from c. 600 barrels per day in 1980 to c. 2,200 in 2003 and now accounts for c. 30 percent of India's energy consumption. Inevitably, such increases are accompanied by additional pollution and carbon emissions to the atmosphere. An analysis of carbon emissions is available from the Energy Information Administration (IEA, 2004b) and Table 6.13 gives carbon emission data for 1950 to 2001. It shows the rapid acceleration of emissions since 1950, including a major leap between 1960 and 1970. Current emissions are four times greater than those of the 1950s and 1960s and are increasing substantially year-on-year. Projections for the next 50 years are examined in Chapter 8.

Table 6.13. Carbon emissions from fossil-fuel use, 1970-2001 (data from Worldwatch Institute, 2002)

YEAR	Carbon emissions Gt
1950	1.61
1960	1.75
1970	3.98
1975	4.52
1980	5.18
1985	5.27
1990	5.94
1995	6.18
2000	6.48
2001	6.55

Urbanization has been a phenomenon since the Industrial Revolution and like recent industrialization it has accelerated in the last three decades. It is fuelled by population growth and opportunities for work, and is important in the context of carbon domestication because it facilitates carbon acquisition. First, labour is provided for industry; second, it centralises carbon consumption as food; third, fuel provision for domestic use is facilitated by a concentrated rather than a dispersed population. Most of the population of developed nations is already concentrated in cities (including suburbs) but the process of urbanization has characterized many developing nations for the last 50 years. Cohen (2004) has reviewed these urbanization trends and states that "Today, at the beginning of the 21st century, there are around 400 cities around the world that contain over a million residents, and about three-quarters of these are in low- and middle-income countries". Some of these are large enough to be classified as mega-cities with populations in excess of ten million people. According to the latest projections from the UN (United Nations, 2002), virtually all of the world's population growth over the next 30 years will occur in urban areas. Since 1900, when there were only 16 cities with populations in excess of one million, the world has undergone an urban transition. Today, there are some 400 cities with a population of more than one million and about 75 percent of these cities are in low and middle income countries. Approximately half of the world's population and c. 75 percent of westerners live in cities, a trend set to continue as shown in Table 6.14. Much of the increase will occur in smaller cities and towns through migration from rural areas and internal growth.

Table 6.14. Projected population growth by sector (based on United Nations, 2002)

YEAR	URBAN	RURAL
2000	2.86×10^9	3.19×10^9
2030	4.98×10^9	3.29×10^9
TOTAL	7.84×10^9	6.48×10^9

The highest rates of urbanization are in Africa and Asia and by 2030 there will be 53 and 54 percent of their respective populations in urban areas. In Asia many cities will become mega-cities as industrialization proceeds but in Africa the growth of cities is not linked to industrialization and many cities are described as 'economically marginalized' by Cohen (2004). Overall, the use of carbon will become less spatially uniform as its consumption, mainly as food and fuel, increases as cities grow.

6.4 Measures of human impact

As the sections above illustrate, human impact on the Earth's surface is considerable and increasing. This impact is mainly due to carbon appropriation, notably food and fuel procurement. In the last decade measures to describe this impact quantitatively have been constructed (see review in Mannion, 2002). To some extent such measures are surrogates for a measure of carbon use. The habitat index of Hannah *et al.* (1994) and the living planet index (LPI) devised by the Worldwide Fund for Nature (WWF, 1998 and updates including Living Planet Report, 2004) provide measures of land-cover alteration and environmental quality respectively. The ecological footprint, first suggested by Rees (1992), provides a measure of resource consumption and is now widely used as a measure of sustainability. In addition, the human appropriation of net primary productivity (HANPP) provides a direct measure of carbon flow from nature to society (see Haberl *et al.,* 2004) while it is also possible to construct a carbon index, as suggested by Mannion (2002).

The habitat index (Hannah *et al.,* 1994) is a direct measure of disturbance of the natural land cover as recorded by a variety of techniques such as atlases, maps and remote sensing. The scale of analysis provides results at the continental scale and so the index provides a large-scale snapshot of human-induced change. The measures of land-cover change are given in Table 6.15 which shows that Europe has the lowest habitat index, i.e. it has been changed more extensively than any other continent. In contrast Australia and South America have the highest indices reflecting the highest percentage of undisturbed land area.

Table 6.15. The habitat index of Hannah *et al.* (1994).

	North America	South America	Europe	Africa	Asia	Austr-alasia
% undisturbed (mostly primary vegetation)	56.3	62.5	15.6	48.9	43.5	62.3
% partially disturbed (mostly agriculture)	18.8	22.5	19.6	35.8	27.0	25.8
% human dominated (permanent agriculture, urban areas)	24.9	15.1	64.5	15.1	29.5	12.0
HABITAT INDEX (HI)	**61.0**	**68.1**	**20.4**	**57.9**	**50.3**	**68.8**

HI is calculated by (% undisturbed area + 0.25 x partially disturbed area x 100) ÷ 100

In general, c. 52 percent of the Earth's surface remains undisturbed, a value which declines to only 27 percent if bare rock and ice-covered areas are excluded. The Living Planet Index (LPI) is a more complex measure than the HI and takes into account a variety of quantified data on biodiversity, notably measures of the abundance of 555 vertebrates in forests, 323 vertebrate species in freshwaters, and 267 marine organisms. These indices are used because they represent a large proportion of what is known as natural capital, a measure applicable at scales from the local to the global. The index employs a baseline date of 1970 and thus provides a comparative measure for the last 30 years. It can be calculated for regions or nations with higher values representing greater disturbance. LPI for the Earth as a whole is shown in Figure 6.13. It shows that between 1970 and 2000 significant losses in natural capital occurred, with the greatest losses occurring in freshwater environments. WWF also calculated consumption pressure based on grain, marine fish, forest products, freshwater consumption, carbon emissions and cement production for regions and nations. All but freshwater consumption are carbon based so this measure is almost a surrogate for carbon manipulation.

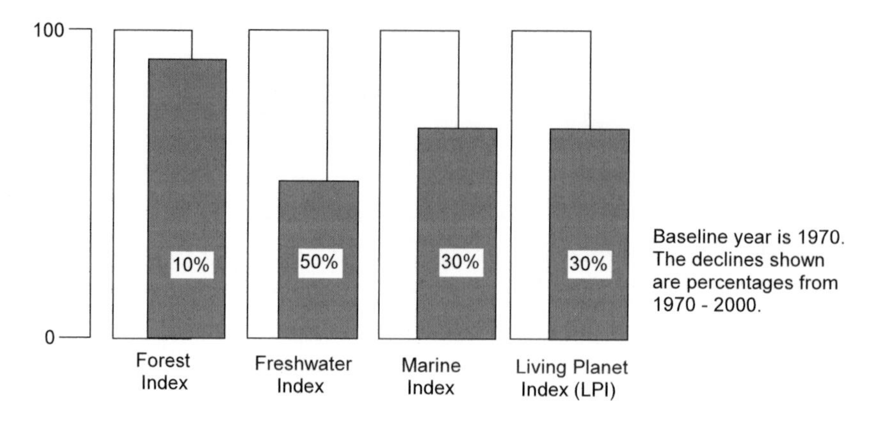

Figure 6.13. LPI for the earth as a whole (based on WWF, 2004a)

Table 6.16 gives examples of regional and national consumption pressure. The disparity between the developed and developing worlds is evident but significant unevenness occurs within these categories, as reflected in the data for the USA and the UK and the differences between Africa and Asia. Such differences have been reiterated by the Living Planet report of 2004 (WWF, 2004). The WWF also promotes the ecological footprint (EF) approach advocated by Rees (1992; see also Chambers *et al.*, 2000). The EF, like the consumption pressure referred to above, focuses on resource consumption and the area of land necessary to provide those resources and to deal with waste; it calculates the area of land required to support the population of a given area on a *per capita* basis. For areas of high population density such as cities (see Section 6.2 above) the area required to support the population will greatly exceed the area occupied by the city. The sustainability or unsustainability of a population will depend on the capacity of its hinterland to provide adequate resources and waste disposal. Wackernagel *et al.* (1997) have calculated the EF on a national basis, some examples of which are given in Table 6.17. Where this exceeds the biocapacity, i.e. the ability of supporting ecosystems in the hinterland, there is an ecological deficit, a situation that is unsustainable in the long term. As the examples in Table 6.16 show, many nations have negative ecological deficits because their EF exceeds their biocapacity. Moreover, Wackernagel *et al.* (1997) show that globally there are 2 ha of biologically productive land per person which is exceeded to produce an ecological deficit. The implication is that current levels of consumption and waste disposal are unsustainable.

Table 6.16. Selected WWF consumption pressure data (based on WWF, 1998).

1. REGIONAL VALUES	**Consumption units** **1 = world average per capita 1995**
Africa	0.55
North America	2.70
Europe	1.72
Asia	0.83
Middle East/Central Asia	1.11
2 NATIONAL VALUES	**Per capita**
USA	2.74
UK	1.43
China	0.85
India	0.47
Niger	0.34

Table 6.17. Examples of the ecological footprint, biocapacity and ecological deficit (based on Wackernagel *et al.,* 1997).

	Ecological footprint ha *per capita*	**Biocapacity ha *per capita***	**Ecological deficit ha *per capita***
EUROPE			
Germany	5.3	1.9	-3.4
Norway	6.2	6.3	0.1
Russia	6.0	3.7	-2.3
UK	5.2	1.7	-3.5
ASIA			
Bangladesh	0.5	0.3	-0.2
China	1.2	0.8	-0.4
Hong Kong	5.1	0.0	-5.1
India	0.8	0.5	-0.3
Japan	4.3	0.9	-3.4
AMERICAS			
Canada	7.7	9.7	1.9
USA	10.3	6.7	-3.6
Chile	2.5	3.2	0.7
Peru	1.6	7.7	6.1
AUSTRALASIA			
Australia	9.0	14.0	5.0
New Zealand	7.6	20.4	12.8
WORLD average	2.8	2.1	-0.7

Overall, developed nations have greater ecological deficits than developing nations. This is because they consume more resources, such as fuel, and may have limited hinterlands. In Europe, for example, biocapacity may be exceeded because the relative lack of forests limits the uptake of carbon dioxide from fossil-fuel consumption. Consequently these nations export a proportion of their waste to be absorbed by forests elsewhere. Such calculations underpin the notion of carbon trading which may become a means of managing the carbon cycle as part of the Kyoto protocol on climatic change. This is discussed in Chapter 7. The EF approach can also be used to examine the sustainability of individual cities or settlements or of specific activities such as fish farming or tourism. For example, Holden (2004) determined that the EFs of Greater Oslo and Førde in Norway indicate that the least impact comes from small compact cities while McDonald and Patterson (2004) have shown that Auckland has the largest EF of any New Zealand city and is especially dependent on the province of Waikato. Recently, Wackernagel *et al.* (2004) have demonstrated the use of the EF to monitor changing consumption patterns. For example, they have shown that South Korea and the Philippines have become ecological debtors as their economies have expanded between 1960 and 1990; prior to this development both were ecological creditors with 'spare' biocapacity.

The indices of sustainability so far examined are based on patterns of consumption and/or waste. However, a direct measure of carbon flow from nature to humans is that of the human appropriation of net primary productivity (HANPP). Vitousek *et al.* (1986) first drew attention to the scale of this appropriation in the mid-1980s when they determined that humans used c. 40 percent of net primary production on land. Recent work on HANPP includes that of Haberl *et al.* (2004) who point out that the HANPP reflects how much of the total energy available in a given area is diverted to humans and is a measure of the intensity of land use re the energetics of the producing system. The disadvantages of this approach include the difficulty of determining thresholds and its failure to take into account the wider impact of human consumption of natural capital which impinge in areas beyond that of production and consumption, e.g. carbon dioxide emissions. However, since the commodities consumed by humans are carbon based, a reflection of the degree of carbon domestication, a carbon index could also prove to be a useful tool for environmental management, as proposed by Mannion (2002). Table 6.18 gives details of how such an index could be calculated. It includes, agricultural, wood and paper production/consumption and carbon dioxide emissions, the combination of which give a carbon index. This can be divided by the number in the population to give a *per capita* carbon index, as shown in Table 6.19.

Table 6.18. Examples of carbon indices (based on Mannion, 2002).

	CO_2 emission Gt	Agricultural production Gt	Wood production Gt	Paper production t x 10^3	Total Carbon Index Mt
Ethiopia	3.5	9.8	46	8	59
China	3192	585	326	25468	4129
Singapore	64	NIL	0.145	93	64
UK	542	29.4	10	5777	588
USA	5468	347	606	85261	6508
World	22714	2649	3805	272082	29440

Table 6.19. Per capita carbon indices (based on Mannion, 2002).

	Total Carbon Index Mt	Population x 10^6	Carbon Index *per capita*
Ethiopia	59	55	1.07
China	4129	1221	3.38
Singapore	64	2.8	22.44
UK	588	58	10.08
USA	6508	263	24.72
World	29440	5716	5.15

The data of Table 6.19 show that the *per capita* world average for 1995 is 5.15, a value exceeded by most developed nations. These data reflect the importance of a nation's impact on the global carbon cycle. As with the ecological footprint, the carbon index could be applied at local, regional and national scales.

Chapter 7

7 THE POLITICIZATION OF CARBON

Carbon appropriation has always been political and most of politics concerns carbon! This is because resource manipulation, especially that of food, wood and fuel energy, has been strategic in the development, internal operation and international influence of any tribe, group or nation. The availability and abundance of food, wood and fuel has always conferred considerable advantage to human groups throughout prehistory and history. The command of carbon has generated and continues to generate political pre-eminence; carbon security thus tends to equate with political security, power and influence. In modern times the importance of carbon-based resources has led to institutionalization and politicization worldwide. This has occurred at national and international levels, a complete appraisal of which is beyond the scope of this book. What follows is a brief survey of some of the major instruments of carbon control beginning with the establishment of the concept of national parks in the late 1800s and emphasizing the 'globalization' of carbon politics in the post World War II era.

While little can be stated with certainty about the carbon politics of human groups in the distant past, it is axiomatic that carbon availability, as food and wood, was a major advantage; it almost certainly featured in the development of group acquisition and consumption. As Lewin and Foley (2004) have discussed, scavenging, hunting and food sharing have loomed large in the debate about what it is to be human. The ramifications of early hominid food acquisition/sharing, especially in relation to meat, include roles in human evolution and the social organization of human groups. Indeed, food procurement was so important that it culminated in the domestication of animals and plants and the beginning of agriculture c. 10,000 years ago. Why it came about remains enigmatic but it must have been expedient politically for reasons which include population maintenance and the generation of commodities for trade (see comments in Section 5.2).

The expansion of Europe beginning in the 1500s was undoubtedly political; power and resource acquisition underpinned a movement of people and resources that changed the world. Much of that change involved carbon appropriation as biomass resources and the exchange of crops, animals and agricultural practices. Carbon flows to Europe constituted a powerful force in European development. That there were concerns about environmental change and degradation is illustrated by the efforts of William the Conqueror (1066-1087) and his successors to enforce protection for designated 'forest' land and the planting of forests by the Tudor kings. Such concerns were also voiced in the publications of the English countryman John Evelyn (1620-1706) in the mid-1600s who was also concerned about the dearth of trees. Two centuries later, John Muir (1838-1914) was extolling the virtues of Nature in North America, especially in his beloved Rocky Mountain environments. By 1872, against a backdrop of predictions of a timber famine in the USA based on rapid clearance rates in the 1800s, the first national park, that of Yellowstone, was created; by 1907 the number of forest reserves had increased to 159 (see review in Mannion, 1997a). Not only did this promote conservation, and thus maintenance of the carbon store, but it also preserved the wildlife and environment for the emerging and now burgeoning industry of tourism. Similar concerns about timber resources were being aired in Britain and forest protection organizations were established in some colonies, e.g. Dehra Dun in India. Following World War I, when the increased need for wood had highlighted the dearth of forest resources, the Forestry Commission was established in 1919 to manage forests and forestry at home. Thus a policy of afforestation commenced. These developments reflect the increasing role of government, and hence of politics, in environmental management. Subsequently, national parks or their equivalent have been established worldwide and forest management at national level is now generally controlled by governments.

However, the political engagement of the modern conservation/environmental movement has more recent origins. In the years immediately following World War II two major institutions were created: the Food and Agriculture Organization in 1945, an agency of the United Nations, and the World Conservation Union (IUCN, originally the International Union for the Protection of Nature – IUPN). Both were concerned with assessing resources, improving management practices worldwide and monitoring change. However, it was during the 1970s that more substantial change in relation to the role of the environment as a political issue occurred. First, there was the founding of formal Green political parties in the early 1970s. Thus the environment, as a primary focus, entered mainstream politics just before the oil crisis of 1973 which highlighted the significance of so-called petropolitics, one of the many components of globalization. Second, two internationally important pressure groups came into existence: Friends of the Earth (FoE) and Greenpeace (GP), both formed in 1971 in the UK and Canada respectively.

The 1970s also witnessed further involvement of the United Nations in environmental issues with the establishment of the Man (*sic*) and the Biosphere (MAB) programme of the United Nations Education Scientific and Cultural Organization (UNESCO) in 1971 and the United Nations Environment Programme (UNEP) in 1973. The World Conservation Monitoring Centre (WCMC) was established soon after, in 1979, to provide information and technical assistance to the UNEP. Many international agreements have resulted from initiatives of the UNEP, e.g. United Nations Convention on the Law of the Sea established in 1982.

In 1983 the United Nations established the World Commission on Environment and Development (WCED) chaired by the former prime minister of Norway, Gro Harlem Brundtland. The commission's report was published in 1987 (WCED, 1987) and is a landmark in environmental politics because it established the concept of sustainable development. One year later, the Intergovernmental Panel on Climatic Change (IPCC) was created to address the science and implications of climatic change and in 1997 the Kyoto protocol was signed with the major objective of reducing carbon dioxide emissions. Subsequently, political agendas for this and biodiversity loss have evolved. Conventions on Climate Change and Biodiversity emerged from the Earth Summit of 1992 (in Rio de Janeiro) which also spawned Agenda 21. This is an action plan for sustainable development, the buzz words of the environmental movement, which recognizes the roles of local, regional and national directives and the role of the individual in the achievement of economic advancement without compromising ecological integrity. Since then sustainable development at all scales has become a major objective of governments, industry and institutions. Moreover, in the last twenty years major initiatives on environmental conservation at national and continental scales have been inaugurated, e.g. Natura 2000, a product of the European Union's Habitat Directive aimed at establishing a network of key wildlife habitats.

Many non-governmental organizations (NGOs) have also been established; some are essentially lobbyists and politically activist while others are research-based organizations. The World Wide Fund for Nature (formerly known as the World Wildlife Fund) was one of the earliest NGOs to be formed. It was established in 1961, has close links with the IUCN and is based in Switzerland. Other examples include the Worldwatch Institute (WWI) and World Resources Institute (WRI); both are based in the USA. The former, founded in 1974, is another product of the growing environmental movement of the 1970s. The WRI was founded in 1982. Both are independently financed, produce publications with a statistical basis of wide appeal and have educational programmes.

Transcending all these organizations is trade, an important component of mainstream politics. Trade, past and present, has exerted major influences over carbon flows, as illustrated in Chapter 6, especially through agriculture, forestry and fuel resources. Trade agreements are contentious because of the advantages and disadvantages they confer. The General Agreement on Tariffs and Trade (GATT) was established at the end of World War II to oversee the rules of trade between nations. In 1995 it was superseded by the World Trade Organization (WTO), an organization of most trading nations based in Switzerland whose agreements regulate trade between countries. Many regional trading blocs have also been established to liberalize trade. Not all trade involves carbon but since a great deal of it does the agreements of the WTO are major players in contemporary carbon flows.

Interestingly, in 2004 the Nobel Prize for Peace was awarded to Wangari Maathai, a Kenyan veterinarian and academic for 'her contribution to sustainable development, democracy and peace'. She founded, and for more than 30 years has been involved with, the Green Belt Movement which has focused on encouraging poor women to plant more than 30 million trees in an effort to curb desertification and provide resources.

7.1 Initiatives 1860 to 1939

A major reason for focussing on the ecological history of the USA and its relationship with conservation is because it is well documented and because it reflects the far-sightedness of its founding people. The transition from exploitation to management of forests and woodlands in the USA (see Section 6.2.2 for general comments on forest transition) occurred relatively rapidly when compared with Europe. This is because of the speed of colonization in the early nineteenth century from east to west, not least because it was more or less contemporaneous with the advent of the railways (see Mannion 2005) which contributed to the spread of agriculture and facilitated the transport of lumber. Certainly, the forests of the eastern seaboard suffered onslaught as initial colonization occurred and, in keeping with contemporary philosophies, this peopling was consciously or unconsciously an attempt at conquering or subjugating Nature for human advancement. Given the rapid rate of forest destruction which accompanied this wave of settlement, fears were expressed at government level about the future of the new nation's forests which were perceived primarily as economic commodities, notably as sources of much needed lumber and fuel.

Precedents were set, not with Yellowstone which is accorded the status of the first US National Park, but with Yosemite in California. According to Mackintosh (2000) various public figures persuaded the government to designate the area for public use. Accordingly, an act of congress, signed by President Abraham Lincoln was passed in 1864 to give the state of California control over the area; the land was to "be held for public use, resort, and recreation.....inalienable for all time". In contrast Yellowstone was designated as a national park under federal legislation because the administrative units in which it is located, Wyoming and Montana, had not yet received statehood. President Ulysses S. Grant signed the act in 1872 and some 2 million acres were "dedicated and set apart as a public park or pleasuring-ground for the benefit and enjoyment of the people". Control of the park was placed in the hands of the Secretary of State for the Interior and thus the federal system of national parks was born. Today there are 388 national parks, some of which are monuments or historic sites rather than true parks; they are administered by the National

Park Service (NPS) which is a federal agency. The role of the NPS is to manage the parks, especially to preserve wildlife habitats, to ensure responsible recreation and tourism and to provide an educational resource. The service also manages any construction or resource extraction and engages in conservation work. In terms of carbon, perhaps the greatest role of the national parks has been the preservation of vast areas of forests by restricting farming and settlement.

Not all the public forests of the USA were or are managed by the NPS. In 1905 congress passed the Transfer Act which allowed the transfer of forest reserves from the Department of the Interior to the Department of Agriculture to create the US Forest Service. This occurred mainly through the efforts of Gifford Pinchot who became the first head of the USDA Division of Forestry. The mission of the newly formed service was stated as "In the administration of the forest reserves it must be clearly borne in mind that all land is to be devoted to its most productive use for the permanent good of the whole people; and not for the temporary benefit of individuals or companies" (reproduced from a memo from James Wilson, Secretary of Agriculture but written by Pinchot and his assistant Olmsted (Roth and Williams, 2003)). In 1907 the forest reserves were redesignated the national forests of which there were c. 63 million acres. Pinchot and his successors inaugurated research programmes as an additional dimension to forest management. Today, the Forest Service is responsible for 155 national forests and 20 national grasslands. Its mission is "to sustain the health, diversity, and productivity of the Nation's forests and grasslands to meet the needs of present and future generations", a mission in keeping with the concept of sustainable development (see Section 7.6). The production of wood and wood products is a major role of the Forest Service through their management of public forests which contribute to the USA's substantial lumber industry.

Table 7.1 gives the main objectives of the Forest Service. Although admirable overall in its aspirations, the Forest Service has in the past been accused of sanctioning unsustainable forest exploitation. For example, in the 1980s, the continued cutting of old growth forests in the northwest with leases being given to Japanese companies for felling and export, came in for much criticism. Nevertheless the Forest Service has achieved sustainable yields in many forests and is thus achieving a high success rate for sustainable forestry.

Table 7.1. The objectives of the US Forest Service (from USDA Forest Service, 2004).

- To advocate a conservation ethic in promoting the health, productivity, diversity, and beauty of forests and associated lands.
- To listen to people and to respond to their diverse needs in making decisions.
- To protect and manage the National Forests and Grasslands so they best demonstrate the sustainable multiple-use management concept.
- To provide technical and financial assistance to State and private forest landowners, encouraging them to practice good stewardship and quality land management in meeting their specific objectives.
- To provide technical and financial assistance to cities and communities wishing to improve their natural environment by planting trees and caring for their forests.
- To provide international technical assistance and scientific exchanges to sustain and enhance global resources and to encourage quality land management.
- To assist States and communities to use the forests wisely in the promotion of rural economic development and a quality rural environment.
- To develop and provide scientific and technical knowledge aimed at improving the ability to protect, manage, and use forests and rangelands.
- To provide work, training, and education to the unemployed, underemployed, elderly, youth, and disadvantaged in pursuit of the Forest Service mission.

Canada's Forest Service (CFS) has a similar history. It was established before the European settlers could inflict irreversible damage on the nation's vast forest reserves (see Section 6.3.2). From c. 1780 uncontrolled logging took place, beginning in the Atlantic Provinces as waves of settlers arrived and carved out farms; by the mid 1800s logging was making inroads into the forests of the pacific coast. Concerns about this and the ravages of forest fires resulted in legislation at local and province level to control fires and by 1899 a federal approach was adopted with the formation of the Dominion Forest Branch, the forerunner of the Canadian Forest Service. Its objectives were "conservation and propagation through fire fighting, tree planting and forest reserves" (Canadian Forest Service, 2004). Today the CFS considers itself to be a leader of global forest sustainability, a role which reflects not only the importance of forests to Canada's economy but its prominence in research and its willingness to co-operate with external bodies. Canada houses c. 10 percent of the world's forests (see Figure 6.1) and is thus a major carbon store. It is the world's largest exporter of wood and wood products and this sector is the greatest source of Canada's external income. The mission of the CFS is "To promote the sustainable development of Canada's forests and the competitiveness of the Canadian forest sector for the well-being of present and future generations of Canadians". Like its US counterpart the CFS engages in conservation, restoration, afforestation, education, recreation, research and forest technology.

Given that wood was such an important resource throughout the history of the UK, it is surprising that formalised control or investigation into forests and forestry did not begin until the early twentieth century. Indeed, this did not begin in the UK but in India where concerns were expressed about declining forests and associated landscape change. So great was this concern that the Indian Forest Service was founded in 1864 and the first Forest Research Institute was established at Dehra Dun in 1906. Subsequently courses on forestry were established at Oxford University with the establishment of the Institute of Forestry. However, during the years of World War I the paucity of national wood resources were highlighted and the need for institutionalised control of forests became evident. In 1919 the Forestry Act was approved by parliament and so the Forestry Commission was set up. Its objectives focussed on promoting forestry, developing afforestation, producing timber and making grants to private landowners for forestry enterprise (Forestry Commission, 2004a). Its mission was to "rebuild and maintain a strategic timber reserve". Limited resources only allowed the purchase of land in marginal upland areas for afforestation and the facility to offer grants to individuals for 25 percent of the costs of tree planting. However, by 1929 some 600,000 acres were being managed in 152 forests and 138,000 acres had been planted, with a further 54,000 acres having been planted in the private sector using grant aid.

By 1934 the total avreage had reached 900,000 and almost doubled again by 1970. In the intervening period forests were heavily exploited during the years of World War II. Thereafter the Forestry Commission acquired derelict and depleted woodlands nationwide to continue to improve the UK's forest resources with an additional impetus deriving from tax incentives available to the private sector. Although commendable, the FC's work attracted criticism. First, there was reliance on conifers, especially alien species such as Sitka spruce, Norway spruce and lodgepole pine, which were planted in vast monocultural stands. This is far from being in character with the UK's native mixed woodlands but the relatively fast-growing conifers provided a harvest within 25 to 30 years. Moreover, in both positive and negative terms these conifers will grow well in inhospitable terrain; the positive aspect is the capacity to produce a crop from land which has little productive value in the conventional sense but the negative aspect is the loss of habitats to afforestation and thus the loss of wildlife habitats such as peatlands. Here, one store of carbon is sacrificed by drainage to produce another store of carbon! More recently, the Forestry Commission has broadened its remit. It now actively plants hardwoods, encourages practices that favour wildlife, engages in recreational and educational activities and has a research programme. Table 7.2 gives data on forest type and extent of Forestry Commission land. The Forestry Commission is responsible for contributing to the rebuilding of the UK's forest resource and thus its carbon store; there is now considerably more carbon in UK forests than in 1919.

Table 7.2. Data on UK forests in 2004 (from the Forestry Commission, 2004b).

Forestry Commission Land	**Conifers x 10^3 ha**	**Broadleaves x 10^3 ha**	**Total Woodland x 10^3 ha**
England	154	52	206
Wales	98	11	110
Scotland	440	25	781
Non Forestry Commission Land			
England	217	693	910
Scotland	64	112	176
Wales	611	254	865
Total UK Woodland	1583	1148	2731

While the beginning of the National Park movement in the USA and the establishment of the Forestry Commission in the UK are examples of primarily national institutions, the 1930s witnessed the beginning of international organizations. The International Council for Science (ICSU) was founded in 1931 and was one of the first international NGOs. Its objective was, and remains, the promotion of scientific activity and collaboration internationally for the benefit of society (see ICSU, 2004). Its current mission statement is given in Table 7.3. In relation to the politicization/institutionalization of carbon the ICSU has instigated global initiatives (often with scientific partners) and acts, on invitation, as scientific adviser to the UN and its international conferences which focus on environment (see Section 7.6). This focus is reflected in its past and current projects such as the International Biological Programme (IBP) which was in operation between 1964 and 1974. Current projects include the International Geosphere-Biosphere Programme (IGBP), the World Climate Research Programme (WCRP), DIVERSITAS which is an integrated programme of biodiversity science and the International Human Dimensions Programme on Global Environmental Change (IHDP). Brief details are given in Figure 7.1.

Table 7.3. The ICSU mission statement (from ICSU, 2004)

- In order to strengthen international science for the benefit of society, ICSU mobilizes the knowledge and resources of the international science community to:
  - Identify and address major issues of importance to science and society.
  - Facilitate interaction amongst scientists across all disciplines and from all countries.
  - Promote the participation of all scientists—regardless of race, citizenship, language, political stance, or gender—in the international scientific endeavour.
  - Provide independent, authoritative advice to stimulate constructive dialogue between the scientific community and governments, civil society, and the private sector.

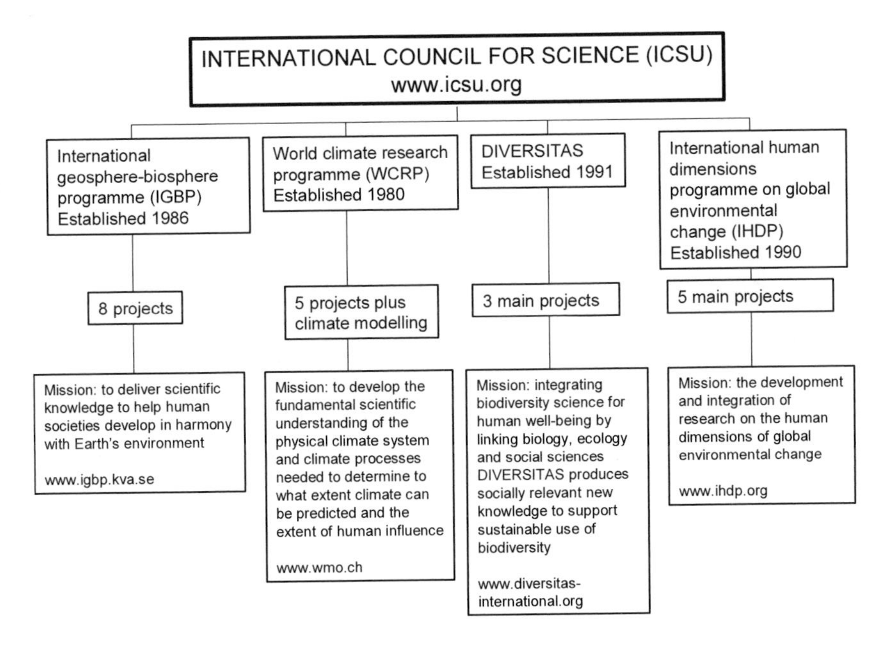

Figure 7.1. Programmes of the ICSU (from ICSU, 2004).

The IGBP was established in 1986 and is an ongoing project with the objective of understanding the dynamics, i.e. the related physics, chemistry and biology, of the biosphere. It currently comprises eight linked projects with six focusing on the ocean, land and atmosphere and two dealing with the temporal element of global change. The WCRP focuses on the dynamics of the climate system and climatic change, including human-induced change. DIVERSITAS comprises three main projects which focus on the assessment and monitoring of biodiversity, the examination of ecosystem services and the formulation of strategies for accommodating sustainable biodiversity use and development. The aim of the IHDP is to describe, analyse and understand the human dimensions of global environmental change through research, capacity building and networking.

7.2 Initiatives 1945 to the early 1970s

In the years immediately after World War II, a political form of globalization was started with the formation of the United Nations to maintain world peace and to encourage international co-operation. At the same time came recognition that famine, hunger and poor nutrition were widespread and that the world's flora and fauna were disappearing at an alarming rate. To counter these trends at international level, the Food and Agriculture Organization (FAO) and the International Union for the Protection of Nature (IUPN), now the World Conservation Union (IUCN), were established.

The Food and Agriculture Organization emerged from discussions in 1943 which involved 44 governments. The new agency was inaugurated in 1945 as an organ of the newly created United Nations and in 1951 its headquarters were moved to Rome. The FAO's stated policy is "to raise levels of nutrition and standards of living, to improve agricultural productivity, and to better the condition of rural populations" (FAO, 2004b). It operates "by leading international efforts to defeat hunger. Serving both developed and developing countries, FAO acts as a neutral forum where all nations meet as equals to negotiate agreements and debate policy. FAO is also a source of knowledge and information. We help developing countries and countries in transition modernize and improve agriculture, forestry and fisheries practices and ensure good nutrition for all. Since our founding in 1945, we have focused special attention on developing rural areas, home to 70 percent of the world's poor and hungry people". These objectives are achieved through education programmes, ensuring that information about best practice and policy is disseminated, by providing a vehicle for information exchange and ensuring that knowledge is passed to farmers. FAO has been responsible for many high-profile international conferences, such as the World Food Summit in 1996, and codes of practice, such as the *Codex alimentarius*, which set international food standards in 1962. FAO espouses the concept of sustainable development (see Section 7.6) and fosters good practice in agriculture, fisheries and forestry. It keeps valuable statistics and provides technical assistance in the field as well as educational programmes in the field. FAO also promotes safety in food production and involves itself in international legislation, e.g. pesticide regulation and the use of genetically-modified crops etc.

The IUCN – the World Conservation Centre – originated from an international conference at Fontainbleu, France, in 1948. It was originally named the International Union for the Protection of Nature (IUPN); in 1956 it became the Union for the Conservation of Nature and Natural Resources (IUCN) and in 1990 it was renamed the IUCN – the World Conservation Centre (IUCN, 2004). Its mission is "to influence, encourage and assist societies throughout the world to conserve the integrity and diversity of nature and to ensure that any use of natural resources is equitable and ecologically sustainable". Thus the IUCN's objectives combine conservation with the recognition that many people depend on natural resources for their livelihoods. Finance comes from five sources: Government aid agencies, voluntary contributions e.g. from the World Wildlife Fund (WWF, see Section 7.5) and the US Department of State, bilateral contributions e.g. the Spanish Ministry of Environment and the UK's Department for International Development, multilateral contributions e.g. the Asian Development Bank and the European Union, and foundations such as the Ford Foundation (USA). The IUCN has an international membership comprising members from 140 countries. It is headed by a Director General and a secretariat of IUCN staff who are based in the organization's headquarters in Gland, Switzerland.

A World Conservation Congress is held every three years at which the organization's policy and programme are decided. The practical programme is the responsibility of six commissions which comprise expert volunteers as well as IUCN staff. Details of the commissions are given in Table 7.4 and reflect the diverse range of activity which characterizes the IUCN. Each commission provides guidance, knowledge of conservation matters, policy and technical advice. National, regional offices administer field projects of which some 500 projects are in operation worldwide. This broad remit incorporates what the IUCN describes as a 'green web' to facilitate partnerships, the dissemination of knowledge, innovation and action.

Table 7.4. The commissions of the IUCN (based on IUCN, 2004).

COMMISSION	ROLE
Species Survival Commission (SSC)	7,000 members advise on technical components of species conservation and draw attention to threatened species.
World Commission on Protected Areas (WCPA)	1,300 members promote the establishment of a network of sites representing terrestrial and marine protected areas worldwide.
Commission on Environmental Law (CEL)	800 members formulate new legal concepts and instruments and encourage societies to devise and apply environmental laws.
IUCN Commission on Education and Communication (CEC)	600 members promote the use of communication and education for conservation purposes, including the empowerment of communities through knowledge acquisition.
Commission on Environment, Economic and Social Policy (CEESP)	500 members provide advice on social and economic factors which affect natural resource use and advise on sustainable resource use.
Commission on Ecosystem Management (CEM)	400 members provide guidance on integrated ecosystem approaches to ecosystem management.

The IUCN also stimulated international legislation to curtail trade in endangered species. The outline proposal for the Convention on International Trade in Endangered Species of Wild Fauna and Flora (CITES) was established at a meeting of IUCN members in 1963. Following lengthy deliberations the text was finally agreed in 1973 and CITES came into force in 1975. According to CITES (2004), the convention is an international but voluntary agreement between Governments to ensure that trade in specimens of wild animals and plants does not threaten their survival. The convention operates through the provision of licenses for specimens of selected species before they can be traded, i.e. imported, exported or re-exported. The species safeguarded by CITES are listed in three appendices which reflect the degree of protection needed. Appendix I includes species threatened with extinction. Trade is permitted only in exceptional circumstances; Appendix II includes species not necessarily threatened with extinction, but in which trade must be controlled in order not to compromise their survival; and

Appendix III contains species which are protected in at least one country and for which other CITES Parties have been asked for assistance. The IUCN and UNEP-WCMC (see Section 7.6) continue to have an input to CITES, including international conferences and the updating of appendices.

Linked with the IUCN and the International Wildfowl Research Bureau (IWRB), that was founded in 1954, is the Ramsar Convention on Wetlands. This was an international agreement signed in Ramsar, Iran, in 1971 following an international conference on wetlands. The convention came into operation in 1975 and its business is handled by the Ramsar Bureau based in Switzerland. Ramsar was the first multinational convention and is the sole convention to address a specific ecosystem type. It provides the framework for national action and international cooperation for the conservation and wise use of wetlands and their resources. Originally signed by 18 countries, there are currently 142 Contracting Parties to the Convention, with 1398 wetland sites, totalling 122.8 million hectares, designated for inclusion in the Ramsar List of Wetlands of International Importance. Its mission is given below (see Ramsar, 2004).

- The Ramsar Convention's mission is the conservation and wise use of all wetlands through local, regional and national actions and international cooperation, as a contribution towards achieving sustainable development throughout the world.

The IUCN also helped to establish the World Conservation Monitoring Centre (WCMC) which originated as an office of IUCN in 1979 in Cambridge, UK, to monitor endangered species. In 1988 this was formally established as the WCMC by the IUCN, WWF and UNEP. In 2000 it became an integral part of UNEP as its vehicle for biodiversity assessment and policy implementation; it is an important intergovernmental environmental organization (see UNEP-WCMC, 2004). UNEP-WCMC has five major roles. It assesses and analyses global biodiversity, determines trends and threats to biodiversity and informs relevant government bodies and non-governmental organizations. It provides support for policy, action plans and agreements at national and international levels as well as facilitation of conservation through the provision of expertise and information. It also makes information widely available and encourages information sharing through networks etc.

7.3 Green politics

The 1960s and 1970s were decades of considerable importance for so-called green politics. This was a period of growing awareness about environmental issues and a period of political activism generally, especially by student bodies in the developed world. Support for the environment, a major component of carbon politics, led to the formation of political parties which focussed on the environment as a central theme. Mainstream green parties adopt principles called the four pillars. These are: ecology (ecological sustainability), justice (social responsibility), democracy (appropriate decision-making) and peace (non violence).

Such parties promoted themselves as alternatives to mainstream political parties in New Zealand, Australia, Canada, UK and Germany. At the same time many mainstream political parties began to adopt 'green issues' in their manifestos. What follows is an introduction to the Green parties and does not include, due to limitations of space, reference to the 'green' aspects of established political parties.

Several green parties were founded in 1972. These included the United Tasmania Group which formed in Hobart, Australia and the Small Party which formed in the Atlantic provinces of Canada. The first national green party, the Values Party, was inaugurated in May 1972 in Wellington, New Zealand and that in the UK was founded in 1973 with the title of People. It was subsequently renamed as the Ecology Party and in 1985 it became the Green Party. It was the first national green party in Europe though it was the German Green Party which first adopted the word 'green' when it was formed in 1980 as '*Die Grünen*'. The Green party in the USA is a relative newcomer having been founded in 2001 from the older Association of State Green Parties which was established in 1996. There are many other green parties worldwide as listed by Global Greens (2004). Some examples are given in Table 7.5.

National green parties are also linked in federations, such as the European Federation of Green Parties, and participate in international conferences. These are called 'global gatherings', the first of which was in Canberra in 2001 when a 'green charter' was established. Their tenets are all based on the four pillars (see above) but the principles in each manifesto vary. For example the US Green Party lists ten values: grassroots democracy, social justice and equal opportunity, ecological wisdom, non-violence, decentralization, community-based economics and economic justice, feminism and gender equality, respect for diversity, personal and global responsibility, and future focus and sustainability. These values reflect equality amongst people and respect for environment.

Table 7.5. Examples of green parties (based on Global Greens, 2004, and European Green Parties, 2004).

COUNTRY	PARTY NAME	DATE FOUNDED	COMMENTS
Japan	Rainbow & Greens	1999	138 politicians in local government
Mongolia	Mongolian Green Party	1990	6000 members; has been in coalition with governing party
Somalia	Somalia Green Party	1990	
Benin	Les Verts du Benin	?	
Taiwan	Green Party Taiwan	1996	Has representatives in National Assembly
Bulgaria	Bulgarian Green Party	1989	Has representatives in National Assembly
Malta	Alternattiva Demokratika Malta	Late 1980s	Strong in European Union elections
Belgium	Ecolo Groen	1980	First green party to have members elected to national parliament

As discussed by Burchell (2002) some green parties have participated in national politics through direct election to national parliaments or assemblies or through the forging of coalitions with traditional political parties. The best known example of the latter is that of the German Green Party which was founded in 1983 and which has been allied with the Social Democratic Society. This alliance was instrumental in the government's decision to end reliance on nuclear power. Both green parties of Belgium (Groen and Ecolo) and the Green Party of Finland have taken part in national government. Currently (2004), there are 169 members of European national parliaments who represent green parties and 33 Green MEPs out of a possible total of 624 members in the European Union parliament (European Green Parties, 2004). Greater success has been achieved in local politics but overall green influence is increasing, not least through campaigning, lobbying and setting agendas which influence the green components of traditional parties. It has been argued that implementation of sustainable practices is most important at local level (O'Riordan, 2004) and thus Green politicians may play a vital role in these 'grassroots' constituencies with which they are most familiar given that most green parties originated at local level.

7.4 Petropolitics and oil crises

Petropolitics is concerned with the interplay between politics and oil/petroleum. It is thus an aspect of politics concerned almost entirely with carbon. The term was coined in the early 1970s when a major international oil crisis occurred and which highlighted the close, almost perilous, relationship between oil availability and international politics. The ownership of oil deposits confers power and advantage in an oil-fuelled world; lack of it confers disadvantage. Since oil is necessary for industrial/economic development it becomes a direct political instrument. The other factors to which oil is related, such as war, climate and poverty, also contribute to petropolitics. Indeed, the issue of global climatic change has rapidly developed a focus in international politics, as discussed in Section 7.6 below. According to Wikipedia (2004), four distinct oil crises can be identified as discussed below.

In 1973 the Arab members of the Organization of Petroleum Exporting Countries (OPEC) decided to discontinue oil exports to nations which supported Israel in its conflict with Egypt, i.e. the Yom Kippur War. Such nations included the USA and much of Western Europe. To counteract the falling value of the dollar, following devaluation and as a show of strength, OPEC members also caused the price of oil to rise dramatically to four times its 1970 value. Many OPEC nations also resented the influence of multinational companies, many based in the USA, Japan and Western Europe, over their resources and indeed OPEC itself was formed to provide a united front against such exploitation. Thus OPEC nations wielded considerable advantage over industrialized nations requiring oil imports. The effects were substantial, including inflation, job losses, fuel shortages and significant declines in share prices on stock markets. Petrol/gasoline was rationed in some countries and people were forced to queue to purchase it. The immediate crisis was short lived and following the Washington Conference in March 1974 OPEC members lifted the embargo. They too needed income. The effects were, however, longer term through inflation and unemployment with some positive aspects involving the instigation of research into alternative fuels and fuel efficiency. Since the mid-1970s, OPEC's influence has declined; this is partly due to internal disagreements and partly because new oil producers have emerged.

Several more recent oil crises have occurred which have had notable economic and political effects. In 1979 the change of government in Iran following the overthrow of the Shah and the installation of a theocratic Islamic government resulted in a deliberate reduction in oil production. Oil shortages once again occurred in the USA. Eventually price controls on oil were removed by the US government. Oil production from Iran and Iraq was also curtailed in 1980 due to the Iran – Iraq war and this too served to increase world oil prices. Yet another crisis occurred in 1990 because of the invasion of Kuwait by Iraq which triggered the Gulf War. A coalition led by the USA ensured that the war was short lived but oil prices rose due to fears of shortages not least because of the deliberate release of 11 million barrels of oil into the Persian Gulf and the deliberate firing of some 700 Kuwati oil wells as an act of sabotage or ecoterrorism (see Mannion, 2002 for details of the environmental impact).

Another oil crisis is occurring at present; once again this is the direct result of conflict due to the Iraq war of 2003 and subsequent unrest in Iraq which involves sabotage of oil pipelines etc. Oil prices have risen dramatically as oil output from Iraq has dropped and fears of wider sabotage of oil installations in the Middle East have intensified. Other factors have also conspired to cause rising oil prices which have exceeded $55 per barrel at times. These include increasing demands by industrializing nations, especially China and India, hurricane damage in oil-producing Caribbean nations, and social unrest in oil-producers in West Africa e.g. Equatorial Guinea. The impact has been considerable, including rising inflation and cost-of-living indices caused by increased costs of production and transport of goods and food. Petrol/gasoline prices for individual motorists are also at an all-time high. This crisis is ongoing.

The problems that fluctuating oil supplies generate at scales from the international to the local or individual person are well exemplified by the situations referred to above. In the event of a true shortage of oil as geological reserves are depleted, an inevitability sometime in the future (predictions vary widely) as discussed in Chapter 8, the crises of the past provide some indication of the impact on economic activity that such shortages would provoke if alternative energy sources were not to be available.

7.5 NGOs

Government or government-sponsored institutions have not been the only influences on carbon and its worldwide flows. Since the end of World War II many non-governmental organizations (NGOs) with environmental and holistic (involving a combination of people-environment-development) concerns have been founded. All are funded independently of governments and political parties; they embrace concerns about the environment and most not only provide information through research but also attempt to influence the environmental policies of governments worldwide. Examples include the World Wide Fund for Nature (WWF) founded in 1961 and the Club of Rome founded in 1968. These were followed in the early 1970s by two activist/lobbyist organizations, notably Friends of the Earth (FoE) and

Greenpeace, while less confrontational organizations such as the World Watch Institute (WWI) and the World Resources Institute (WRI) were established for research, data compilation and education.

The WWF (see WWF, 2004b) was founded in 1961 as the World Wildlife Fund which was changed to the Worldwide Fund for Nature in 1986 but with the retention of its well-known acronym and panda motif. Julian Huxley, the British biologist and the first Director General of UNESCO who also helped to found the IUCN (see above), went to Africa in 1960 on UNESCO business and discovered that habitat destruction and hunting were destroying landscapes and livelihoods. Huxley enlisted the help of Max Nicholson, Director General of the British Nature Conservancy, and with funding from the businessman Victor Stolon, the WWF was established as an international organization with its headquarters in Switzerland which it shared with IUCN (see Section 7.2). A fund-raising campaign was established at national level through national offices which are linked to the international centre. The WWF supports institutions, such as IUCN and the Charles Darwin Research Centre in the Galapagos Islands, and many large and small conservation projects worldwide. For example, it sponsors 350 projects under its forest programme.

The WWF also has a lobbying function insofar as it attempts to educate and influence government and public opinion about conservation issues. One resulting institution is that of the Trade Records Analysis of Fauna and Flora (TRAFFIC) which was set up to regulate and monitor trade in plants, animals and substances such as ivory. Today there are 17 TRAFFIC offices worldwide. In 1980 collaboration between the IUCN, WWF and the United Nations Environment Programme (UNEP, see Section 7.6) resulted in publication of the 'World Conservation Strategy' which advocated the integration of development and conservation. Many nations have used this as a template for national conservation strategies. Another example of WWF's lobbying is the debt-for-nature swaps whereby a proportion of a nation's debt is converted into funds specifically for conservation. Such swaps began in the late 1980s and have benefited many developing countries. By 1990 the WWF had reformulated its mission with three objectives: "the preservation of biological diversity, promoting the concept of sustainable use of resources, and reducing wasteful consumption and pollution". The commitment to work closely with local people was renewed and in 1991 another landmark publication, 'Caring for the Earth: A Strategy for Sustainable Living' was launched. Its focus was on the actions that individuals could take to improve their lives and the environment. In the following year WWF, in its role as a pressure group, pressed governments to subscribe to the treaties agreed at the United Nations Conference on Environment and Development (UNCED, also known as the 'Earth Summit') in Rio de Janeiro in 1992. In more recent years WWF has further publicized the state of the Earth and its rapid changes due to human action in its 'Living Planet Reports', as referred to in Section 6.4.

Another influential NGO was formed in 1968. This was the Club of Rome which describes itself as a global think tank with a mission "to act as a global catalyst of change that is free of any political, ideological or business interest" (Club of Rome, 2004). It is concerned with innovation for the common good and thus its activities are not solely focussed on the environment or carbon. The Club of Rome states that "it brings together scientists, economists, businessmen, international high civil servants, heads of state and former heads of state from all five continents who are convinced that the future of humankind is not determined once and for all and that each human being can contribute to the improvement of our societies". Many of the Club of Rome's reports reflect an holistic approach to development and thus place importance on environment and sustainable resource use. Like the WWF, it comprises national associations which are linked through the international headquarters in Hamburg, Germany. The Club of Rome contributed to the environmental debates in the 1970s and continues to promote environmental sustainability.

Perhaps the most high-profile NGOs with environmental and thus carbon-based foci are the Friends of the Earth (FoE) and Greenpeace (Gp), both founded in 1971. The FoE was started by David Brower who left the Sierra Club, an influential environmental group in the USA which was founded in 1892 with John Muir, the famous naturalist, as its first president (Sierra Club, 2004). This occurred in 1969 and by 1971 FoE had become an international organization. It is primarily a campaigning organization (see FoE International, 2004) which is a federation of national groups co-ordinated through FoE International and an international secretariat; the national groups themselves comprise local 'grassroots' groups. Efforts focus on local and national issues specific to members' interests but the mission reflects global sustainability. The overall mission objectives are listed in Table 7.6. These reflect the political and lobbying roles of FoE though the organization does commission research to boost these activities. There are currently 68 national groups and a worldwide membership of more that one million people.

Table 7.6. Objectives of the Friends of the Earth

- To protect the Earth against further deterioration and repair damage inflicted upon the environment by human activities and negligence;
- To preserve the Earth's ecological, cultural and ethnic diversity;
- To increase public participation and democratic decision-making. Greater democracy is both an end in itself and is vital to the protection of the environment and the sound management of natural resources;
- To achieve social, economic and political justice and equal access to resources and opportunities for men and women on the local, national, regional and international levels;
- To promote environmentally sustainable development on the local, national regional and global levels.

Greenpeace is another activist organization whose main objective is to publicize environmental injustices (see Weyler, 2001 for a brief history and Greenpeace, 2004). It was founded in 1971 and its first campaign was to draw attention to US nuclear underground testing in Amchitka, an island off the Alaskan coast. A group of activists set sail from Vancouver and headed north in the belief that individuals could indeed make a difference.

- Greenpeace is an independent, campaigning organization that uses non-violent, creative confrontation to expose global environmental problems, and force solutions for a green and peaceful future. Greenpeace's goal is to ensure the ability of the Earth to nurture life in all its diversity.

More than thirty years later, Greenpeace is well known and an effective campaigning organization. Its mission statement is given above. Now international, Greenpeace has its headquarters in Amsterdam, the Netherlands, offices in 41 countries and a worldwide membership of 2.8 million. These supporters and various foundations provide its finance so it remains entirely independent of governments, industry etc. in order to maintain objectivity and credibility. Its ship, the *Rainbow Warrior*, has become a symbol of protest, especially against nuclear testing, whaling and sealing. The current ship is named after the original *Rainbow Warrior* which gained notoriety because it was sunk in Auckland Harbour, New Zealand, by the French Secret Service in 1985 following protests about nuclear testing in the Pacific. FoE also developed the concept of environmental space whereby every nation would have equitable access to resources which include water, food etc. Both Friends of the Earth and Greenpeace have waged many successful campaigns and have been responsible for raising the general environmental awareness of people everywhere. They not only campaign about current environmental issues such as biodiversity loss and climatic change (see Section 7.6) but also against globalization.

There are many other NGOs engaged in environmental education, information dissemination and publicity. The Worldwatch Institute and the World Resources Institute are two such organizations. The former was established in 1974 by Lester Brown (World Watch Institute, 2004). It is funded through the sale of its publications, such as the series entitled *State of the World, World Watch, Signposts* and *Vital Signs,* as well as by donations from individuals and foundations.

- The Worldwatch Institute is an independent research organization that works for an environmentally sustainable and socially just society, in which the needs of all people are met without threatening the health of the natural environment or the well-being of future generations.
- By providing compelling, accessible, and fact-based analysis of critical global issues, Worldwatch informs people around the world about the complex interactions between people, nature, and economies. Worldwatch focuses on the underlying causes of and practical solutions to the world's problems, in order to inspire people to demand new policies, investment patterns and lifestyle choices.

Its mission is given above as one focussed on the compilation of data on environmental characteristics and their analysis, with an emphasis on people/environment relationships through an examination of ecological/economic relationships. These publications are widely read and quoted, reflecting a belief in the work of the WWI which is based in Washington DC, USA. The World Resources Institute (WRI) provides a similar service to that of the WWI insofar as it collates and disseminates information about the state of the environment. Its publications, such as the *World Resources* series, provide data on resource production and consumption e.g. fossil-fuel energy, food, fisheries, water, as well as population data. WRI has also established '*Earth Trends*', an online data base comprising freely accessible environmental data in various formats (World Resources Institute, 2004). Its mission is given below.

- World Resources Institute is an independent non-profit organization with a staff of more than 100 scientists, economists, policy experts, business analysts, statistical analysts, mapmakers, and communicators working to protect the Earth and improve people's lives.

WRI has four main objectives: to protect Earth's living systems, to increase access to information, to create sustainable enterprise and opportunity and to reverse global warming. WRI was founded in 1984 as an environmental charity funded by individual donations and foundations. Based in Washington DC, a major aim of the WRI is to influence environmental policy through the provision and analysis of information.

7.6 The UN, UNEP, international agreements and the Earth Charter

The United Nations (UN) was created in 1945 with the objective of securing world security and peace following the upheaval of two World Wars between 1914 and 1945. One of its first agencies was the United Nations Educational Scientific and Cultural Organization (UNESCO) which was founded in 1946. Its objective was, and remains, the promotion of collaboration, and thus peace, between nations through education, science and culture. However, in the first two decades of its existence, environmental issues were of little consequence in UN activities, but increasing awareness of such issues in the late 1960s and early 1970s (see Section 7.3) led the UN to recognize their role in the achievement of security and peace. Although concerns since the 1980s about UNESCO's administration and possible political bias have caused rifts within the organization and the withdrawal of several nations, including the USA and the UK, it has been instrumental in conservation through its establishment of the Man (*sic*) and Biosphere (MAB) in 1971; this was the result of deliberations between FAO, the World Health Organization (WHO), the World Meteorological Organization (WMO) and the International Council for Science (ICSU) in 1968. The MAB programme was charged with promoting an interdisciplinary approach to environmental investigation and conservation and people – environment relationships. In its 33-year history the MAB programme has undergone many changes but its prime objective remains the reconciliation of resource use and conservation. In this respect it is a major advocate of the 'ecosystem approach' adopted by the Convention on Biological Diversity (CBD) at the UN Conference on Environment and Development in 1992 (see below and Table 7.7).

To achieve these objectives MAB has established a series of Biosphere Reserves worldwide which are defined (see UNESCO, 2004) as "areas of terrestrial and coastal ecosystems promoting solutions to reconcile the conservation of biodiversity with its sustainable use". They are internationally recognized, nominated by national governments and remain under sovereign jurisdiction of the states where they are located. Biosphere reserves serve in some ways as 'living laboratories' for testing out and demonstrating integrated management of land, water and biodiversity. Each biosphere reserve is intended to fulfil three basic functions, which are complementary and mutually reinforcing, as given in Table 7.7. The designation of biosphere reserves is ongoing and in November 2004 there were 459 reserves in 97 countries (for a complete list see UNESCO, 2004). Some of these reserves are classified as World Heritage Sites. The MAB programme also publishes a bulletin and organizes workshops and conferences.

Table 7.7. The functions of MAB biosphere reserves (based on UNESCO, 2004).

CONSERVATION	to contribute to the conservation of landscapes, ecosystems, species and genetic variation.
DEVELOPMENT	to encourage economic and human development which is socially, culturally and ecologically sustainable.
LOGISTICS	to provide support for research, monitoring, education and information exchange on scientific and socio-cultural aspects of conservation and development at scales from the local to the global.

A year after MAB was founded, the United Nations Environment Programme (UNEP; see UNEP, 2004a)) was established in 1972 as a component of the United Nations (UN) following the International Conference on the Human Environment in Stockholm, Sweden, in 1972. Its mission is given below.

- To provide leadership and encourage partnership in caring for the environment by inspiring, informing, and enabling nations and peoples to improve their quality of life without compromising that of future generations.

Based in Nairobi, UNEP is 'the voice of the environment' in the UN. It has an executive director who works with a senior management group, regional directors and representatives and national committees, and who liaise with ministries of environment worldwide and international organizations such as the WWF (see Section 7.5). UNEP produces publications, web-based information and technical advice as well as organizing international conferences. The information it makes available includes the Global Resource Information Database (GRID), the International Register of Potentially Toxic Chemicals (IRPTC). The former comprises a network of co-operating centres worldwide whose objective is to make environmental data accessible; a major resource is a range of digital maps, including thematic maps. The latter lists details of potentially toxic chemicals and their mode of action. UNEP also runs an environment network through a web site (www.unep.net) which provides environmental information through links with other web sites. There is also a regular magazine entitled *Our Planet,* a series of reports entitled *Global Environmental Outlook (GEO)* as well as a range of other services to governments, environmental groups etc.

The UN, independently of UNEP, also commissioned the World Conference on Environment and Development in 1983 under the chairpersonship of Gro Harlem Brundtland, former prime minister of Norway. The object was to devise a plan to bring about development within the bounds of maintaining environmental integrity. The outcome of this conference, published in 1987 as *Our Common Future* (WCED, 1987), advocated the concept of sustainable development. This concept, formally proposed as "Development that meets the needs of the present without compromising the ability of future generations to meet their own needs" has been widely adopted by governments worldwide. Now more sentiment than objective, the concept of sustainable development has become clichéd, and much debate surrounds its true meaning and relevance (see Williams and Millington, 2004, for a discussion and Lumley and Armstrong, 2004, for historical background).

However, UNEP has embraced sustainable development as a focus in its activities. This is reflected in UNEP's most high-profile activities, notably its international conferences. Apart from the founding conference in Stockholm in 1972 which established the duality of environment and development, other conferences include the so-called Earth summits of Rio de Janeiro in 1992 and Johannesburg in 2002. The former (the International Conference on Environment and Development) was influential insofar as it achieved several major agreements relating to the environment and which all concern carbon in one form or another. Brief details of the agreements on climatic change, biodiversity, forests and action plans are given in Table 7.8. The principles (see O'Riordan, 2000, for details) established in these agreements have become significant nationally, regionally and locally through their adoption by government agencies.

Table 7.8. Major agreements of the 1992 UN Conference on Environment and Development (UN, 1997).

Framework Convention on Climate Change The aim is to provide an international framework within which future actions can be taken. The main climate problem was recognized as global warming and discussions ensued concerning reductions in carbon dioxide emissions, i.e. to reduce them to 1990 levels by 2000. The UN Framework Convention on Climate Change was drawn up. The EU declined to ratify the treaty; individual countries set their own targets and lengthened the time available to reach them. Developing countries expected developed countries to set the pace and to fund the cuts for both groups of countries. This treaty, although being diluted by ambiguous composition, has at least established the value of the precautionary principle since, in 1992, the evidence for anthropogenic global warming was equivocal. 150+ nations signed this agreement.
The Convention on Biological Diversity The focus of this treaty, which came into force in 1994, is the conservation of the Earth's biota through the protection of species and habitats or ecosystems. The debate polarized developing and developed nations insofar as the developing countries house more of the world's biodiversity (where it is experiencing many threats) than the developed countries which are skilled in biotechnology. The convention was signed by 153 countries, a major exception being the USA because of fears that US biotechnology companies could be disadvantaged. The USA has since signed.
Statement on Forest Principles This represents a diluted (and legally unenforceable) consensus on forest management which resulted from earlier (1990-1991) attempts to formulate a global forest convention. The principles assert the sovereign right of individual countries to profit from their forest resources but advocate that this should take place within a framework of forest protection, conservation and management.
The Rio Declaration on Environment and Development This has its origins in the UN Conference on the Human Environment, Stockholm in 1972. It is a statement of 27 principles upon which the nations have agreed to base their actions in dealing with environment and development issues.

Agenda 21

This is a 40-chapter action plan for sustainable development. It established tasks, targets, costs, etc., for actions which combine economic development and environmental protection. In many nations, e.g. the UK, the suggestions of Agenda 21 have been incorporated into local, regional and national development strategies. It led to the establishment of the UN Commission on Sustainable Development in 1992.

The UN Commission on Sustainable Development, established in 1992 (see Table 7.8), is charged with promoting, implementing and monitoring development focussed on sustainability, including the addressing of social issues such as poverty and grass-roots decision making. This commission, as part of UNEP, lacks authority and its various reports indicate that some, but not nearly enough, progress has been made and that environmental degradation continues. In 2002 a further 'Earth summit', the World Summit on Sustainable Development, was held in Johannesburg. Its objectives were to determine progress in sustainable development, a measure of the success of Agenda 21 which was promoted at the UNCED in 1992 (see Table 7.8); it confirmed earlier reports from the UN Commission on Sustainable Development but also highlighted the partnership approach. This is defined as voluntary, multi-stakeholder initiatives aimed at implementing sustainable development, and governments were urged to promote such liaisons as complementary approaches to those of government agencies (see Johannesburg Summit, 2002). It culminated in the 'Johannesburg Declaration' which restated and renewed commitments on the part of governments to sustainable development.

Its objectives encompass the eradication of poverty, changing unsustainable patterns of consumption and production, protecting and managing the natural resource base of economic and social development, sustainable development in a globalizing world, health and sustainable development, sustainable development of small island developing states, sustainable development for Africa and other regional initiatives, the means of implementing sustainable development, the institutional framework for sustainable development, strengthening the institutional framework for sustainable development at the international level, and the roles of international agencies such as the UN as well as those of national governments in promoting sustainable development. In addition, UNEP was responsible for the First Global Ministerial Environmental Forum which was held in 2000. It reflects the 'globalization' of environmental politics and the need for governments to confer and exchange ideas.

Other initiatives under the auspices of the UN/UNEP have focussed on climatic change and are examined in Section 7.7. Another formal development from the UNCED in Rio de Janeiro is the Earth Charter. Intergovernmental efforts to generate a formal charter at Rio failed but interested parties from civic society, whose interest stemmed from the 1983 World Conference on Environment and Development (WCED), continued to press for it and it came into existence in 1994 under the direction of an international, but non-governmental, Earth Charter Commission. Its history and objectives are described by d'Evie and Glass (2002). The Earth Charter's focus is the recognition of close people-environment relationships and the necessity of sustainable development. It promotes just ecological and social integrity (see papers in Miller and Westra, 2002) and thus embraces equity in resource use as well as peace and democracy. Its mission is given below (see The Earth Charter Initiative; Earth Charter, 2004).

- The mission of the Earth Charter is to establish a sound ethical foundation for the emerging global society and to help build a sustainable world based on respect for nature, universal human rights, economic justice, and a culture of peace.
- The goals of the Earth Charter Initiative are:
    - To promote the dissemination, endorsement, and implementation of the Earth Charter by civil society, business, and government.
    - To encourage and support the educational use of the Earth Charter.
    - To seek endorsement of the Earth Charter by the UN.

The Earth Charter did not gain official UN acceptance at the World Summit on Sustainable Development in Johannesburg in 2002 but its supporters continue to press for recognition and it was endorsed by the IUCN in November 2004. The web site provides material for educational purposes.

7.7 International initiatives on atmosphere/climate

Throughout the 1970s and 1980s concerns were intensifying in the scientific community about the prospect of atmospheric and climatic change. Three main problems can be identified: acid rain, stratospheric ozone depletion and global warming. All are linked to the domestication of carbon.

First so-called 'acid rain' is produced through fossil-fuel burning, especially coal. Its production, impact and mitigation is discussed in Mannion (1997a), Middleton (2003) and Kemp (2004), and referred to in Section 5.5. When carbon dioxide, nitrous oxides and sulphur dioxide are released into the atmosphere they combine with water to produce acids, notably carbonic, nitric and sulphuric acids, and also possibly combine with heavy metals such as lead and cadmium. These are deposited from precipitation either locally where they are produced or at a distance from the source if they are transported by winds. The result is soil acidification and the corrosion of buildings in cities. The former can impair vegetation and cause acidification of lakes and streams which, if severe, may result in the loss of flora and fauna. In cities, the etching of buildings, erosion of material from limestone-clad walls and loss of definition of sculpted façades cause eye-sores, and poor atmospheric quality is a health hazard, especially for those with respiratory problems. Mitigation is, however, possible as is demonstrated by the considerable success of many industrialized nations in the northern hemisphere to curb emissions, a process which has led to much recovery from acidification in vulnerable environments.

The strategies employed include flue gas desulphurization requiring the employment of scrubbers in power stations to remove sulphur dioxide. Catalytic reduction processes have been developed to remove nitrous oxides though they are not as widely applied due to high costs, and little effort has been made to remove carbon dioxide prior to emission (see below in relation to global warming). However, these mitigating strategies are linked with international agreements to reduce the spread of acidification and reverse its impact. Such agreements have come into existence under the umbrella of the Geneva Convention on Long-range Transboundary Air Pollution (LRTAP) of 1979. According to the United Nations Economic Commission for Europe (UNECE, 2004), this convention has its origins in the 1960s when it was first recognized that pollution could be transported between nations; research indicated that Scandinavian lakes were being acidified by emissions from other European nations. This issue was highlighted at the UN Conference on the Human Environment in 1972 and following collaborative studies which highlighted long-distance transport of pollutants (see Section 5.5) the LRTAP was established and came into force in 1983. There are now eight protocols as listed in Table 7.9. The 1985 protocol is often referred to as the 30 percent club because of the requirement of signatories to reduce sulphur emissions by at least 30 percent and it was the first major protocol, and indeed international agreement, to tackle 'acid rain'. The other protocols deal with other aspects of atmospheric pollution.

Table 7.9. Protocols to the Convention on Long-Range Transboundary Air Pollution (LRTAP), (based on UNECE, 2004)

Year	Protocol
1999	Protocol to Abate Acidification, Eutrophication and Ground-level Ozone; 31 Signatories and 14 ratifications. Not yet in force. (Guidance documents to Protocol adopted by decision 1999/1).
1998	Protocol on Persistent Organic Pollutants (POPs); 21 ratifications parties. Entered into force on 23 October 2003.
1998	Protocol on Heavy Metals; 24 ratifications parties. Entered into force on 29 December 2003.
1994	Protocol on Further Reduction of Sulphur Emissions; 25 Parties. Entered into force 5 August 1998.
1991	Protocol concerning the Control of Emissions of Volatile Organic Compounds or their Transboundary Fluxes; 21 Parties. Entered into force 29 September 1997.
1988	Protocol concerning the Control of Nitrogen Oxides or their Transboundary Fluxes; 29 Parties. Entered into force 14 February 1991.
1985	Protocol on the Reduction of Sulphur Emissions or their Transboundary Fluxes by at least 30 percent; 22 Parties. Entered into force 2 September 1987.
1984	Protocol on Long-term Financing of the Cooperative Programme for Monitoring and Evaluation of the Long-range Transmission of Air Pollutants in Europe (EMEP); 41 Parties. Entered into force 28 January 1988.

Today 'acid rain' is much less of a problem then it was 20 years ago in Europe and North America but it is now spreading in industrializing countries in Asia and Latin America (see Section 6.3.3). In order to help prevent acidification in these regions the Regional Acidification Information and Simulation (RAINS) programme has been devised by EU researchers as 'early warning' system for Asian governments. Details for RAINS for Europe and for Asia are documented by the International Institute for Applied Systems Analysis (IIASS, 2004).

Another aspect of atmospheric pollution involving carbon is that of chlorofluorocarbons (CFCs). As discussed in Section 2.2.1 (see also Table 2.2), these are molecules containing carbon combined with halides such as chlorine. They do not occur in nature and were first created artificially in the late 1920s; widespread use as refrigerants and propellants in aerosols began in the 1950s and as cleaning agents in the electronics industry in the early 1980s. Unfortunately, these compounds are now known to damage the ozone layer in the stratosphere, a layer of the upper atmosphere between 12 and 50 km above the Earth's surface. The ozone is important because it absorbs ultraviolet light and thus limits the amount reaching the Earth's surface where high concentrations can cause skin cancer and genetic alterations in organisms. Moreover, CFCs are themselves potent heat-trapping gases and thus contribute to global climatic change. The problem of ozone depletion went unrecognized until the 1970s when Molina and Rowland (1974) produced evidence that CFCs cause stratospheric ozone to break down and in the mid-1980s an 'ozone hole' was first identified over the Antarctic by Farman *et al.* (1985). By 1979 the governments of Canada, Norway, Sweden and the USA had banned CFC use for aerosols and by 1987 concern had become international. This culminated in the UN-sponsored Montreal Protocol on Substances that Deplete the Ozone Layer (see UNEP, 2004b for details) which was widely adopted, with subsequent amendments in 1990 and 1992. This protocol prescribes that the production and use of compounds with the capacity to deplete ozone in the stratosphere, i.e. chlorofluorocarbons, halons, carbon tetrachloride, and methyl chloroform, should be phased out by 2000, or by 2005 for methyl chloroform. The protocol has been widely though not universally accepted, but it is anticipated that the stratospheric ozone layer will return to its pre-CFC condition by c. 2050.

However, while the issues of 'acid rain' and stratospheric ozone depletion remain significant, and provide evidence that internationally agreed strategies can be effective, the most pressing problem of carbon dioxide emissions remains. This has spawned another round of international protocols under the auspices of the UN. In 1988 the Intergovernmental Panel on Climate Change (IPCC) was established as a joint venture between the World Meteorological Organization (WMO) and UNEP. Its role "is to assess on a comprehensive, objective, open and transparent basis the scientific, technical and socio-economic information relevant to understanding the scientific basis of risk of human-induced climate change, its potential impacts and options for adaptation and mitigation" (IPCC, 2004). IPCC is managed by a secretariat and is based in Geneva at the WMO. As Table 7.10 shows, it comprises three working groups and a task force whose objective is to appraise the international community on all aspects of climatic change through the analysis of published literature. The first report was produced in 1990 and provided impetus for the formation of an Intergovernmental

Negotiating Committee which documented the UN Framework Convention on Climate Change (UNFCC) in 1992 which was signed at the UNCED (Earth Summit) in Rio de Janeiro (see Table 7.8); it came into force in 1994.

Table 7.10. The structure of the IPCC (based on IPCC, 2004)

Working Group I	assesses the scientific aspects of the climate system and climate change
Working Group II	assesses the vulnerability of socio-economic and natural systems to climate change, negative and positive consequences of climate change, and options for adapting to it
Working Group III	assesses options for limiting greenhouse gas emissions and otherwise mitigating climate change
Task Force	is responsible for the IPCC National Greenhouse Gas Inventories Programme

The second report of the IPCC was published in 1995 and was instrumental in the formulation of the Kyoto Protocol to the UNFCC in 1997. It highlighted the significance of human impact on global climate. Its third report was in 2001 and is referred to in Chapters 2, 4 and 6; a fourth report is scheduled for 2007.

The Kyoto Protocol was signed initially by 84 nations who accepted the principle that emissions of heat-trapping (greenhouse) gases should be reduced by at least 5 percent of levels in 1990 in the commitment period of 2008 -2012 (UNFCCC, 2004). The gases are listed in Table 7.11.

Table 7.11. The major heat-trapping (greenhouse) gases

Carbon dioxide (CO_2)
Methane (CH_4)
Nitrous oxide (N_2O)
Hydrofluorocarbons (HFCs)
Perfluorocarbons (PFCs)
Sulphur hexafluoride (SF_6)

How this was to be achieved was decided at a later meeting in Marrakech in 2001, with details documented as the Marrakech Accords. Some nations have not signed the Kyoto Protocol, notably the USA which is one of the world's major producers of heat-trapping gases. The EU has a block target of an 8 percent reduction and an agreed redistribution within EU nations will take place to meet it. This began in January, 2005, and as pointed out by Pearce (2005a) it involves the allocation of carbon dioxide quotas to specific companies; those with insufficient quotas to match their real emissions will have to purchase additional quotas from those companies which produce less carbon dioxide than their permits allow. A record or permit allocation and exchange is to be effected through an electronic registration system. All signatories are required to establish domestic policies and measures to achieve their targets though not all measures have to be concerned with actual emissions. Other actions involving carbon sinks may be developed to offset emissions. Possibilities include afforestation, reforestation and deforestation (as defined in the Kyoto Protocol) and forest management, cropland management, grazing land management and land revegetation. The Protocol also establishes three 'mechanisms' known as joint implementation, the clean development mechanism and emissions trading to assist nations in reducing emissions. Joint implementation, which is essentially a technology transfer, involves reducing emissions or providing sinks in another country which is also a signatory of the same status i.e. Annex 1 countries. These are defined as "the industrialized countries listed in this annex to the Convention who sought to return their greenhouse-gas emissions to 1990 levels by the year 2000.... They have also accepted emissions targets for the period 2008-12. They include the 24 original OECD members, the European Union, and 14 countries with economies in transition (e.g. Croatia, Liechtenstein, Monaco, and Slovenia, with the Czech Republic and Slovakia replacing Czechoslovakia)". The 'clean development mechanism' allows credit for emission reductions and/or the establishment of carbon sinks in another nation not in Annex I. Emissions trading allows an Annex I Party to transfer some of the emissions under its assigned amount, as assigned amount units (AAUs), to another Annex I Party that finds it relatively more difficult to meet its emissions restrictions. Trading in gains made under the auspices of joint implementation and the clean development mechanism may also be traded. An overview of carbon trading etc. has been compiled by Hasselknippe (2003).

On the world stage, carbon trading began in 2004. According to CO_2e (2004) the first legally-binding transaction in October involved the purchase of Clean Development Mechanism (CDM) Certified Emission Reductions (CERs) by a European company. The collective CERs from three Brazilian electricity generating stations run on waste from sugar cane, i.e. a renewable resource, were purchased by an unnamed but leading European country. CO_2e were also involved in an earlier arrangement between Chilean, Japanese and Canadian companies. The Chilean company developed a new treatment for pig manure which facilitates the efficient capture and use of methane, generating CDM credits which were traded with two leading electricity generation companies in Canada and Japan. Already several companies have been established to facilitate carbon trading, e.g. Point Carbon, which describes itself as "the leading global provider of independent analysis, news, market intelligence and forecasting for the emerging carbon emission markets" (Point Carbon, 2004).

7.8 The politics of biodiversity

The use, misuse, monitoring and preservation of biodiversity have become mainstream foci of political interest since the UNCED in 1992 though biodiversity has always been political because it is a resource which has facilitated human development. Moreover many countries had formal systems for nature conservation in place long before UNCED in 1992. However, concerns about the loss of natural ecosystems, especially forests (see Section 6.3), high rates of species extinction and the recognition that climate and world ecosystems are closely linked culminated formally in the Convention on Biological Diversity (CBD) signed at UNCED in 1992 (see Table 7.7). This was the first international treaty to address conservation of biological diversity worldwide. It was signed by 150 government leaders with the prime objective being the implementation of Agenda 21. It stresses that to stem the losses of biodiversity three major factors must be considered: the conservation of biodiversity, sustainable use of its components and equitable sharing of benefits derived from its use (for details see Convention on Biodiversity, 2004). An important element of the CBD is the Protocol on Biosafety which was signed in 200 at Cartagena. Its objective is to protect biological diversity from the potential risks posed by living modified organisms created through applications of modern biotechnology. One aspect of the protective framework involves the adoption of the precautionary principle through the use of the 'advance informed agreement' (AIA) procedure for ensuring that countries are provided with the information necessary to make informed decisions before agreeing to the import of such organisms. A further aspect is the establishment of a Biosafety Clearing-House as a forum for the exchange of information on living modified organisms and to facilitate the implementation of the Protocol through the provision of information.

The Conference of the Parties (COP) has established seven thematic programmes with 16 cross-cutting issues, as listed in Table 7.12. These reflect endangered environments and their organisms in conjunction with issues which influence most, if not all, of these environments and which reflect connections between them. In terms of action, programmes operate at national level with responsible bodies filing reports to the CBD (these are available at CBD, 2004). The goals are stated in the strategic plan, the primary objective of which is to curtail the loss of biodiversity.

Table 7.12. Details of the Convention on Biological Diversity (based on CBD, 2004).

THEMATIC PROGRAMMES:
1. Agricultural biodiversity
2. Dry and sub-humid lands biodiversity
3. Forest biodiversity
4. Inland waters biodiversity
5. Island biodiversity
6. Marine and coastal biodiversity
7. Mountain biodiversity
CROSS-CUTTING ISSUES:
1. Access to genetic resources & benefit sharing
2. Alien species
3. Traditional knowledge, innovations and practices
4. Biological diversity & tourism
5. Climate change & biodiversity
6. Economics, trade & incentive measures
7. Ecosystem approach
8. Global strategy for plant conservation
9. Global taxonomy initiative
10. Impact assessment
11. Indicators
12. Liability & redress
13. Protected areas
14. Public education & awareness
15. Sustainable use & biodiversity
16. Technology transfer & co-operation

The mission statement is given below. The targets and indicators identified in the CBD are based on material prepared by the Subsidiary Body on Scientific, Technical and Technological Advice (SBSTTA) and the COP , including the deliberations of a meeting on the '2010 Global Biodiversity Challenge' in 2003 in London.

- Parties (are required to) commit to a more effective and coherent implementation of the three objectives of the Convention, to achieve by 2010 a significant reduction of the current rate of biodiversity loss at the global, regional and national level as a contribution to poverty alleviation and to the benefit of all life on Earth.

The UNCED in 1992 also prompted new initiatives on conservation by individual governments. In the EU, for example, the UNCED principles of sustainable development were adopted within existing conservation strategies which focus on the Birds Directive of 1979 and the Habitat Directive of 1992, and new goals were established through Natura 2000. This is the creation of an EU-wide network of sites for protection, called Special Protection Sites (SPAs) and Special Areas for Conservation (SACs) which are considered to be 'type' sites for specific natural features. In the UK, for example, some 237 SPAs have been designated, amounting to 1.25 million hectares, along with 567 SACs, amounting to 2.16 million hectares (Department for Environment Food and Rural Affairs (DEFRA), 2004). In the EU seven biogeographic zones have been designated with the Continental and Atlantic zones being the most extensive. According to Europa (2004) recent developments involve "the decision to include more than 7,000 nature sites in the Atlantic and Continental regions of the EU in the network. The 197 animal species, 89 plant species and 205 habitats covered are scientifically considered of European importance. This means that their protection must be enhanced to preserve valuable bio-diversity in Europe. Species such as the wolf, the otter, the salmon as well as certain coastal lagoons and river systems are part of the lists, which cover most of the EU's territory (France, Germany, Belgium, the Netherlands, Italy, UK, Sweden, Austria and Denmark)".

The Statement on Forest Principles made at the UNCED in 1992 (see Table 7.8) also concerns biodiversity. While this statement (see UN, 1992, for the full text) recognizes the right of individual nations to exploit their own forest resources they are urged to develop sustainable use, trade and conservation policies. Such policies should include afforestation and tree planting. The statement also recognizes the universal value of forests in environmental processes, heritage, history and culture and that the sustainable use of forests will require sustainable patterns of production and consumption at a global level. However admirable these principles are, forest demise continues at alarming levels (see Section 6.3) and the UNCED declaration is more a statement of good will or advice than a mechanism for conservation.

7.9 Trade

Since prehistoric communities exchanged resources, trade has always been important as an instrument of carbon domestication. Moreover, the spread of agriculture and the expansion of Europe (see Chapter 5) stimulated trade, much of which involved carbon-based commodities such as crops and animal products. More recently, trade in coal, oil and natural gas represents substantial flows of carbon worldwide. Consequently, the economics of trade influence carbon flows and the economics of trade are determined to a large extent by trading agreements at regional, national or international level. While a full appraisal of past and present trade arrangements and agreements is beyond the scope of this book, it is nevertheless important to consider some of the practices in operation today. Not only do such practices influence carbon flows worldwide but they also influence overall development through the placing of controls on the flows of cash. Amongst the most recent and important trade agreements are the General Agreement on Tariffs and Trade (GATT) and the World Trade Organization (WTO). There are also a number of regional trading arrangements such as those of the European Union and the North American Free Trade Agreement (NAFTA).

The GATT came into existence in the years immediately after World War II through the desire of governments to liberalize trade. It was first signed in 1947 and came into effect in 1948. Efforts to create an International Trade Organization (ITO) from the more informal GATT in the 1960s and 1970s failed, leaving GATT as the major international agreement on trade. It provides a forum for the promotion of trade and a mechanism for resolving disputes between nations about trade. The former emphasizes free trade through the regulation and reduction of tariffs on traded goods. Since its inception there have been eight rounds of talks, the last and biggest being the Uruguay round between 1986 and 1994. The outcome of these, often acrimonious, deliberations included the reduction of many tariffs by 40 percent. At the last meeting, which involved 123 countries, the WTO came into existence at the beginning of 1995 to take the place of GATT (see WTO, 2004). Whereas GATT was concerned with trade in goods, the WTO also covers trade in services and intellectual property. The WTO, which is based in Geneva, has adopted rules drawn up under the GATT and since 1994 it has negotiated further agreements. The WTO states that "At its (WTO's) heart are the WTO agreements, negotiated and signed by the bulk of the world's trading nations. These documents provide the legal ground rules for international commerce. They are essentially contracts, binding governments to keep their trade policies within agreed limits. Although negotiated and signed by governments, the goal is to help producers of goods and services, exporters, and importers conduct their business, while allowing governments to meet their social and environmental objectives". Today there are 148 members, of which almost 100 are developing nations.

The scope of the WTO is immense since it deals with trade in all carbon-based commodities to telecommunications and finance. However, there are several underpinning principles on which the WTO operates. These are given in Table 7.13.

Table 7.13. The underpinning principles governing the World Trade Organization (based on WTO, 2004).

TRADE WITHOUT DISCRIMINATION This comprises two components: most-favoured-nation and national treatment. The most-favoured-nation concept requires equal treatment among trading partners e.g. one customs duty for all. National treatment: Imported and locally-produced goods/services should receive equal treatment once the former have entered a country.
FREER TRADE THROUGH NEGOTIATION Lowering trade barriers, such as duties/tariffs and quotas or bans, is a priority.
PREDICTABILITY THROUGH BINDING AND TRANSPARENCY The clear statement and maintenance of trade conditions is important because it provides a stable and predictable trading environment. WTO negotiates 'bindings' i.e. commitments to trade specific commodities under clearly stated conditions such as ceilings on customs tariffs. The Uruguay talks increased considerably the number of binding agreements. In agriculture 100% of products have bound tariffs and thus a more secure trading environment for producers and consumers is created.
PROMOTING FAIR COMPETITION WTO's rules promote fair trade which may not always be free trade (see most-favoured-nation etc above). They also cover dumping , i.e. low cost export to win markets, for which compensation may be required.
ENCOURAGING DEVELOPMENT AND ECONOMIC REFORM The stability and predictability of trade encourage liberalization which contributes to development. For developing nations there is a transitional period to implement WTO rules and 'special' trading conditions may be implemented to assist developing nations.

The WTO has received much criticism, and attempts to disrupt its meetings have been made by protesters against globalization. It is accused of not encouraging global justice because of the special pleading and lobbying by powerful nations to operate protectionist policies.

There are also a number of regional trading blocs which are essentially customs unions. The objective, like that of the WTO, is to stimulate trade and thus assist development. There are four types of trading blocs, brief descriptions of which are given in Table 7.14.

Table 7.14. Types of trading blocs (based on BizEd, 2004).

FREE TRADE AREAS Constituent sovereign nations trade freely with each other, so all have favoured-nation-status (see text on WTO) but they exercise individual trade barriers with nations outside the free trade area. Example: North American Free Trade Agreement (NAFTA)
CUSTOMS UNIONS Constituent nations do not have complete control over trade policy but are bound by some level of common policy. They have a common external tariff with nations outside the agreement. The European Union (EU) is a customs union.
COMMON MARKETS Common markets are customs unions plus unrestricted movement of capital and labour between members. Mercosur, comprising Argentina, Brazil, Paraguay and Uruguay is an example.
ECONOMIC UNIONS Economic unions are common markets with more integration and common policies such as the Common Agricultural Policy (CAP) of the EU. There is also a common currency and exchange rate for most members.

There are many different trade blocs, as documented by the World Bank (2001) but the largest and most influential are the European Union (EU), the North American Free Trade Agreement (NAFTA), the Mercado Comun del Cono Sur (MERCOSUR) also known as Southern Common Markets (SCCM), and the Association of Southeast Asian Nations (ASEAN). A more comprehensive list is given in Table 7.15.

Table 7.15. The major trade bloc groups (based on World Bank, 2001).

Trade bloc groups
Andean Community (CAN)
Asia-Pacific Economic Cooperation (APEC)
Association of Southeast Asian Nations Free Trade Area (AFTA)
ASEAN + 1 (Mainland China) Free Trade Area
Caribbean Community and Common Market (CARICOM)
Central European Free Trade Agreement (CEFTA)
Closer Economic Partnership Arrangement (CEPA) between Hong Kong, Macau and Mainland of the People's Republic of China
East African Economic Community (EAEC)
Economic Community of West African States (ECOWAS)
European Economic Area (EEA)
European Free Trade Association (EFTA)
European Union (EU), formerly the European Community
Mercado Comun del Sur (MERCOSUR or MERCOSUL)
Mercado Comun Centro Americano (MCCA)
North American Free Trade Agreement (NAFTA)
South Asian Association for Regional Cooperation (SAARC)
Southern African Development Community (SADC)

There are also proposals for others; for example the former prime minister of Spain has advocated a formal trade agreement, to be known as the Atlantic Economic Association, between Atlantic nations (Aznar, 2004/5). World trade and thus carbon flows are complicated by the financial and development prospects generated by such associations. Add to this the various governmental and non-governmental institutions which influence the use, conservation and trade of all types of biological resources and the picture of carbon movement worldwide is indeed complex.

Chapter 8

8 CONCLUSION AND PROSPECT

It is an obvious but unequivocal and necessary statement that carbon in all its combinations is ubiquitous and a fundamental component of planet Earth. In view of its pivotal role in the past, present and future of the Earth, any highlighting of its qualities, distribution and importance is invidious, not least because carbon as the element is relatively rare and unreactive and because it is carbon's role in mobilizing other elements that is the all important link within and between environmental systems. Alone it is innocuous; in combination it is a dynamic force. The versatility of carbon chemically is but one of its strengths and this, in turn, is linked to the highlight of carbon's central role in all life forms. From bacteria to plants and animals, the wonder of carbon as life and its history is daunting. The organic and inorganic are united through carbon and the continuous interplay between the two is the backbone of Earth's geological history through the intermediary of atmospheric change. This relationship is the essence of Earth's past, present and future. It hinges on the biogeochemical cycling and sequestration of carbon in the Earth's crust and biosphere through geological time and the efforts of one of carbon's manifestations, modern humans, to reverse that process via the unquenchable thirst of society for biomass-based resources and energy.

This latter constitutes the domestication of carbon. As discussed in Chapter 5, this can be considered as a developmental process with several thresholds: the domestication of fire involving the use of wood for fuel, the harnessing of specific plants and animals for agriculture with attendant loss of natural ecosystems, the exploitation of fossil fuels, and modern biotechnology. Such developments have occurred at different times and at different intensities throughout the world and have involved carbon flows within and between regions/countries. This domestication has altered the Earth's surface to such an extent that there is little manifestation of its pre-human-intervention state. It has also led to considerable inequities in access to wealth over time. There are two ways to assess this condition as it is manifest in the opening decade of the twenty-first century. Either it has led to an interplay of natural, semi-natural and cultural landscapes within which almost 6.4×10^9 people (data from Population Reference Bureau, 2004) live. The achievements, over and above that of the sustenance of so many of a single species, include the urban-industrial landscapes of the past and present. Archaeological sites attest to the aspirations and achievements of former civilizations and their cultural apogees, while modern cities, many with an ancient past, bear witness to the industriousness, in its broadest sense, of society in the process of wealth generation, much of which is the economic face of contemporary carbon flows worldwide. Spin-offs to such past and present activities include art, theatre, literature, music, sport, etc. Alternatively, the loss of so much 'Nature' and the transformation of the Earth's surface, which is associated with the appropriation by humans of such a large proportion of the biosphere's primary productivity (see Section 6.4), is considered a matter of grave concern; rather than representing the achievement of human endeavour, it represents the profligacy and poor stewardship of their environment by human communities. Rather than constituting the sacrifices necessary for humans to manipulate or subjugate nature in order to achieve industrial and cultural apogees, the outcomes of this profligacy are the many environmental problems of today, notably global climatic change, biodiversity loss, pollution etc. These two scenarios are examined below; though both are indisputable, the final analysis rests on philosophical viewpoints and whether or not humans reign supreme (within limits); a key question is whether they can continue to reign supreme by maintaining the adequacy of biogeochemical services for such supremacy.

So what of the future prospect? How can development continue with little or no further impairment of the biosphere and atmosphere? How can the intimately-coupled society and carbon-based resources be uncoupled? How can such bonds be broken? And how can technology be brought to bear on environmental problems created by the domestication of carbon? It is unlikely that any major shifts in energy sources will occur in the near future so it remains to be seen what affect carbon taxes, trading and international agreements curbing carbon emissions have on limiting global warming. In the longer term, breaking the bond with fossil and probably most carbon-based fuels will become essential; renewable fuels such as wind, solar and tidal power require research and development because they are less polluting, and even the much-debated nuclear energy might prove to be more acceptable than fossil fuels. Given that world population will continue to increase in the next 20 years, how will these people be adequately fed? That carbon-processing system which is agriculture will need to adjust. Should existing agricultural systems be intensified, or should more of the biosphere be converted into agricultural land? Moreover, new developments in biotechnology offer promise for improving agricultural productivity and for environmental remediation. Is this, however, too little too late? Moreover, uneven development, i.e. uneven carbon domestication which has given rise to uneven access to and control of carbon, has generated a world pattern of wealth and poverty. Can and should this situation persist and what bearing does this pattern have on carbon flows? These are just some of the issues facing the world community today.

8.1 Conclusion

The versatility of carbon is a product of its chemistry. The abundance of inorganic and organic carbon compounds is a result not only of carbon's chemistry but also of the temporal changes which have occurred throughout the Earth's history. Carbon's versatility is especially reflected in its biology; the structure, composition and abundance of life forms and their changes through geological time to which the fossil record bears spectacular, yet still limited, witness; not all organisms are readily fossilized, especially those with few hard parts or those living in environments with little deposition of materials in which preservation could occur. Not least is the complexity of human biology and the creativity which is conferred by the human brain, a factor which has rendered carbon even more versatile and invented a whole new vista of synthetic carbon compounds. In terms of composition, the abundance of organic chemicals in living organisms is amazing; that one element could be so mercurial and at the same time be taken so much for granted is a paradox. Apart from its fossil content, the geological record also attests to the operation of the biogeochemical cycle of carbon since the Earth came into existence. As with the biological characteristics of carbon, its geological/environmental aspects reflect versatility and dynamism. Moreover, the interplay between life and environment provides the theatre, the plays and the actors in a continuous reappraisal that is Gaia; this is rather like the action of a kaleidoscope within which the component parts remain constant but are continuously reassembling. And then came Nature's more recent achievement: the evolution of the hominids only c. 5×10^6 years ago culminated in modern humans and a new age. Modern humans set about building on the achievements of their ancestors and have eventually changed the face of the Earth through their domestication of carbon.

Until the opening of the present interglacial period c. 12,000 years ago, humans and their hominid ancestors had little impact on the environment. Nevertheless, as scavengers and then hunter-gatherers they initiated the process of carbon domestication through the harnessing of fire and use of biomass resources; the latter began the process of manipulating plant and animal communities for food. These, along with stone tool-making were the earliest technologies. Some 2×10^6 years ago *Homo erectus* evolved, the first hominid species to expand beyond its continent of origin, Africa, into Europe and Asia. Modern humans (*Homo sapiens)* evolved c. 250,000 years ago and subsequently colonized every continent, except Antarctica which remained a true wilderness until the early 1900s. They became skilled hunter-gatherers and as the last ice age drew to a close, they initiated a shift from this relatively opportunist food-procurement to a more predictable procurement system based on permanent agriculture. This set in train a degree, scale and rate of environmental change unprecedented in nature, excepting volcanic eruptions and earthquakes. Few parts of the Earth's surface remain unaffected by agriculture which remains the most important agent of environmental change today. Of particular significance is the onset of deforestation and thus the start of the impairment of one of the biosphere's main stores of carbon. Humans began to reject their status as integral components of the biosphere and to emerge as dominants and directors of ecosystems and the biosphere. Perhaps this shift in status and responsibility is the manifestation of the parable of the biblical Adam and Eve. A major threshold in carbon domestication, agriculture fostered co-operation between humans, and provided sufficient food that a proportion of the population could withdraw from food production and develop other skills and other resources. Carbon, as the biomass of natural vegetation and soil, was sacrificed in order to create a controllable carbon-processing system with enhanced edible carbon. Today, industrialized agriculture utilizes the carbon of fossil fuels to do exactly the same. These are the tradeoffs, the compromises or drawbacks which characterize all forms of agriculture. Moreover, such tradeoffs etc are characteristic of all technologies; no technology is entirely beneficial, a concept sometimes difficult to accept in what has become a risk-averse society and which is epitomized by the current debate about genetically-modified crops in Europe.

Agriculture is amazingly varied worldwide. It is characteristic of all but the most hostile parts of the Earth, notably the most extreme hot and cold desert areas. It ranges in style from extensive herding of animals, in environments which can only supply products consumable by humans through animal digestion of biomass, to the intensive industrialized agriculture of North America and Europe which produces an abundance of grain/cereal, fibre, meat and dairy products. Other agricultural products include beer, wine, oils, beverages, flowers, perfumes, some medicines, illegal and legal pharmaceuticals and biomass fuels. Such commodities are both the essence of subsistence and the basis of considerable wealth generation through trade; the carbon flows they generate also constitute international links. The production of a food surplus confers advantage economically and politically whether at local, regional or national scales, and has done so in the past just as it does at present. Food security contributes to the development of other resources and cultural attributes such as art, communication, science, and promotes innovation in general. Today, agriculture feeds more than 6×10^9 people worldwide. That in itself is a remarkable achievement, and would be even more remarkable if the food was more evenly distributed. Moreover, food availability has supported cultural development and one of the wonders of the world must be the huge cultural diversity and heritage which exists. That humans have been able to fashion the world in which they live within the environmental limits, e.g. geology, soil type, climatic characteristics, of that world is a matter for celebration and wonderment. Such mutuality is the essence of Gaia and the manifestations of the relationship, the heterogeneity of people, environment and culture, attest to the versatility of people and the resilience of the environment.

The creation of landscapes from wildscapes involves a host of human activities other than agriculture. They include forestry, mineral extraction, industrialization and urbanization; all generate land-cover change and cause carbon to flow internationally and from the rural to the urban environment. A summary is given in Figure 8.1.

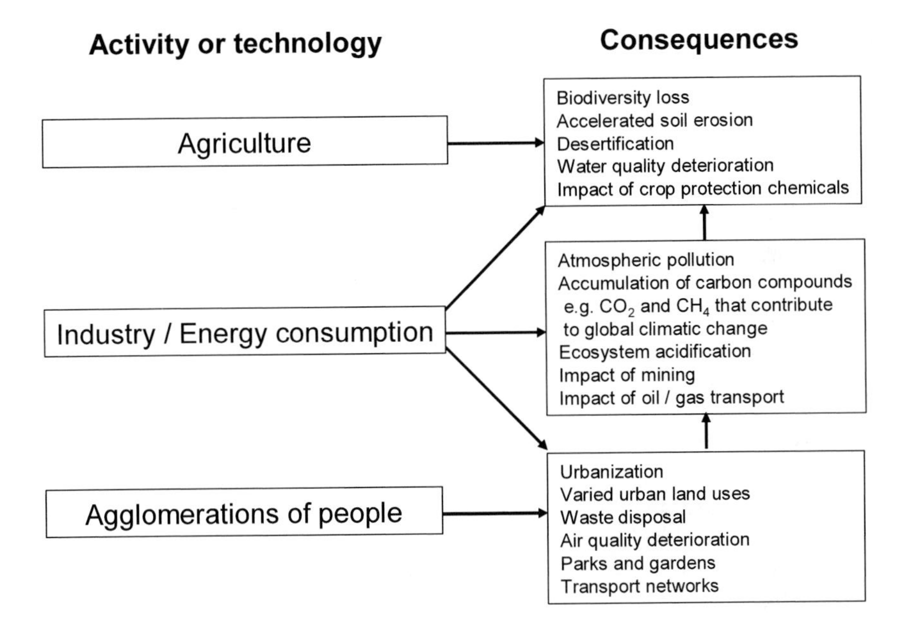

Figure 8.1 Anthropogenic factors in landscape formation

These factors facilitate the flow of carbon to the cities which are the major consumers of food and fuel energy and raw materials such as wood, fibre, and the products of petrochemicals. Although cities are the major producers of carbon dioxide, a cause of climatic change, they also attest to the artistic, architectural, technological and scientific capacities of nations and house a wealth-generating legacy through tourism. The overall impact of carbon domestication can be classified into four separate but linked categories involving land transformation, a process which will undoubtedly continue as socio-economic factors drive change through their impact on agriculture, industry and urbanization. These activities generate environmental problems which can be summarized under four major headings: ecosystem and biodiversity loss, impairment of water quality, global climatic change, and uneven development and (wealth distribution.)

8.1.1 Ecosystem and biodiversity loss

Biodiversity loss and accelerated extinction rates began with the earliest agriculture; its onslaught on the Earth's cover of natural vegetation, especially its forests; agriculture and other activities have subsequently been unrelenting. The many uses to which living organisms can be put also places them in jeopardy; the inherent versatility of carbon threatens its existence in usable forms. No subject better illustrates this condition than trees. They not only succour a huge range of other living organisms but they provide fuel, construction material, paper, gums/resins, fruits, nuts, beverages and oils. In addition, trees provide on-going functions including habitats, shade, water cycling and biogeochemical cycling, not least of carbon. As Figure 8.2 shows, deforestation continues to be widespread with the highest rates in Indonesia, tropical Africa and Central America.

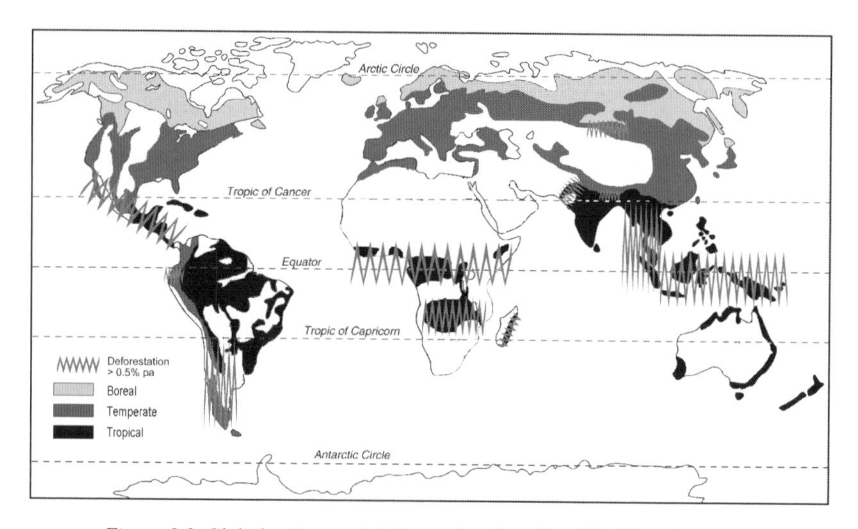

Figure 8.2 Global patterns of deforestation (based on World Bank, 2004a)

The products and functions of other vegetation types are also as vital as those of trees. Any loss of the natural vegetation cover impairs, possibly even eradicates, these products and functions as well as adversely affecting dependent organisms such as animals, insects, fungi and other micro-organisms. Of particular importance is the impairment of ecosystems to store carbon. This may occur due to the extinction of species but it also involves the loss of unique genetic characteristics and thus lost opportunities for biotechnology. These two issues reflect the importance of carbon at macro- and micro-scales.

When under pressure whole ecosystems may be lost. The data on land-cover alteration given in Section 6.4 reflect a considerable anthropogenic impact; approximately 52 percent of the Earth's surface remains undisturbed, a value which declines to only 27 percent if bare rock and ice-covered areas are excluded (based on Hannah *et al.,* 1994). Moreover, the Living Planet Index (WWF, 2004a) shows that in the last 30 years significant losses in natural capital have occurred in forests, freshwater and marine environments. The impact is not, however uniform; the domestication of carbon has been greatest in developed countries whose environmental impact has been most intense. The ecological footprint (Section 6.4) deepens and broadens as resource consumption increases and in most cases the ecological footprint extends beyond the country in which the resource consumption occurs. Consequently, many nations consume not only their own natural capital but also that of other nations. This is another element of globalization and it also reflects uneven development and degree of wealth or poverty (see below). Given this massive alteration of the Earth's ecosystems, it follows that many organisms have been driven to extinction. It is impossible to determine the rate of extinction because there are many unknown factors, not least being any estimate of numbers of plants, animals, fungi etc. prior to anthropogenic disturbance, the inadequate understanding of the status of many organisms currently, and adequate data on numbers of organisms extant today. Nevertheless, estimates range from 1,000 to 10,000 times the natural rate of extinction; estimates at the higher end of this scale equate with estimates for extinction phases recorded in the geological record, e.g. at the Permian – Triassic and Cretaceous – Triassic boundaries (see Sections 4.1.4.6, 4.1.5.1, 4.1.5.3 and 4.1.6.1). Apart from the fact that this represents a major loss of life forms, it means that a huge genetic resource is being lost just at the time when science has provided the wherewithal to comprehend genetic codes and engineer them to advantage. The loss of both species and ecosystems must also have a bearing on the capacity of the biosphere to provide biogeochemical services which are vital to the maintenance of atmospheric composition and life. Although the relationship is complex and research on it relatively new (see Tilman and Downing, 1994), available information indicates that ecosystem stability and productivity are linked with biodiversity. Thus any losses through disturbance and extinction have a detrimental, though as yet unquantifiable, impact (see Naeem and Wright, 2003 and references therein). Once again, this reflects the intricate interplay between biogeochemical cycles, global temperatures and the character of the biosphere. As Figure 2.17 shows, only a relatively small proportion of the global biogeochemical cycle of carbon has been appropriated by humans but already profound effects are becoming manifest (see below).

8.1.2 Impairment of water quality

Most human activities require water and, according to the LPI water quality and aquatic ecosystems have declined substantially in the last 30 years. Much of this is related to carbon domestication. The LPI is concerned primarily with the organisms in given ecosystems and there is abundant evidence to show that the range and populations of many aquatic species have declined. The plight of many whale species worldwide and the imperilled cod fisheries in the waters of the North Atlantic are a case in point; on a smaller scale but nonetheless important are the decline in sea cucumbers in the waters around the Galapagos Islands and the decline in sea-horse populations in the waters of Southeast Asia. Over-exploitation of such marine resources has resulted in scarcity and attendant threats to livelihoods. Declines in the fish stocks of freshwaters have also occurred. Freshwater is a vital component of appropriating carbon. According to UNEP (2002) c. 75 percent of total global water consumption is for agriculture, chiefly for crop irrigation, while industrial use accounts for about 20 percent, and the remaining 5 percent is used for domestic purposes.

Irrigation for agriculture is a major cause of declining water quality; in the quest for increased yield, poorly-managed irrigation systems have resulted in salinization of soils and water and sometime to land abandonment.

Excessive use of phosphate and artificial nitrogen fertilizer, again part of the quest to increase yields, is another cause of water-quality decline. It may result in cultural eutrophication, notably a substantial increase in the nutrient content of fresh and salt water, which encourages algal growth and an eventual overall decline in biodiversity (see Section 6.3.1). The problem is widespread in countries where intensive industrialized crop production is prevalent and is a major problem in enclosed seas such as the Mediterranean. Industry and domestic uses of water also exact a toll. For example, in parts of the USA, Canada and Europe 'acid rain', caused by the burning of sulphur-rich fossil fuels, has acidified catchments and reduced water pH below the levels tolerated by most fish species; algal and microbial populations have also altered. Acidification is likely to spread substantially as industrialization continues in China, India etc. The issue is discussed in Sections 5.5 and 7.7.

Concerns about the accessibility of fresh water have led Kofi Annan, Secretary General of the United Nations, to state "Global freshwater consumption rose sixfold between 1900 and 1995 - more than twice the rate of population growth. About one third of the world's population already lives in countries considered to be 'water stressed' - that is, where consumption exceeds 10% of total supply. If present trends continue, two out of every three people on Earth will live in that condition by 2025". The availability of clean water remains a problem in many parts of the world; 20

percent of the world's 6.2 x 10^9 people still do not have access to safe drinking water (UNEP, 2002). Safe sanitation is also a problem; the contamination of water supplies by faecal material remains a major source of disease in parts of the least developed world and UNEP states that "Polluted water is estimated to affect the health of 1.2 billion people, and contributes to the death of 15 million children annually". These latter problems are especially acute in parts of the developing world.

8.1.3 Global climatic change and its impact

By the start of the new millennium, global climatic change had become a fact of life. Despite continued wrangles about the Kyoto Protocol (Section 7.7) governments generally had been convinced that global warming was occurring. Figure 5.5 shows that the last few decades have been the warmest on record, with 1998 being the warmest at 0.58°C above the 1961-1990 mean, and the years 2002 and 2003 jointly being the second warmest. Although it is possible that the temperature increases of the last few decades simply represent a minor warming trend rather than a major shift in global temperatures, other evidence from palaeoenvironmental archives supports, though does not prove, the latter view. As noted in Section 2.3, data from polar ice cores indicate that there is at least a c. 25 percent increase in the average concentration of carbon dioxide and a doubling of that of methane in warm (interglacial) stages when compared with cold (glacial) stages. These cores also show that in the last c. 200 years there has been further increase in carbon dioxide concentrations by c. 25 percent and of methane by c. 100 percent, an increase unprecedented in nature and due to human appropriation of the carbon sequestered in the Earth's crust and, in the case of methane, a huge increase in the number of farm animals. According to the Energy Information Administration (EIA, 2004b) world carbon dioxide emissions grew at an average annual rate of 1.2 percent between 1980 and 2001, from 18,651 x 10^6 t to 24,082 x 10^6 t. During this period, the average annual growth rate of OECD and non-OECD carbon dioxide emissions was 0.7 percent and 1.8 percent, respectively. Moreover, total OECD carbon dioxide emissions increased from 10,202 x 10^6 t in 1980 to 11,690 x 10^6 t in 2001. Non-OECD emissions increased from 8,448 x 10^6 t in 1980 to 12,392 x 10^6 t in 2001. The industrial production of other heat-trapping gases, notably CFCs, has also contributed to this enhanced 'greenhouse effect'. Moreover, it is acknowledged that atmospheric composition has been a major factor in determining global temperatures during the Earth's geological history (see Section 4.1). Thus it is reasonable to conclude that such substantial anthropogenic perturbations to global carbon dynamics could not proceed without any impact. Moreover, Stott et al. (2004) have proposed a methodology for determining the proportion of global warming due to human activity. Taking data for 2003, when summer temperatures in Europe were at their hottest since c. 1500 AD, they estimate that human influence has doubled the risk of similar heat wave conditions. Actual temperatures were 5 to 7°C higher than average in France, Switzerland and Germany where death rates were also higher than average. In addition, climate models predict that such conditions are likely to increase by a factor of one hundred in the next 80 years with implications for heat-related deaths. What remains problematic, however, is how to predict the impact of global warming on ecosystems, hydrology, the oceans, agricultural ecosystems and society in terms of health, housing etc., and how to cope with these changes.

What has occurred in the last decade which reflects global warming and might be a portent of things to come? Table 8.1 gives a summary of the findings of the IPCC (2001a) in relation to climatic conditions in the 1990s. It reflects global variation and widespread variation of a magnitude that points to change rather than oscillation. More recently, a warning from the World Meteorological Organization (WMO, 2004) based on weather events in 2003 draws attention to a possible increase in extreme weather events. For example, the USA experienced 562 tornados in May 2003 as compared with the previous high record of 399 in June 1992; such an increase resulted in 41 deaths and considerable destruction of property. Heat waves in India, with temperatures as high as 45 to 49°C which were 2 to 5°C higher than average, caused some 1,400 deaths and in neighbouring Sri Lanka unusually high rainfall caused flooding and landslides.

There is also a growing body of evidence for the effect of increasing temperatures of flora and fauna. A survey by the Worldwide Fund for Nature (WWF, 2005) records that with the exception of a few, glaciers worldwide have retreated due to melting. In Alaska most glaciers are melting, and thinning rates in the last 5 to 7 years are more than twice those of earlier years. This melting contributes about 50 percent of the water flowing into the oceans globally. Almost all glaciers in the Andes are retreating and melting has accelerated in the 1990s. In the European Alps between 10 and 20 percent of glacier ice has melted in the last 20 years and since 1850 Europe's Alpine glaciers has lost half their volume; c. 50 percent of those remaining will have disappeared by 2100. Tropical glaciers in Africa have decreased in area by 60 to 70 percent on average since the early 1900s while most Himalayan glaciers have diminished in length and volume since 1990. In the pacific region, glacier retreat is evident except for some New Zealand glaciers where increased snowfall has resulted in growth. This has also occurred in Scandinavia and Iceland but in Greenland, which houses c. 12 percent of the world's ice, ice sheets have thinned and begun sliding towards the sea. Similar thinning and sliding has occurred on the periphery of Antarctica which houses some 95 percent of the world's freshwater. All of these changes are contributing to rising sea levels.

Table 8.1. Aspects of climatic change during the 1990s (adapted from IPCC, 2001a)

A. Global average temperature of the lowest 8 km of atmosphere has changed by + 0.05 ± 0.10 °C per decade but the increase in global average surface temperature has been greater, i.e. by 0.15 ± 0.05 °C per decade. The difference occurs mostly over tropical and sub-tropical regions.
B. There have been decreases of c. 10 % in the extent and duration of lake ice cover in the mid- and high-latitudes of the northern hemisphere during the 1990s.
C. Mountain glaciers in non-polar regions have retreated during the 1990s.
D. Spring and summer sea-ice in the northern hemisphere has decreased in extent by c. 10-15 % since the 1950s. Some evidence suggests that Arctic sea-ice thickness in late summer and early autumn has declined by c. 40 % with a more limited decline in winter sea-ice thickness.
E. On average global sea level rose between 0.1 and 0.2 m during the 1990s.
F. Global ocean heat content has increased since the late 1950s.
G. There have been changes inprecipitation: (i) between 0.5 and 1.0% per decade increase in continental mid- and high-latitudes of the northern hemisphere in the 1990s; (ii) an increase of 0.2 to 0.3% per decade over tropical land areas in the 1990s.
H. It is probable that the mid- and high-latitudes of the northern hemisphere have experienced c. 2-4% increase in the frequency of heavy precipitation events.
I. There is the possibilityof a 2% increase in cloud cover over mid- high-latitude continental regions in the 1990s.
J. There appears to have been a reduction in the frequency of extreme low temperatures and a small incfease in the occurrence of extreme high temperatures.
K. Warm episodes, e.g. of El Niño – Southern Oscillation (ENSO) have been more frequent and longer lasting since the mid-1970s.
L. There have been some increases in the occurrence of drought and severe wetness.
M. The frequency and intensity of droughts have increased in the last few decades.

Apart from an abundance of hearsay reports of fish, birds and insects expanding beyond their normal ranges a quantified assessment of shifts in range boundaries and phenology (seasonally-linked characteristics such as time date of flowering, awakening from hibernation and arrival of migrants) for a range of species has been undertaken by Parmesan and Yohe

(2003). They have examined the recorded range boundaries of many species of birds, butterflies and alpine herbs and found that there has been a shift northward, or upward in the case of alpine communities, of 6.1 km per decade. Analysis of the phenological characteristics of a wide range of herbs, trees, shrubs, birds, butterflies and amphibians also shows that change has occurred, notably an earlier timing of spring by 2.3 days. Thus, global warming is already affecting many species. In a similar study, Root *et al.* (2003) examined the changes in the phenology of 143 species of plants and animals and found that more than 80 percent had shifted in a manner consistent with climatic warming; in species of the temperate zone a shift in spring phenology of 5.1 days was apparent. Such results reflect the close relationship between climate, the capacity of organisms to respond and the possibility that ecosystems will alter substantially in the future as organisms react at species level and new assemblages of species emerge.

8.1.4 Uneven development and wealth distribution

Carbon resource development and acquisition as well as carbon flows vary in space and time. The rise and fall of civilizations throughout prehistory and history have been linked in some way to the control of these (and other) resources because food and fuel are so fundamental in human societies. The patterns of carbon control in the past *per se* have not, or possibly cannot, be determined. Present day (2004-2005) patterns of development and degree of wealth and poverty can be measured in various ways and there are important links with other attributes of society such as levels of education, health care, longevity, type of employment etc.

Figure 8.3 is a simple measure of poverty; it shows the proportion of people by region who live on less than $1 per day and how that proportion has changed in the last 20 years. It also shows how uneven development is globally. Approximately 1.1 x 10^9 people, i.e. 20 percent of the world's population, live on less than $1 a day; a further 20 percent live on less than $2 per day (United Nations, 2004). The greatest decline of numbers of people in the former situation in the last 20 years has been in China, East and South Asia where industrialization is occurring rapidly. This also reflects a substantial increase in carbon harnessing and consumption. In contrast, the situation in Sub-Saharan Africa has remained largely unchanged. The World Bank (2004b) makes the following observation "Growth in India and China is expected to reduce the proportion of people living on less than $1 a day from 28.3 percent in 1990 to 12.5 percent by 2015. But many countries will be left behind – particularly those in Sub-Saharan Africa, where poverty rates are already very high and the situation is unlikely to improve if current trends continue. Sub-Saharan Africa is expected to account for half of the world's poor by 2015, compared to 19 percent in 1990". Indeed by 2000 the number of people living on less than

$1 a day in Sub-Saharan Africa actually rose to 314 million, i.e. 47 percent of the population. Thus the gap between those who have and those who have not appears set to continue for some time to come. Mechanisms to redress this unevenness will require efforts on the part of all nations, including those who are less wealthy (see Section 8.2.6).

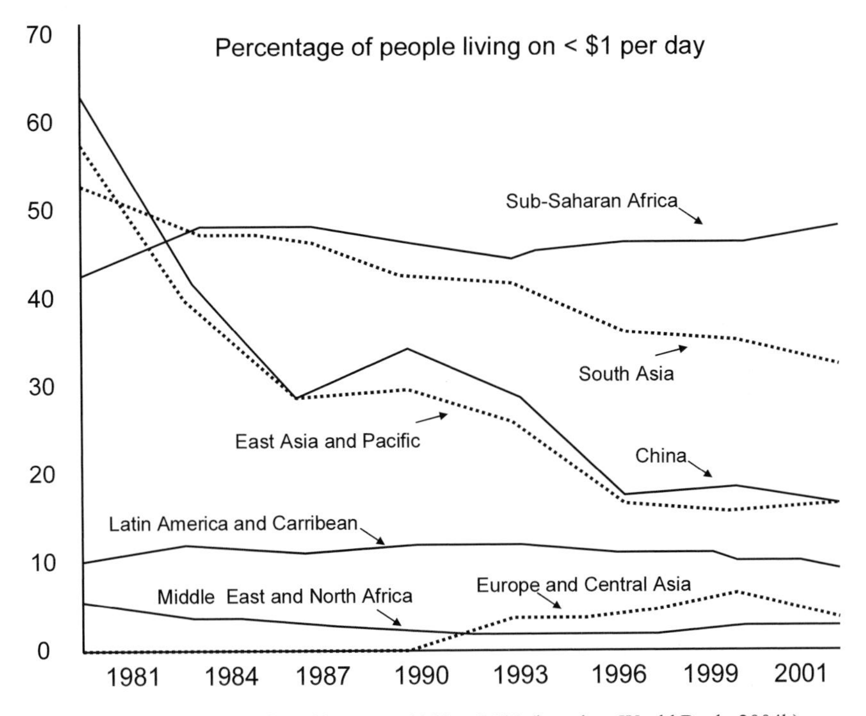

Figure 8.3. Patterns of world poverty 1980 to 2001 (based on World Bank, 2004b)

Another means of measuring development involves resource consumption and relates to the indices, such as the ecological footprint, discussed in Section 6.4. Overall, the global ecological footprint of humanity is not only increasing but is also in excess of the Earth's natural capital. Consequently, the current situation is unsustainable in the longer term. This situation has recently been examined by Diamond (2005a) who states that "On average, each citizen of the UK, Western Europe, the US, and Japan consumes 32 times more resources such as fossil fuels, and puts out 32 times more waste, than do inhabitants of the Third World". These nations have domesticated more carbon than any others; they not only consume their own carbon but also much of that produced elsewhere; they also produce more carbon emissions as carbon flows predominantly from the poor to the rich. For example, the USA has less than 5 percent of the world's population but it uses 26 percent of the oil, 25 percent of the coal, and 27 percent of the natural gas consumed worldwide.

Historically, Europe developed by using its own carbon resources, i.e. its trees and agricultural production; carbon flows were essentially internal until the 1500s. During the next three centuries Europe's development was greatly facilitated by exporting its people and importing resources from its colonies. Such activities are not open to those nations aspiring to Europe's standards of living today. Interestingly, however, many naturally carbon-rich nations have been slow to develop and their *per capita* levels of wealth continue to be low. These are oil-rich nations such as Nigeria and those in the Middle East whose oil wealth brings in high revenues. Other factors must be at work to curtail the spread of wealth, such as the concentration of oil revenues in the hands of a few, corruption and war.

Table 8.2. GDP *per capita* for selected countries (data from Central Intelligence Agency, 2005)

COUNTRY	**GDP** *per capita*	**COUNTRY**	**GDP** *per capita*
USA	$37800	India	$2900
Norway	$37700	Vietnam	$2500
UK	$27700	Iran	$2000
Singapore	$23700	Zimbabwe	$1900
Greece	$19900	Sudan	$1900
Oman	$13400	Cambodia	$1700
Saudi Arabia	$11800	Iraq	$1500
Argentina	$11200	Rwanda	$1300
Russia	$8900	Chad	$1200
Bulgaria	$7600	Kenya	$1000
Romania	$6900	Ethiopia	$700
Peru	$5200	Burundi	$600
China	$5000	Somalia	$500
Venezuela	$4800	Sierra Leone	$500
Albania	$4500	East Timor	$500

Data on Gross Domestic Product (GDP) *per capita* provide another measure of development. GDP is defined as a measure of the amount of the economic production, i.e. the total value of goods and services of a particular territory or nation per unit time (usually per year); it may be expressed on a *per capita* basis which facilitates comparison. It reflects wealth generation and is an accepted measure of national income and output. Table 8.2 gives the GDP for selected nations and shows that there is difference of c. 750 percent between the richest and poorest nations. The data also show that the richest nations are very rich and that the poorest nations are very poor. Many Sub-Saharan countries are amongst the poorest. Calorie intake is a direct measure of carbon consumption, as is oil consumption. Brief details of both are given in Table 8.3 which highlights the low food and fuel consumption of Africa and Asia and the high consumption of North America and Europe.

Table 8.3. Approximate kilo calorie and fuel consumption *per capita*

CONTINENT/REGION	Kilo calories per day for 2000 (data from FAO, 2004)	Fossil-fuel use, 1999 (tonne oil equiv.) (data from WRI, 2004)
Africa	2424	Not available
Sub-Saharan Africa	Not available	> 0.7
North Africa and Middle East	Not available	1.3
Asia	2696	0.87
Latin America and Caribbean	2860	1.2
Oceania	2961	5.7
Europe	3332	3.5
North America	3756	3.5
Developing world	2656	0.77
Developed world	3281	4.55
World Average	2792	1.62

Within each group there are substantial variations which mask even greater inequities. For example, the fuel consumption data for Sub-Saharan Africa include c. 5.2 tonne oil equivalent for South Africa while many other countries in the region rely heavily on fuel wood and biomass and use very little fossil-fuel energy. Similarly there is much variation in kilocalorie consumption; for example, in many Sub-Saharan the number of kilo calories available for consumption is less than 2300 while in South Africa the number is 2886. Moreover, Table 8.3 shows, the overall difference between the developed and developing worlds is marked for both fuel and fossil-fuel consumption being different by c. 23 percent for food and a huge c. 600 percent for fuel. The latter also means that carbon emissions are c. 600 percent higher from the latter. These are beginning to cause environmental change through the alteration of climate which has an impact globally and is not confined to the producing nations. So too do the wealth, high standards of living including food security, stability and political influence which accompany this command of carbon.

8.2 Prospect

In the short term, i.e. the next 50 years, carbon domestication will intensify. In the longer term, i.e. the next 150 years, there could and should be some decoupling of society and carbon. Population trends for the world as a whole and the regional differences which will occur in this millennium are central to any evaluation of future carbon domestication and thus to the prospects for people-environment relationships. The related issue of rising standards of living is equally important because enhanced material wealth involves additional pressure on all types of resources, not least on carbon. Fuel and food energy are key factors to consider, though they are not entirely separate as fuel energy is an important and increasing input into food production. In the light of the relatively limited understanding of global carbon cycle dynamics and the growing recognition that the more human intervention there is in this cycle the more susceptible global climate is to alteration, it is axiomatic that some semblance of sustainable development can only be achieved if the bonds between humans and carbon are relaxed. Breaking such bonds will not be easy and indeed alternatives may themselves pose other problems. The most obvious target could and should be the reliance on fossil fuels. This bond *should* be a target because fossil-fuel use is a major cause of perturbation to the carbon cycle and *can* be a target because there are alternatives which involve little or no carbon, e.g. solar, wind and nuclear energies. On the other hand, food production will always involve carbon manipulation and that bond cannot be broken; agriculture could, however, be made more efficient not least through technologies which involve carbon manipulation such as genetic engineering. Other technologies, notably information technologies, could also be employed to improve the efficiency of carbon-based resource use and of resources whose manipulation depends on carbon; examples include wood, paper, fertilizers. Another aspect of sustainable resource use and energy conservation is recycling, e.g. of plastics, glass, paper and metals, some of which may be undertaken using organisms such as bacteria and fungi. Future, as yet indefinable, developments in technology may also contribute to relaxing the bonds between people and carbon, though such developments will need to be undertaken in the context of the precautionary principle (see O'Riordan, 2000) which requires risk assessment etc. to pre-empt possible problems. As history relates, all technologies carry some level of risk; pre-emption is essential to minimize adverse impact but even the most benign technologies have some impact. However, none of the bonds with carbon will be relaxed or severed unless carbon management becomes a paramount political issue; global warming is now widely accepted politically but carbon management requires a globally holistic approach with cascading effects at regional, national and local scales.

8.2.1 Population change

As Figure 8.4 shows, there is not full agreement for the period 2000 to 2150 between the main agencies which generate population projections. However, there is general agreement at changes until c. 2050, beyond which conjecture is in any case unreliable due to the many unknown factors which may operate. Figure 8.4 shows that population is set to rise to c. 10-12 x 10^9 by 2050; this is almost a doubling of the world's present population. This is despite a reduction in the rate of increase which reached a peak of c. 2.04 percent per year in the late 1960s and which is now 1.35 percent and likely to decrease to c. 1.1 percent in 2010 to 2015 (United Nations, 2001). At the very least this will mean a doubling of pressure on resources and agricultural systems in the next 30 to 50 years (see Kemp, 2004 for a review).

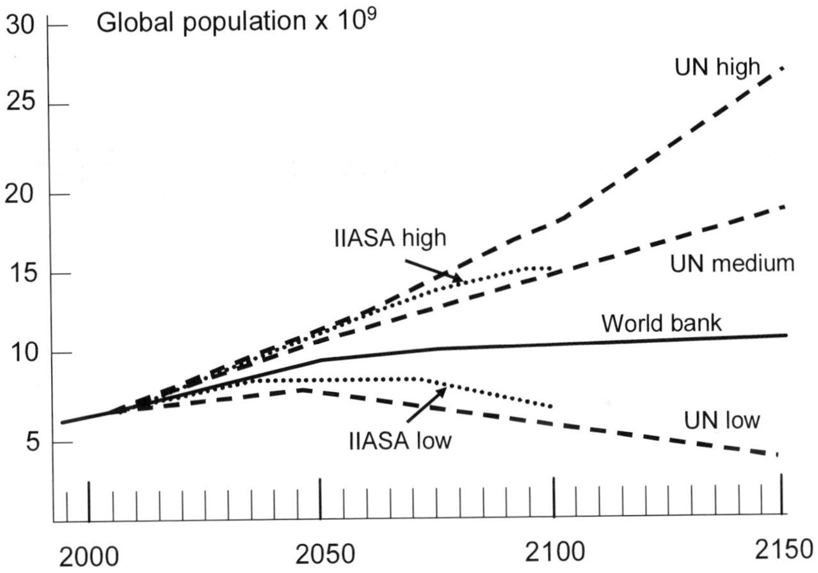

Key: UN=United Nations; IIASA= International Institute of Applied Systems Analysis; World Bank=World Bank

Figure 8.4. World population projections by the major demographic agencies (based on IPCC, 2000).

However, this growth will not be uniform as the populations of developing countries will increase substantially while those of developed countries will increase only slightly or, in some cases, actually decrease. Some details are given in Table 8.4. Of particular note is a substantial increase in Sub-Saharan Africa due to growth rates more than twice the world average. This region is already the poorest in the world (see Section 8.1.4 and Table 8.2); poverty and unemployment are high and prospects are not good. Moreover, this region is likely to experience the impact of global warming and thus basic food production may become vulnerable in the absence of intervention. North Africa and the Near East also have high growth rates and a high incidence of poverty. Although growth rates are relatively low, and declining, the next three decades will witness a large increment in Asia; China and India alone account for much of this increment. However, as stated in Section 8.1.4, both nations are industrializing rapidly and standards of living are rising as additional wealth is generated. The implication is considerably enhanced pressure on resources. This latter is a worldwide issue. The developed world already commands more resources, and uses more energy, than the developing world and within the latter there is a widening gap between nations, and even within nations, in relation to *per capita* resource consumption. Further analyses reveal differences in age structure, gender, education, health etc. (see Kemp, 2004 and United Nations Population Fund, 2004); whilst important, such comparisons are beyond the scope of this book but in general high standards in the latter two go hand-in-hand with wealth which, in turn, is an outcome of carbon domestication.

Table 8.4. Projections of population change (based on United Nations, 2001)

	POPULATION x 10^9		INCREMENT pa x 10^6		GROWTH RATE % pa	
Years	**2015**	**2030**	**2000-2015**	**2015-2030**	**2000-2015**	**2015-2030**
WORLD	7.207	8.207	76	67	1.2	0.9
Developing World	5.827	6.869	74	66	1.4	1.1
Sub-Saharan Africa	0.883	1.229	20	24	2.6	2.2
Near East/North Africa	0.520	0.651	9	9	1.9	1.5
Latin America + Caribbean	0.624	0.717	7	6	1.3	0.9
South Asia	1.672	1.969	22	19	1.6	1.1
East Asia	2.128	2.303	16	9	0.9	0.5
Industrial countries	0.951	0.979	2	1	0.4	0.2
Transition countries	0.398	0.381	-1	-1	-0.2	-0.3

Overall, population increase under current technological and political conditions will at least double the pressure on resources but rising standards of living in some regions will intensify this pressure further. It is difficult to envisage how this can be sustained as further domestication of carbon will cause serious environmental problems.

8.2.2 Climatic change

As stated above (Section 8.1.3), there is now general acceptance that climatic change is occurring. As Houghton (2004) discusses, the precise prediction of the magnitude of change, notably of global warming, and its impact is impossible because of the many factors involved, including possible political action, and the difficulties of modelling climate. One problem is the estimation of greenhouse gas emissions (see Section 8.2.4) and their changes as energy consumption patterns alter; another problem is the emissions of aerosols, such as the droplets associated with 'acid rain' (see Section 5.5) which actually offset global warming by reradiating heat back into space. Generally, climate models use scenarios which consider 'business as usual' conditions i.e. emissions in 2000, and conditions with various percentage increases or decreases of aerosols.

Table 8.5. Likely regional alterations caused by global change (adapted from IPCC, 2001b)

A. AFRICA
• Low adaptability generally • Decrease in grain yields; reduction in food security • Reduction in water availability; increase in desertification • Increased droughts and floods will affect agriculture • Increased extinctions of plants and animals
B. ASIA
• Low adaptability in some areas; average adaptability overall • Increased floods, droughts, forest fires, cyclones will cause land-cover alteration • Decreases in agricultural productivity will compromise food security in some areas • Exacerbation of threats to biodiversity • Reduction in permafrost extent in the north
C. AUSTRALIA / NEW ZEALAND
• Some positive advantages re crop productivity, but only in short term • Water resources may decline in quantity • Heavy rains and tropical cyclones may increase infrequency • Biodiversity will decline in vulnerable areas, e.g. alpine, wetlands, semi-arid habitats and coral reefs

D. EUROPE
• Adaptive capacity is high • Increased differences between north and south, especially re water • Large reduction in alpine glaciers and permafrost; increased river flooding • Some positive effects on agriculture in the north but decreases in productivity are likely in southern and eastern Europe • Altitudinal and latitudinal shifts of vegetation communities are likely, though some habitats, e.g. tundra and wetlands, will be reduced • Altered temperature and precipitation regimes may affect tourism
E. LATIN AMERICA
• Adaptive capacity is generally low • Loss and retreat of alpine glaciers will reduce water availability • Increased frequency of floods and droughts; increased frequency of tropical cyclones • Crop yields are likely to decrease; subsistence farming is particularly vulnerable • Increased biodiversity loss especially in the Amazon basin • Coastal flooding of major cities
F. NORTH AMERICA
• High adaptive capacity • Some benefits in Canada for crop production and forestry • Increased drought in southern areas • Long-term benefits may be poor • Loss of glaciers, ice caps, permafrost; varied impact on water resources • Coastal cities are prone to flooding; weather-related property damage may increase • Loss of habitats such as prairie wetlands and tundra
G. POLAR REGIONS
• These regions are particularly vulnerable to change, especially the Arctic, Antarctic Peninsula and Southern Oceans • Ice-sheet thickness and extent will decline, as will permafrost • Invasion by more warmth-demanding species • Ice melting will affect all regions

The deliberations of the IPCC, assuming no significant mitigation, are given in Table 8.5. This shows that high latitudes are likely to warm more significantly than low latitudes and that some regions will experience more extreme weather events, such as storms, droughts, floods etc., than at present. The entire globe will be affected; predictions of temperature change include those of the IPCC (2001b), which envisages a rise in temperature of between 1.4 and 5.8 °C between 1990 and 2100, and the Hadley centre of the UK Meteorological Office which predicts rises in the range 2.0 to 4.0 °C. The ranges are broad because of assumed variations in greenhouse gas emissions. Recently, results from a project (climateprediction.net) designed to co-ordinate the additional computer capacity of individually-owned computers to assist with the modelling of global climate have generated results ranging from under 2°C to more than 11°C. These data, reported by Stainforth et al. (2005), are averages for the Earth as a whole with a doubling of atmospheric carbon dioxide concentrations. Figure 8.5 gives the likely temperature increases if greenhouse gas emissions continue under the 'business as usual' scenario. This means that the atmospheric concentration of carbon dioxide will more than double by 2100, assuming mid-range economic growth but with no measures to reduce greenhouse-gas emissions. It shows substantial warming at high latitudes generally and some low latitudes such as Amazonia. Clearly, planning for such change is as difficult as prevention or limitation! Even if major reductions are achieved, an unlikely possibility as discussed in Section 8.2.4, some global warming is inevitable. Moreover, no matter how much energy conservation ensues, human impact on the carbon cycle is already of sufficient magnitude to affect carbon stores in the biosphere. On the one hand, species extinction due to warming is likely, as discussed by Thomas et al. (2005). Using data on 1103 plants and animals from sample regions which cover c. 20 percent of the Earth's surface, they estimate that 18, 24 and 35 percent of the species investigated will become extinct by 2050 under conditions of the minimum, mid-range and maximum expected impact of climatic change. Coupled with the likely, even higher, incidence of extinction due to disturbance and clearance, these figures forewarn of considerable species loss. Inevitably, this will impair carbon storage in the biosphere. On the other hand, the peatlands and wetlands of the world, which are major biospheric stores of carbon, are likely to become producers rather than consumers and thus reinforce global warming. Such mechanisms reflect the holistic nature of the planet and its carbon cycle. Moreover, the draining of wetlands, and the removal of forests and other types of vegetation for agriculture (see Section 8.2.3) all reduce carbon storage and thus compound the effects of global warming.

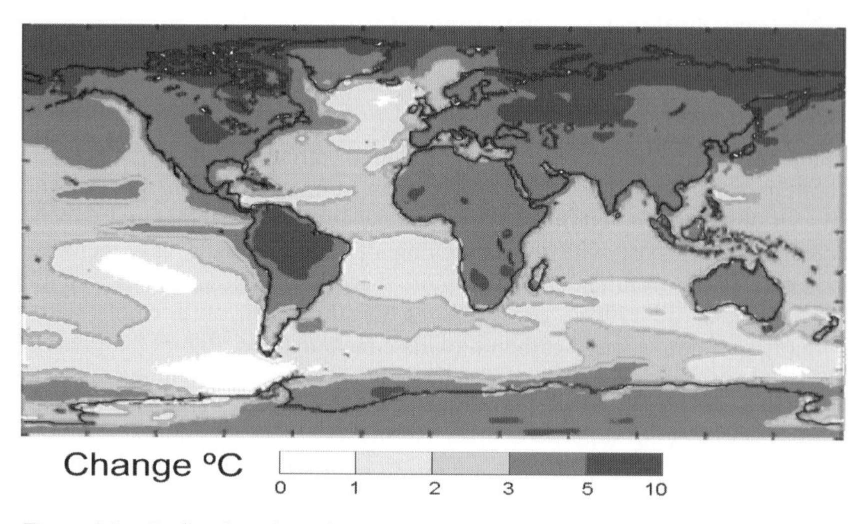

Figure 8.5. Predicted surface air temperature change from present to 2100 under the 'business as usual' scenario (based on Hadley Centre, 2005).

Global warming will affect carbon reservoirs of carbon and their interactions in many ways. The impacts will be biological, such as species extinction as discussed above, and include alterations in the composition and extent of the world's remaining natural ecosystems. The impact of increased temperatures and any alterations in precipitation that this causes will be felt at the species level. Consequently, the ecosystems of 2100 will be different from those of 2000, just as those of the last cold stage of the Earth c. 120,000 years ago were different from those of the present interglacial c. 5,000 years ago. It is impossible to determine just how different they will be, especially as other factors, such as the spread and /or intensification of agriculture, will also be operating. The possible impact of global warming on micro-organisms remains to be investigated but change is inevitable given change in flora, fauna and soils. Equally important is the effect global warming will have on other biogeochemical cycles as all are in some way linked to that of carbon and in which micro-organisms play a significant but less well understood role. The global hydrological cycle will also be affected, creating opportunities for some and adverse conditions for others. Many of the most vulnerable regions are those already suffering water stress, such as the Middle East and North Africa where hydropolitics are fraught due to shared rivers and irrigation needs. There are also implications for agriculture worldwide, as discussed in Section 8.2.3.

The communities of the oceans will also change; as Figure 8.5 shows. Ocean temperatures are likely to rise almost everywhere and the oceans are likely to become increasingly acidic as they absorb carbon dioxide; organisms will respond at the species level, altering food chains and webs and competition. As in terrestrial ecosystems increased extinction is highly likely. Thus climatic change will exacerbate problems due to marine pollution and over fishing. Sea levels will rise worldwide with implications for coastal and island ecosystems as well as for people. How much rise will occur depends on two factors: thermal expansion of the oceans and the melting of ice caps and glaciers. The magnitude of both will relate to the degree of warming, and estimates vary considerably. The IPCC (2001b) suggest a range of 9 to 88 cm, with a c. 40 cm rise by 2080 if mid-range climatic change occurs. The environmental impact would be considerable and include flooding, accelerated erosion, loss of protective coastal vegetation and the fisheries they support, species extinction and incursions of salt water into freshwater aquifers. All of these factors would affect human communities in coastal regions, including some of the world's largest cities. Amongst the most vulnerable nations are some of the world's poorest such as Bangladesh, Egypt, Indonesia and Pakistan while some of the island nations, e.g. the Maldives and Seychelles could disappear altogether.

Concerns are also being voiced about possible 'environmental refugees'. According to Townsend (2002), 150×10^6 people could be rendered homeless or without means of support due to climatic change and environmental degradation by 2050; he states that "in Bangladesh alone a one-metre rise would uproot 20 million people". The developing nations will be hardest hit and have inadequate resources and infrastructure to cope. Developed nations will be affected directly, as in low-lying nations such as The Netherlands and many coastal areas of the USA, and indirectly as displaced and disadvantaged people attempt to find a better life in less vulnerable regions. Add to these issues prospects of invasion by undesirable plants and animals which compete with indigenous species, and the likelihood of increased incidence of plant and animal diseases as well as diseases, such as malaria, which affect humans, then the potential of global warming for disrupting environmental and socio-economic systems is immense. No person and no place will remain unaltered; even if there are no direct effects, repercussions of effects elsewhere will require adjustments environmentally and socially.

8.2.3 Agriculture

Likely future change in world agriculture by 2015/2030 has been addressed in depth by the FAO (Bruinsma, 2003). The issue is complex but integral to any discourse on carbon domestication and its impact on the carbon biogeochemical cycle. Two factors are paramount: climatic change and population increase; neither of these can be predicted with certainty though both, as discussed above (Sections 8.2.1 and 8.2.2), will instigate change in carbon acquisition.

As discussed in Section 8.2.1, global population is likely to double by 2050, with most growth occurring in developing nations. In many developed nations low or negative population growth is already causing a contraction in agriculture and the removal of land from agriculture, i.e. set aside, through deliberate policies and the encouragement of farmers to find alternative land uses such as recreation and tourism. This contrasts with the situation in the developing world where pressure on agriculture will intensify; as well as substantial population growth standards of living will rise in industrializing nations, such as China and India, more varied foodstuffs are sought, such as increased dairy products and meat. Bruinsma (2003) notes that many developing nations are no longer producing a surplus of agricultural goods and have become net importers. He also notes that meat consumption in developing countries is currently increasing at a rate of 5 to 6 percent per year and that of dairy products at a rate of 3.4 to 3.8 percent per year.

How will sufficient food for these additional people be produced? Bruinsma (2003) states that "By 2030, crop production in the developing countries is projected to be 67 percent higher than in 1997/1999". Such growth can be achieved in only two ways. One possibility is to turn land currently occupied with natural vegetation into agricultural land and another possibility is to intensify production on existing agricultural land. In reality FAO predicts (Bruinsma, 2003) that c. 80 percent of the increase will be achieved through intensification and 20 percent through expansion. The first possibility is not desirable because it would result in further loss of ecosystems and biodiversity and thus further impairment of the biosphere's capacity to store carbon and provide the environmental services associated with all biogeochemical cycling. Intensification through the application of technology is far from ideal but is preferable (Avery, 2000). The disadvantages include problems associated with pesticides, artificial fertilizers, soil erosion and irrigation, though with good management these problems can be minimized.

Genetically modified (GM) crops and animals hold considerable promise for increasing productivity though there are disadvantages as well as advantages. These are summarized in Figure 8.6. On the positive side, genetically modified crops and animals should lead to a reduction in fossil-fuel energy inputs to agricultural systems but on the negative side they may encourage the spread of agriculture into areas where it is currently not viable and thus compromise natural ecosystems.

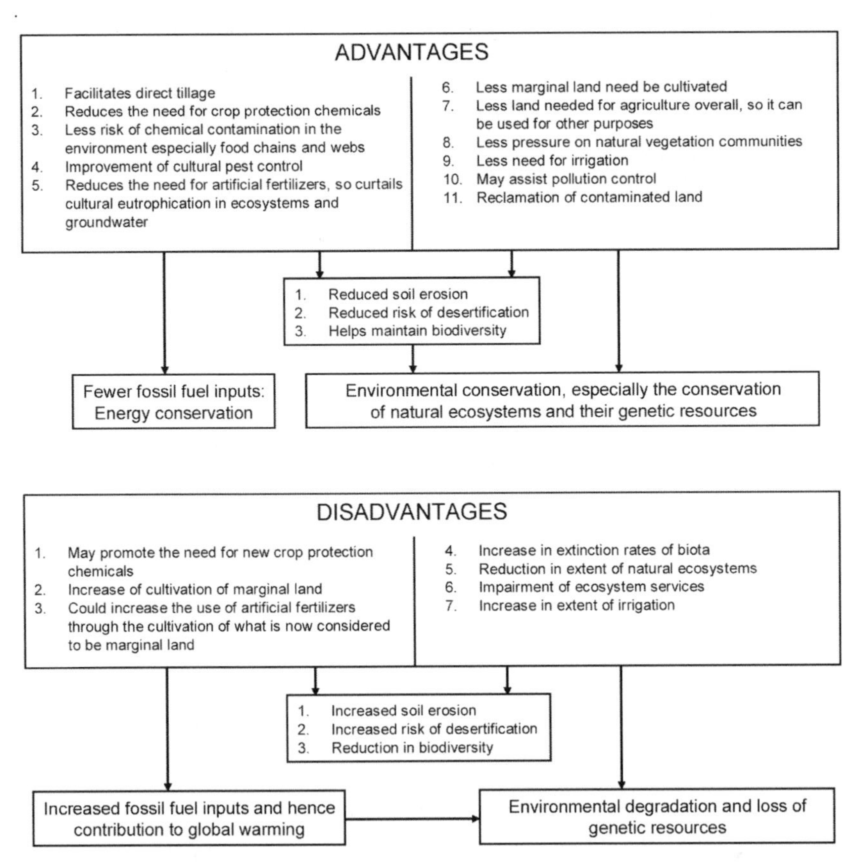

Figure 8.6. Advantages and disadvantages of GM crops (from Mannion, 2002)

In Europe there has been widespread antagonism to genetically modified crops; some fears have been fuelled by the contamination of the food chain with prions, the cause of bovine spongiform encephalopathy (BSE) and Creutzfeldt-Jakob disease. The possibility that similar contamination could occur through genetic modification, plus other fears concerning the crops' environmental impact has led to their rejection (see Figure 8.6). Indeed, Europe does not need GM food because it already produces a food surplus and population increase is likely to be relatively low in the next few decades. However, the technology has been embraced by many countries, notably the USA, as detailed in Table 5.10. Worldwide, cotton, canola, soybean, and maize occupy 1.5 x 10^9 ha of cultivable crop land, of which c. 5 percent now comprises GM, as detailed in Table. 8.6.

Table 8.6. The extent of genetically modified crops worldwide in 2003 and 2004 (data from James, 2004).

	2003 (x 10^6 ha)	**2004 (x 10^6 ha)**
Soybean	41.4	48.4
Maize	15.5	19.3
Cotton	7.2	9.0
Canola	3.6	4.3

According to James (2004) approximately 8.25 million farmers in 17 countries planted biotech crops in 2004 representing 1.25 million more farmers than planted them in 18 countries in 2003. Moreover, there was a substantial increase in developing countries in 2004; for the first time, the growth in biotech crop area was higher in developing countries, where it increased by 7.2 x 10^6 ha, than in industrial countries where it increased by 6.1 x 10^6 ha. Details are given in Table 5.10. GM crops also have the potential for improving the nutrient intake of people; for example the vitamin E content of rice can be improved.

The impact of global warming on agriculture is debateable and some possibilities are outlined in Table 8.5. Alterations in average temperatures, temperature ranges and precipitation patterns as well as an increased incidence of extreme events and unpredictability will cause problems for some regions and open up opportunities for others. The FAO (Bruinsma, 2003) assessment of the impact of climatic change on agriculture states "The main impacts of climatic change on global food production are not projected to occur until after 2030, but thereafter they could become increasingly serious. Up to 2030 the impact may be broadly neutral or even positive at the global level. Food production in higher latitudes will generally benefit from climate change, whereas it may suffer in large areas of the tropics". That food production is likely to increase in some regions is to be welcomed but many nations currently subject to food shortages and slow development may find that conditions deteriorate, as in Sub-Saharan Africa and possibly in other semi-arid regions where rainfall becomes increasingly erratic, in the next few decades. Coupled with high rates of population increase, this will cause environmental and social degradation. Beyond 2030, FAO believes that the impact of climate change will intensify with adverse affects and recommends that governments etc. begin now to consider ways and means to counteract this trend.

8.2.4 Energy issues

Energy issues are crucial in any debate about human appropriation of carbon, not least because this is one factor in the people-environment debate that can be tackled via radical change. There is no equivocation about energy requirements in the next 50 years; a substantial increase is necessary to underpin development in many parts of the world despite efforts to reduce fossil-fuel energy use in many industrialized societies. Fossil-fuel consumption cannot, however, be maintained at current levels because of diminishing resources and the climatic impact of increasing carbon dioxide in the atmosphere. Energy is vital to socio-economic progress but it does not have to be carbon-based energy. As stated above, this particular bond between carbon and society could and should be broken via the use of non-carbon based energy sources. In a recent meeting (February, 2005) of climate scientists in Exeter, UK (see report by Pearce, 2005b), there was emphatic support for the widespread evidence for imminent global warming; the question is no longer when but how much and what measures can be taken in mitigation. The participants concluded that the aim should be to keep atmospheric carbon dioxide concentrations below 400 ppm in order to limit global temperature rise to less than 2 °C. Given that atmospheric carbon dioxide concentrations are already c. 380 ppm and rising rapidly this target is unlikely to be met.

8.2.4.1 Energy projections

According to the Energy Information Administration (2004), world energy consumption is predicted to increase by 54 percent of 2001 volumes by 2025. Energy consumption in 2001 stood at 404×10^{15} Btu (British thermal units) and is predicted to reach 623×10^{15} Btu. Some 40 percent of this increase will be in Asia, including the rapidly industrializing countries of India and China. This is linked with Gross Domestic product (GDP) which is set to increase at 5.1 percent as compared with a world average of 3.0 percent. The developing world overall will account for c. 70 percent of the energy increase. As Figure 8.7 shows, the greatest growth is likely to occur in oil, natural gas and coal consumption with much lower increases in renewable and nuclear energies. Thus the domestication of carbon, as an energy source, is set to accelerate substantially under current socio-economic and political conditions. This will inevitably generate huge increases in carbon dioxide emissions and the enhanced greenhouse effect will be reinforced. The world is already getting warmer; it could get even hotter and more unpredictable quite soon!

What remains to be seen, however, is how governments rise to the challenge of global warming in the next few years. The robust application of restrictions on carbon emissions, as agreed through the Kyoto Protocol (Section 7.7), is, if parties adhere to it, likely to alter patterns of energy use and influence land-cover characteristics through the encouragement of carbon-sequestering land use. It came into force on 16th February 2005 but several nations which emit vast quantities of carbon dioxide, notably the USA, have not signed the protocol. There is also pressure on China and India to participate.

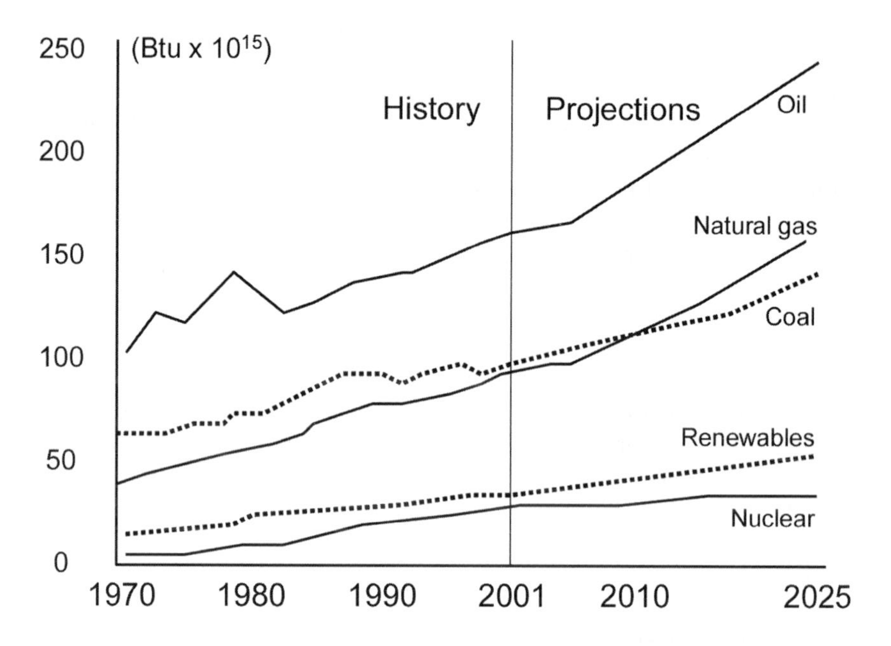

Figure 8.7. Energy use by fuel 1970 to 2025 (based on Energy Information Administration, 2004b)

Other negative and positive factors may also come into play. The former are essentially reactive and include the necessity of combating problems, such as those associated with increased incidence of extreme weather events, reduced food production, and human health problems. In contrast, pre-emptive or prescriptive actions may include investment in, and promotion of, the science and technology of energy conservation at all stages from extraction to electricity production and energy-efficiency in buildings as well as the development of non-carbon-based energy sources, as discussed below. Adopting a bleak view, the situation will deteriorate further if fossil-fuel use continues unabated and indeed it could be worse if coal-rich nations, such as the rapidly industrializing China and India, choose to use such energy resources without substantial emissions controls. Without such controls, not only will carbon dioxide concentrations rise rapidly in the atmosphere but so too will other pollutants, namely the substances involved in 'acid rain'. Although such pollutants deflect global warming to some extent, they have an adverse impact on many ecosystems which results in an impairment of their capacity to absorb carbon. Such interactions once again highlight the holistic character of the biosphere.

8.2.4.2 Breaking the bonds

There are many technologies which provide an alternative source of energy to fossil fuels. Not all of these are carbon free and none are without problems. If society is serious, as it must be, about breaking the bonds with carbon, then the most appropriate alternative energies are those in which carbon is not at all involved. These include nuclear fission and fusion, the harnessing of wind and water energy and the direct harvesting of solar energy (see Morgan, 2002, and Boyle, 2004, for reviews). None are especially easy to implement given the technology of the early twenty-first century and all have drawbacks. As Figure 8.7 shows, only about 20 percent of current energy needs derives from non-fossil fuels, and much of that also derives from carbon-based sources. These include biomass fuels which range from wood to biofuels such as those from oil crops. On a positive note, biomass fuels are renewable and they do not add carbon dioxide to the atmosphere because there is a balance between the carbon dioxide they absorb and that which they emit when burnt. Such fuels are advantageous for those nations without reserves of fossil fuels and they can often be produced as a secondary product from crops. The production of fuel alcohol from sugar cane in Brazil is a case in point. On a negative note, the commercial production of biomass fuels may be energy intensive if artificial fertilizers are necessary to grow the crops and so these fuels will result in a net increase of carbon dioxide in the atmosphere. However, the harnessing of methane from waste tips and farms would make economic sense and at the same time be environmentally sensible to prevent it entering the atmosphere.

Of the non-carbon-based fuels currently used, nuclear energy is the most abundant. First established in the 1950s and with significant plant construction until the 1970s, nuclear power plants were considered as the most important means for the provision of cheap and plentiful electricity. During this period some 54, 59 and 104 plants were constructed in Japan, France, and the USA respectively (International Atomic Energy Agency, 2005). Indeed, the serious consequences of a fossil-fuel economy have led some scientists and politicians to call for a revival of investment in nuclear technologies. The advantage is that nuclear fission, whereby energy is released when atoms of heavy metals, such as uranium, are split, does not produce heat-trapping gases and is thus not a threat to the integrity of the atmosphere. However, there are also disadvantages: problems of waste treatment, high costs of running and decommissioning, the potential for devastating accidents such as that of Chernobyl in 1986, and the potential for developing nuclear weapons. Consequently, the nuclear option is not in favour and it is likely that its contribution to world energy production will decline from c. 16 percent in 2002 to c. 12 percent in 2015; there are currently 441 units in 30 countries (IAEA, 2005). Nuclear fusion is a possible alternative to fission and would indeed represent a major breakthrough in energy availability. It involves the fusion of the nuclei of hydrogen atoms to create helium, a reaction which takes place in the sun and which releases heat. However, the technology to harness the heat has not yet been developed so fusion is not a viable option in the short term. Despite the drawbacks of fission, it could be more widely used as a short-term measure to replace fossil fuels until other, safer and cleaner technology, can be developed.

Wind, hydroelectric, tidal and geothermal power are all relatively clean but can only be developed where the geology and topography allow. Wind energy is the most widespread and comprises wind turbines, often within a wind farm, which drive a generator. This is increasing in extent worldwide as data from the EU reflect: "Cumulative wind power capacity increased by 20 percent to 34,205 MW at the end of 2004, up from 28,568 MW at the end of 2003. 5,703 MW of wind power capacity were installed last year, representing a wind turbine manufacturing turnover of some €5.7 billion" (European Wind Energy Association, 2005). There are disadvantages, not least being the occupation of land, intermittent energy production dependent on the weather, unattractive visual impact, and hazards to birds. On the positive side, it does not produce heat-trapping gases. Hydroelectric power harnesses the energy of fast-flowing rivers which drive turbines. It does not produce greenhouse gases but requires dam construction which can be destructive of landscapes and communities, as is occurring in the Three Gorges region of China's Yangtze River where the world's largest dam is located. This will displace nearly 2×10^6 people. Tidal power works on the same principle: dam construction across a bay causes sea water to flow through turbines and thus generate power. The disadvantages include disruption to wildlife and unsightliness, and a substantial difference between low and high tide is essential. Geothermal energy is also location specific insofar as it capitalizes on the occurrence of heat from the Earth's core which is brought close to the surface as molten rock. This heats groundwater and can be circulated to heat buildings or to drive turbogenerators. It is feasible only in volcanic regions and is in wide use in Iceland where energy consumption *per capita* is among the highest in the world and where c. 87 percent of all housing is heated with geothermal energy (Orkustofnun, 2005).

Perhaps the best alternative to fossil fuels is solar energy. Developing this and encouraging its use worldwide would really break the bonds with carbon-based fuels. Wright and Nebel (2002) state that "Just 40 minutes of sunlight striking the land surface of the United States yields the equivalent energy of a year's expenditure of fossil fuel. If less than one tenth of 1 percent of the Earth's surface were dedicated to collecting solar energy, that area alone could supply the electricity needs of the world". The promise is, thus, considerable. The major drawback is the technology required to capture a diffuse source of energy and concentrate it into a usable and transportable form. Direct heating of water and/or space can be achieved using flat-plate solar collectors which convert sunlight to heat. Alternatively, solar energy may be used to produce electricity; currently this is possible through the use of photovoltaic cells and solar-trough collectors. The former comprises a small (c. 5 cm by 8 cm) sheet of material made from silicon; costs of production are low and each cell has a life-span of c. 20 years. Solar-trough collectors comprise large curved (trough-shaped) reflectors which focus sunlight onto an oil-bearing pipe; this heats up and can be used to produce steam for a turbogenerator. The reasons why solar energy is not more widely used is because the technology is expensive when compared with other fuel types and it can be unreliable insofar as it works best in climates which are sunny all year round and does not work during hours of darkness. However, there is considerable potential and a real opportunity to break the bond with carbon.

Other possible fuels include hydrogen gas but inventing a system for powering vehicles this way, i.e. based on electrolysis, is very difficult and costs are prohibitive. Research has, however, advanced in relation to the use of fuel cells based on hydrogen. Hydrogen is recombined with oxygen to produce an electrical potential which is the source of power. Vehicles with this technology are now being produced but there are problems of refuelling and the storage of hydrogen which is explosive in air. These developments are highlighted by the International Association for Hydrogen Energy (2004).

8.2.5 Other technologies

All technologies have drawbacks and none offer a universal panacea for the world's problems. Before the implementation of any technology the degree of risk should be assessed under the precautionary principle. Many technologies can be used to good effect, including genetic modification which is a branch of biotechnology. It has major applications in agriculture, as discussed in Sections 5.6 and 8.2.3, and its applications will increase as additional genetic research is undertaken and as the growth in world population brings additional pressure to bear on agricultural systems. Other aspects of biotechnology include the use of organisms to treat pollutants and contaminants; this is bioremediation. It has been referred to in Sections 3.2, 3.3 and 3.4.2. Other technologies include information technology, those associated with recycling and nanotechnology.

Information technology is a powerful tool. It can assist in environmental management in three ways: disseminating information rapidly, modelling or predicting impacts and increasing the efficiency of existing technologies. During the last 100 years huge advances in communications have occurred; these include the radio, telephone, television, video recorder, DVD, computers and mobile telephones. Where once communal communication was the norm, i.e. communal newspapers, cinema news etc. now individuals are linked to a global system of news and information via personal computers and telephones. Such developments have 'made the World smaller' as information about individuals and events is disseminated instantaneously. This is part of the globalization process and, of course, not everyone has the same advantage; inevitably, citizens of the developed world have the advantage over those of the developing world. According to Internet World Stats (2005) there are currently more than 605 x 10^6 internet users worldwide. As Table 8.7 shows, the world is rapidly becoming enmeshed through the internet though the development is uneven, with the lowest numbers in Africa and the Middle East when rated as a proportion of the population. This growth is likely to continue in the next few years, with significant growth expected in Latin America and industrializing Asia.

Table 8.7. The number of internet users, by region (based on Internet World Stats, 2005).

World regions	Population (2005 est.) x 10^3	Internet usage x 10^3	% usage growth 2000-2005	% using 2005	% of world users
Africa	900,465	12,937	186.6	1.4	1.6
Asia	3,612,363	266,742	133.4	7.4	32.6
Europe	730,991	230,923	124.0	31.6	28.3
Middle East	259,500	17,326	227.8	6.7	2.1
North America	328,387	218,400	102.0	66.5	26.7
Latin America/Caribbean	546,917	55,280	205.9	10.1	6.8
Oceania / Australia	33,443	15,838	107.9	47.4	1.9
WORLD total	6,412,067	817,447	126.4	12.7	100

This interconnectedness means that all types of data and news about 'good practice' in agriculture, waste management, pollution control, fuel use etc. can be spread quickly, though there must be a political will to allow this to happen (see Section 8.2.6). 'Good practice' derives from experience and modelling. Examples of such use include the control of fertilizer furrow by furrow through the use of probes which relay soil nitrate or phosphate availability to a computer on board a tractor; the same computer can then adjust the amount of fertilizer applied to the soil. Irrigation systems can be similarly controlled so that just enough water is applied and when, so that losses due to evaporation can be reduced and salinization prevented. Areas at risk of global warming, soil erosion, flooding, deforestation, desertification etc. can also be determined using information technology, often linked with satellite remote sensing. Many such applications have been discussed by Lillesand *et al.* (2004) and Treitz (2004). Indeed, any determination of the impact of global warming at any temporal or spatial scales is dependent on the collection of remotely sensed and remotely collected data and computer modelling (see various IPCC reports, 2001).

Increased recycling could also make an additional contribution to energy and resource conservation, and is considered to be a key component of sustainable development. Not all recyling is economic when taken at face value but it is in general when considered in a broad context. On the one hand, recycling can increase the availability of scarce resources and reduce the need for additional primary resources, such as metals. On the other hand, less energy is required to reuse such resources when compared with the energy cost of manufacturing primary resources and less disposal of rubbish is necessary which saves land and/or incineration. The most commonly recyled materials are glass, paper, metals and plastics as well as organic wastes. Each requires a specific technology, as discussed by Wright and Nebel (2002) and Craig *et al.* (2001). Recycling can be encouraged by local authorities charged with waste disposal and this has the advantage of encouraging action at the scale of the individual or the individual household. As Scott (2004) demonstrates, there are many ways in which individuals can contribute to sustainable development.

Looking into the future, the developing field of nanotechnology may also contribute to sustainable resource use and energy conservation. Nanotechnology is a collective term which refers to technological developments on the nanometer scale, usually 0.1-100 nm: one nanometer equals one thousandth of a micrometer or one millionth of a millimetere. The term is sometimes used to describe all types of phenomena at atomic scales. The core tenet is that as materials are scaled down, even to the molecular level, they exhibit novel properties. For example, some metals act as catalysts at nanoscale but not at macroscale. Consequently, the technology opens up possibilities for many industries, such as chemical synthesis, ceramics, powders etc. and in medicine. Energy saving could ensue in manufacturing processes. However, the disadvantages include potential health problems due to the inhalation of such small particles. These and other possibilities associated with nanotechnology are examined by Nanotechnology Now (2005) and by Newton (2002) and Freitas and Merkle (2004).

8.2.6 Politics

From the local to the global, politics has a pivotal role to play in carbon management. 'Twas ever thus'! However, in the past the involvement has been indirect insofar as governments of all descriptions have encouraged all manner of activities to foster development, wealth generation and political power. These activities range from food acquisition in hunter-gatherer communities to the co-operation required for scientific advancement and industrial development. Technology, notably that to capture carbon in various forms, has underpinned socio-economic success for some but lack of it has left others behind to create a broadening divide between the rich and the poor. Redress is now necessary, indeed essential in order to prevent yet more wars and to alleviate poverty. This can only be achieved through political means. Moreover, that harnessing of the carbon resource through technology by a few has created environmental problems for the entire globe; for many the damage has not resulted in high standards of living, longevity, food security and strong political influence.

First, international efforts must focus on breaking the bonds with carbon to stop and possibly reverse global warming. This will not only reduce environmental stress worldwide but also reduce pressure on the socio-economic systems, including agriculture, of the poorest nations which are most vulnerable to environmental change. It will require considerable investment in and commitment to reductions in carbon emissions, preferably by the conservation of fossil fuels through increased efficiency and especially through the promotion of non-carbon-based fuels such as solar energy. Adjuncts to this include sharing the technology of clean energy, and of other technologies to improve the quality of life, with those nations most in need. Such actions will require an enormous change in attitudes politically. Some of the seeds have already beeen set providing, as Diamond (2005) has recently acknowledged, a glimmer of hope that the globalizers are recognizing the impact of globalization in its broadest terms. The United Nations and its environment-related constituent agencies have the know-how to make a difference; the political will is needed to make it happen. What is also needed is the international recognition that standards of living cannot continue to rise within a carbon-rich economy such as that of the present.

Second, poverty alleviation is also linked with world trade. Economic and political considerations are not always conducive to fair trade. Such distortions of global carbon flows, as commodities, need redress. Moreover, it would behoove developed nations to reduce their appetites for exotic but wasteful products of agriculture and horticulture. The issue is complex but the question must be asked: Why do Ecuador and Colombia, to name but two nations, grow flowers for export to Europe and North America when a substantial proportion of their populations rely on Food Aid? Flowers, attractive as they are, provide no calories; they are decorative carbon which requires soil, water, sunshine, labour and crop protection chemicals and fertilizers to produce. Another case in point is tobacco which does much more harm than good, not to mention poppy and other narcotic-producing crops. If there were no markets for such commodities no one would grow them and valuable resources would not be wasted. Poverty is also linked with food production and the likelihood of 12 x 10^9 people to feed by 2050 is challenge enough; food-stressed nations need assistance to improve productivity through technology transfer.

The encouragement and provision of education in general and research into environmental processes and the socio-economic factors which drive change is also in the domain of governments. So too is support for conservation and its strict enforcement through instruments of the law. The regulation of emissions and pollutants, again through legal enforcement, can also contribute to the increased efficiency of carbon use. Corruption, often associated with carbon-based resources such as wood and oil as well as trade, must not be tolerated at any political level; it is conducive to inequity, the persistence of poverty and often wasteful use of resources; some form of international watchdog might help to combat this problem. At the international level, the alleviation of debt for the poorest nations, together with technical assistance, should contribute to the development process and raise standards of living. Neither governments nor the international community should be seduced by the so-called cornucopian or technocentrist condition that has so far, with a few hiccups (see Diamond, 2005, for a range of examples of past societies which have thrived and then collapsed), enabled society to survive, multiply and wield influence. Technology has indeed facilitated development but it may have also brought society to the brink of a threshold, the crossing of which it may not be sufficiently robust to prevent; the so-called ecocentrist voice must be heard. The less extreme variations of these two approaches, which are reflected in the somewhat polarized works of Wilson (2002, the ecocentric view) and Lomborg (2001, the technocentric view), are not entirely incompatible as technocentrism can facilitate ecocentrism through shared goals and shared and recycled resources. The promotion of both should be an objective of governments though a major hurdle is the relative absence of politicians worldwide with a background in geography, environmental or similar studies. It is possible that politicians have not yet recognized what is at stake in terms of environmental change, probably because carbon domestication has facilitated their divorce from the very environment which sustains them.

8.3 Envoi

There are many environmental and socio-economic trends which promote apprehension for the future. First, there is the growing consensus of the scientific community about global warming and the possibilities it will generate in terms of opportunities for some and increased hazards for others, plus implications for ecosystems and societies worldwide. Second, there is the prospect of at least a 60 to 80 percent increase in global population by 2050, and all its implications re agriculture, land use, loss of remaining natural ecosystems and urbanization. All are influences on the global biogeochemical cycle of carbon and whatever happens on planet Earth in the next few centuries, the global biogeochemical cycle of carbon will continue to operate as the life and soul of Gaia. However, there are two regions which will have to cope with rapid population growth as well as enhanced water shortages and increased frequency of extreme weather conditions. These are Sub-Saharan Africa and most of the Middle (Near) East. The people in these regions are already amongst the most impoverished and are especially vulnerable to environmental hazards, and they are also susceptible to disputes and warfare. Somehow there is always sufficient money for armaments but never enough money for the basics of life; there is also continued grumbling about colonialism by European nations, rampant corruption and exploitation as well as environmental degradation. Will this gap between rich and poor widen and can the world community afford to let it widen? Moreover, and a question which is rarely asked, is how the world will cope with limitations on the production of petrochemical-derived substances, e.g. plastics, resins, pharmaceuticals, crop-protection chemicals etc., if or when petrochemicals are no longer available by political decree to curb carbon use or through resource depletion.

Perhaps the environmental problems due to carbon domestication will, in contrast, unite a fractured world; solving the carbon problem will indeed require a united global effort. Nowhere will escape the globalization created by the domestication of carbon; the imbalance created by the intensity and rapidity of this appropriation will undoubtedly have major repercussions for both environment and society. This harnessing of carbon has served well for many people; it has facilitated development, fuelled and fed civilizations, and generated knowledge and wealth, but its apogee has come and is now passed. The tamed beast is shedding its shackles and sending ripples of retribution, remonstration and reprisal; eventually, the tamed may tame the tamer. Beginning to break the society-carbon bonds now, through positive measures, is the best way forward – perhaps the only way forward.

8.4 Addendum

As this book went to press, the Millennium Ecosystem Assessment Committee (2005) issued an interim report on the state of the Earth's ecosystems and the services they provide. This report, based on a four-year study and the collective findings of 1,300 scientists from 95 countries, provides a synthesis of environmental data on an unprecedented scale (see the Living Planet Reports of the World Wide Fund for Nature discussed on p. 202-205) and arrives at four conclusions as summarized below:

1. Since c. 1950 anthropogenic ecosystem change has been more rapid and extensive than at any other time in human history and has caused unprecedented loss of biodiversity and species extinction.
2. Such alterations have provided the wherewithal for development and wealth and health generation for many but at considerable cost in relation to impaired ecosystem services [note that this includes services associated with the carbon biogeochemical cycle] and the continuance of poverty for some groups of people. There is evidence that non-linear alterations in the environment, e.g. accelerating degradation, may increase the risk of future irreversible change with adverse consequences for society. All such change jeopardizes the resources available to future generations and is thus not compatible with sustainable development.
3. There is the likelihood of continued degradation between now and 2050, exacerbating problems of achieving sustainable development.
4. The reversal of degradation and thus the safeguarding ecosystem services is not impossible but success requires major changes in local, regional and international policies, politics and governance. The report highlights various possibilities to achieve such goals.

A full report is due to be published in mid 2005.

While there is no specific reference to the domestication of carbon, it is axiomatic that environmental degradation and the impairment of ecosystem services is precisely the result of this process. Consequently, the Millennium Assessment Report reinforces many of the suggestions and conclusions posed in this book. Breaking the bonds with carbon, along with additional measures to manage the carbon and other biogeochemical cycles, is paramount for future environmental and social integrity.

"I saw Hamlet Prince of Denmark played; but now the old plays begin to disgust this refined age." John Evelyn (1620-1706).

References

Aber, J.D. and Melillo, J.M. 2001. **Terrestrial Ecosystems.** 2nd Edition. Harcourt: San Diego.

Alberts, B, Bray, D., Johnson, A., Lewis, J., Raff, M., Roberts, K. and Walter, P. 1998. **Essential Cell Biology.** Garland Publishing: New York.

Allaby, M. 1999. **Temperate Forests.** Facts on File Inc.: New York.

Allen, H.D. 2001. **Mediterranean Ecogeography.** Prentice Hall: Harlow.

Altman, A. 2004. From plant tissue culture to biotechnology: scientific revolutions, abiotic stress tolerance and forestry. *In Vitro Cellular and Developmental Biology – Plant* 39, 75-84.

American Spice Trade Association. 2004. **Spice History.** www.astaspice.org

Annan, K.A. 2000. **We the Peoples. The Role of the United Nations in the 21st Century.** United Nations: New York. (available at www.un.org)

Archibold, O.W. 1995. **The Ecology of World Vegetation.** Chapman and Hall: London.

Arora, D.K., Bridge, P.D. and Bhatnagar, D. 2003. **Fungal Biotechnology in Agricultural, Food and Environmental Applications.** Marcel Dekker: New York.

Aspartame Information Center 2004. **Aspartame Fact Sheet.** www.aspartame.org

Aspirin Foundation. 2004. **The Chemistry of Aspirin.** www.aspirin-foundation.com

Atjay, G.L., Ketner, P. and Duvigneaud, P. 1979. Terrestrial primary production and phytomass. In: Bolin, B., Degens, E.T., Kempe, S. and Ketner, P. (Eds.) **The Global Carbon Cycle.** John Wiley and Sons: Chichester, pp. 129-181.

Avery, D.T. 2000. **Saving the Planet with Pesticides and Plastic: The Environmental Triumph of High-Yield Farming.** 2nd Edition. Hudson Institute: Washington.

Aznar, J.M. 2004/05. Look West, Europe. *Newsweek* Special Issue, p. 29. www.newsweekinternational.com

Bacteria Museum. 2004. **Bacterial Species.** www.bacteriamuseum.org

Balfour-Park, J. 1998. **Indigo.** British Museum: London.

Barnola, J.-M., Raynaud, D., Korotkevich, Y.S. and Lorius, C, 1987. Vostok ice core provides 160,000-year record of atmospheric CO_2. *Nature* 329, 408-414.

Barnola, J.-M., Raynaud, D., Lorius, C. and Barkov, N.I. 2003. Historical carbon dioxide record from the Vostok ice core. In Oak Ridge Laboratory **Trends: A Compendium of Data on Global Change.** Carbon Dioxide Information Analysis Center, Oak Ridge National Laboratory, US Department of Energy: Oak Ridge, Tennessee. Available at http://cdiac.ornl.gov

Bauer, K., Garbe, D. and Surburg, H. 2001. **Common Fragrance and Flavor Materials.** Culinary and Hospitality Industry Publications: Weimar, Texas.

Behling, H. 2002. Carbon storage increases by major forest ecosystems in tropical South America since the Last Glacial Maximum and the early Holocene. *Global and Planetary Change* 33, 107-116.

Bekker, A., Holland, N.D., Wang, P.-L., Rumble III, D., Stein, H.J., Hannah, J.L., Coetzee, L.L. and Beukes, N.J. 2004. Dating the rise of atmospheric oxygen. *Nature* 427, 117- 120.

Bell, P.R. and Hemsley, A.R. 2000. **Green Plants: Their origin and Diversity.** Cambridge University Press: Cambridge.

Bellis, M. 2004. **History of Aspirin.** www.inventors.about.com

Bellwood, P. 2004. **The First Farmers.** Blackwell: Oxford.

Beilstein, F.K. 1881-1883 *et seq.* **Beilstein Handbook of Organic Chemistry**. Beilstein Institute: Frankfurt/Main.

Ben-Shem, A., Frolow, F. and Nelson, N. 2003. Crystal structure of plant photosystem I. *Nature* 426, 630-635.

Berner, R.A. 2003. The long-term carbon cycle, fossil fuels and atmospheric composition. *Nature* 426, 323-326.

BIODIDAC. 2005. **A Bank of Digital Resources for Teaching Biology.** www.biodidac.bio.uottawa.ca

BizEd. 2004. **The Formulation of Trading Blocs.** www.bized.ac.uk

Bonar, G. 2002. **Ecological Climatology.** Cambridge University Press: Cambridge.

Boyd, M. 2002. Identification of anthropogenic burning in the palaeoecological record of the northern prairies. *Annals of the Association of American Geographers* 92, 471-487.

Boyle, G. (Ed.), 2004. **Renewable Energy.** Oxford University Press: Oxford.

Brain, C.K. and Sillen, A. 1988. Evidence from Swartkrans cave for the earliest use of fire. *Nature* 336, 454-466.

Brasier, M.D., Green, O.R., Jephcoat, A.P., Kleppe, A.K., Van Krankendonk, M.J., Lindsay, J.F., Steel, A. and Grassineau, N.V. 2002. Questioning the evidence for the Earth's oldest fossils. *Nature* 416, 76-81.

Brasier, M., Green, O., Lindsay, J. and Steele, A. 2004. *Origins of Life and Evolution of the Biosphere* 34, 257-269.

British Plastics Federation 2004. **The History of Plastics.** www.bpf.co.uk

Brocks, J.J., Logan, G.A., Buick, R. and Summons, R.E. 1999. Archean molecular fossils and the early rise of eukaryotes. *Science* 285: 1033-1036.

Brown, P., Sutikna, T., Morwood, M.J., Soejono, J., and Saptomo, E.W. and Due, R.A., 2004. A new small-bodied hominin from the Late Pleistocene of Flores, Indonesia. *Nature* 431, 1055-1061.

Brown, R.C. 2003. **Biorenewable Resources: Engineering New Products from Agriculture.** Iowa State Press: Ames, Iowa.

Bruford, M.W., Bradley, D.G. and Luikart, G. 2003. DNA markers reveal the complexity of livestock domestication. *Nature Reviews Genetics* 4, 900 -910.

Bruinsma, J. (Ed.), 2003. **World Agriculture: Towards 2015/2030.** Earthscan: London.

Brunt, A.A., Crabtree, K., Dallwitz, M.J., Gibbs, A.J., Watson, L. and Zurcher, E.J. (Eds.), 1996 onwards. Plant Viruses Online. Descriptions and lists from the VIDE database. www.biology.anu.au.

Burchell, J. 2002. **The Evolution of Green Politics. Development and Change within European Green Parties.** Earthscan: London.

Burnett, J.H. 2003. **Fungal Populations and Species.** Oxford University Press: Oxford.

Calorie Control Council. 2004. Acesulfame. www.caloriecontrol.org

Calow, P. (Ed.), 1998. **The Encyclopedia of Ecology and Environmental Management.** Blackwell: Oxford.

Cambell-Culver, M. 2001. **The Origin of Plants.** Headline: London.

Cameron, D.W. and Groves, C.P. 2004. **Bones, Stones and Molecules.** Elsevier: Amsterdam.

Canadian Forest Service (CFS). 2004. **Over a Century of Innovative Solutions.** www.nrcan.gc.ac

Carlile, M.J., Watkinson, S.C. and Gooday, G.W. 2001. **The Fungi.** Academic Press: San Diego.

Carroll, S.B., Grenier, J.K. and Weatherbee, S.D. 2001. **From DNA to Diversity: Molecular Genetics and the Evolution of Animal Design.** Blackwell: Oxford.

Central Intelligence Agency (CIA). 2005.**The World Factbook.** www.cia.gov

Chambers, N., Simmons, C. and Wackernagel, M. 2000. **Sharing Nature's Interest. Ecological Footprints as an Indicator of Sustainability.** Earthscan: London.

Chevron. 2004. **What is crude oil?** www.chevron.com

Chrispeels, M.J. and Sadava, D.E. 2003. **Plants, Genes and Crop Biotechnology.** Jones and Bartlett: Boston.

Christopherson, R.W. 2005. **Geosystems.** 5th Edition. Pearson: Upper Saddle River, New Jersey.

Clardy, J. and Walsh, C. 2004. Lessons from natural molecules. *Nature* 432, 829-837.

Clark, J. D., Beyene, Y., Woldegabriel, G., Hart, W.K., Renne, P.R., Gilbert, H., Defleur, A., Suwa, G., Katoh, S., Ludwig, K.R., Boisserie, J.-R., Asfaw, B. and White, T.D. 2003. Stratigraphic, chronological and behavioural contexts of Pleistocene *Homo sapiens* from Middle Awash, Ethiopia. *Nature* 423, 747–752.

Clowes, M. 2004. The ediacaran biota. www.palaeos.com

Club of Rome. 2004. **Organization.** www.clubofrome.org

CO_{2e}. 2004. **Carbon Markets**. www.co2e.com

Cohen, B. 2004. Urban growth in developing countries: a review of current trends and a caution regarding existing forecasts. *World Development* 32, 23-51.

Conservation International. 2004. **Comparing Hotspots**. www.biodiversityhotspots.org

Craig, J.R., Vaughan, D.J. and Skinner, B.J. 2001. **Resources of the Earth.** 3rd Edition. Prentice Hall: Upper Saddle River, New Jersey.

Crosby, A.W. 1986. **Ecological Imperialism: The Biological Expansion of Europe.** Cambridge University Press: Cambridge.

Convention on Biological Diversity (CBD). 2004. **Convention on Biological Diversity.** www.biodiv.org

Convention on International Trade in Endangered Species of Wild Fauna and Flora (CITES). 2004. **Discover CITES.** www.cites.org

Crummett, W.B. 2002. **Decades of Dioxin.** Xlibris Corporation: Philadelphia/

Davis, B.G. and Fairbanks, A.J. 2002. **Carbohydrate Chemistry.** Oxford University Press: Oxford.

Department for Environment Food and Rural Affairs. 2004. **European Union Birds and Habitats Directives. Natura 2000.** www.defra.gov.uk

Department of Trade and Industry (UK). 2004. **Supply and Consumption of Coal (Table 2.7).** www.dti.gov.uk

d'Evie, F. and Glass, S.M. 2002. The Earth Charter: an ethical framework for sustainable living. In Miller, P. and Westra, L. (Eds.) **Just Ecological Integrity.** Rowman and Littlefield: Lanham, USA, pp.17-25.

Diamond, J. 2002. Evolution, consequences and the future of plant and animal domestication. *Nature* 418, 700–707.

Diamond, J. 2005. **Collapse. How Societies Choose to Fail or Survive.** Allen Law: London.

Dillehay, T.D. 2003. Tracking the first Americans. *Nature* 425, 23-24.

Dillehay, T.D. and Collins, M.B. 1988. Early cultural evidence from Monte Verde in Chile. *Nature* 332, 150-152.

Dixon, R.K., Brown, S., Houghton, R.A., Solomon, A.M., Trexler, M.C. and Wisniewski, J. 1994. Carbon pools and flux of global forest ecosystems. *Science* 263, 185-190.

Dobson, C.M. 2004. Chemical space and biology. *Nature* 432, 824-828.

Earth Charter. 2004. **The Earth Charter Initiative.** www.earthcharter.org

Edwards, L.E. and Pojeta Jr., J. 1997. **Fossils, Rocks and Time.** United States Geological Survey (USGS), available at www.usgs.gov

El Eid, M. 2005. The process of carbon creation. *Nature* 433, 117-118.

Energy Information Administration (EIA) (USA) 2004a. **Annual Energy Review.** www.eia.doe.gov

EIA. 2004b. **World Energy Use and Carbon Dioxide Emissions.** www.eia.doe.gov

Environment Canada. 2002. **Land-use Change and Forests. Factsheet 6.** www.ec.gc.ca

Environmental Protection Agency (EPA). 2004. **Types of Petroleum Oils.** www.epa.gov

EPICA Community Members. Eight glacial cycles from an Antarctic ice core. *Nature* 429: 623- 628.

Erickson, G.M., Makovicky, P.J., Currie, P.J., Norell, M.A., Yerby, S.A. and Brochu, C.A. 2004. Gigantism and comparative life-history parameters of tyrannosaurid dinosaurs. *Nature* 430, 772-775.

Etheridge, D.M., Steele, L.P., Francey, R.J. and Langenfelds, R.L. 2002. Historical CH_4 records since about 1000AD from ice core data. In Oak Ridge Laboratory **Trends: A Compendium of Data on Global Change.** Carbon Dioxide Information Analysis Center,

Oak Ridge National Laboratory, US Department of Energy: Oak Ridge. Available at http://cdiac.ornl.gov
Europa. 2004. **Nature and Biodiversity.** www.eu.int
European Green Parties. 2004. **Member parties and Observers.** www.europeangreens.org
European Wind Energy Association (EWEA). 2005. **Latest News.** www.ewea.org
Facorellis, Y., Kyparissi-Apostolika, A. and Maniatis, Y. 2001. The cave of Theoptera, Kalambaka: radiocarbon evidence for 50,000 years of human presence. *Radiocarbon* 43, 1029-1043.
Farabee, M.J. 1999. **On-line Biology Book.** www.emc.maricopa.edu
Farman, J.C., Gardiner, B.G.and Shanklin, J.D. 1985. Large losses of total ozone in Antarctica reveal seasonal ClO_x/NO_2. *Nature* 315, 207-210.
Fiedel, S. 2000. The peopling of the New World: present evidence, new theories and future directions. *Journal of Archaeological Research* 8, 39-103.
Field, C. 2003. **Stone age campfire could be Britain's earliest.** www.24hourmuseum.org.uk
Field, C.B., Behrnfeld, M.J., Randerson, J.T. and Falkowski, P. 1998. Primary production of the biosphere: integrating terrestrial and oceanic components. *Science* 281, 237-240.
Finlay, B.J. 2004. Protist taxonomy: an ecological perspective. *Philosophical Transactions of the Royal Society* **B**359: 599-610.
Food and Agriculture Organization. 2000. **Global Network on Integrated Soil Management for Sustainable Use of Salt-Affected Soils.** www.fao.org
Food and Agriculture Organization. 2004a. **FAOSTAT**. www.fao.org
Food and Agriculture Organization. 2004b. **What is FAO?** www.fao.org
Forestry Commission (UK) 2004a. **History of the Forestry Commission.** www.forestry.gov.uk
Forestry Commission (UK) 2004b. **Forestry Facts and Figures.** www.forestry.gov.uk
Fossil Museum. 2004.**Geological History.** www.fossilmuseum.net
Fouquet, R. and Pearson, P.J.G. 1998. A thousand years of energy use in the United Kingdom. *The Energy Journal* 19, 1-41.
Fouquet, R. and Pearson, P.J.G. 2003. Five centuries of energy prices. *World Economics* 4, 93-119.
Frängsmyr, T. (Ed.), 1983. **Linnaeus: The Man and His Work.** University of California Press: Berkeley.
Freese, B. 2003. **Coal: A Human History**. Perseus Publishing: New York.
Friends of the Earth International (FoEI). 2004. **A Short History of FoEI.** www.foei.org
Freitas, Jr., R.A. and Merkle, R.C. 2004. **Kinematic Self-Replicating Machines.** Landes Bioscience: Georgetown, USA.
Fynbo, H.O.U. plus 30 others. 2005. Revised rates for the stellar triple-α process from the measurement of ^{12}C nuclear resonances. *Nature* 433, 136-139.
Garry Laboratory. 2004. **The Big Picture Book of Viruses.** www.virology.net
Gately, I. 2001. **La Diva Nicotina. The Story of How Tobacco Seduced the World.** Simon and Schuster: London.
Gerasimidis, A. and Athanasiadis, N. 1995. Woodland history of northern Greece from the mid-Holocene to recent times based on evidence from peat pollen profiles. *Vegetation History and Archaeobotany* 4, 109-116.
GlaxoSmithKline 2004. **Clavulanic Acid.** www.us-gsk.com
Global Forest Watch 2004. **Interactive Maps.** www.globalforestwatch.org
Global Forest Watch: Canada. 2004. **Canada's Forests.** www.globalforestwatch.org
Global Greens. 2004. **Green Party History.** www.globalgreens.info
Gould, S.J. 1989. **Wonderful Life. The Burgess Shale and the Nature of History.** W.W.Norton: New York.
Goren-Inbar, N., Alperson, N., Kislev, M., Simchoni, O., Melamed, Y., Ben-Nun, A. and Werker, E. 2004. Evidence of hominid control of fire at Gesher Benot Ya'aqov, Israel. *Science* 304, 725-727.

Goudie, A. 2002. **Great Warm Deserts of the World: Landscapes and Evolution.** Oxford University Press: Oxford.
Grace, J. 2004. Understanding and managing the global carbon cycle. *Journal of Ecology* 92: 189-202.
Graham, L. and Wilcox, I. 1999. **Algae.** Prentice Hall: Upper Saddle River, New Jersey.
Greenpeace. 2004. **History.** www.greenpeace.org
Groombridge, B. and Jenkins, M.D. 2002. **World Atlas of Biodiversity.** University of California Press: Berkeley.
Haberl, H., Wackernagel, M., Krausmann, F., Erb, K.-H. and Monfreda, C. 2004. Ecological footprints and human appropriation of net primary productivity: a comparison. *Land Use Policy* 21, 279-288.
Hadley Centre (UK Meteorological Office). 2005. Climate change projections. www.metoffice.com
Hall, C., Tharakan, P., Hallock, J., Cleveland, C. and Jefferson, M. 2003. Hydrocarbons and the evolution of human culture. *Nature* 426, 318-322.
Hannah, L., Lohse, D., Hutchinson, C. Carr, J.L. and Lankerani, A. 1994. preliminary inventory of human disturbance of world ecosystems. *Ambio* 23, 246-250.
Hasselknippe, H. 2003. Systems for carbon trading: an overview. *Climate Policy* 3, Supplement 2, S43-S57.
Hatcher, J. 1993. **The History of the British Coal Industry. Volume I. Before 1700: Towards the Age of Coal.** Clarendon Press: Oxford.
Head, I.M., Jones, M.J. and Larter, S.R. 2003. Biological activity in the deep subsurface and the origin of heavy oil. *Nature* 426, 344-351.
Heaphy, S. 2004. **Prion Diseases.** www.micro.msb.le.ac.uk
Heller, S.R. 1998. The Beilstein system: an introduction. In Heller, S.R. (Ed.), **Beilstein System: Strategies for Effective Searching.** American Chemical Society: Washington D.C., pp. 1-12.
Hillman, G.C. 1989. Late Palaeolithic plant foods from Wadi Kubbaniya in Upper Egypt: dietary diversity, infant weaning, and seasonality in a riverine environment. In Harris D.R. and Hillman G.C. (Eds.), **Foraging and farming; the Evolution of Plant Exploitation.** Unwin Hyman: London, pp.207-239.
Hillman G., Hedges, R., Moore, A., Colledge, S., and Pettitt, P. 2001. New evidence of Lateglacial cereal cultivation at Abu Hureyra on the Euphrates. *The Holocene* 11, 383-393.
Hoffman, P.F. 1999. The breakup of Rodinia, birth of Gondwana, true polar wander and the snowball Earth. *Journal of African Earth Sciences* 28,17-33.
Hoffman, P.F. and Schrag, D.P. 2000. Snowball Earth. *Scientific American* 282, 68-75.
Holden, E. 2004. Ecological footprints and sustainable urban form. *Journal of Housing and the Built Environment* 19, 91-109.
Holdridge, I.R., Grenke, W.C., Hatheway, W.H., Liang, T. and Tosi, J.A. 1971. **Forest Environments in Tropical Life Zones.** Pergamon Press: Oxford.
Holmes, B. 2005. Diamonds in the rough. *New Scientist* 185 (8th January), 36-39.
Hong, S., Candelone, J.P. and Boutron, C.F. 1996. History of ancient copper smelting pollution during Roman and Medieval times recorded in Greenland ice. *Science* 272, 246-249.
Hong, S., Boutron, C.F., Edwards, R. and Morgan, V.I. 1998. Heavy metals in Antarctic ice from Law Dome: initial results. *Environmental Research* Section A, 78, 94-103.
Houghton, J. 2004. **Global Warming: The Complete Briefing.** Cambridge University Press: Cambridge.
Houghton, R.A. 2003. Revised estimates of the annual net flux of carbon to the atmosphere from changes in land use and land management 1850 – 2000. *Tellus* 55B, 378-390.
Hu, Y., Meng, J., Wang, Y. and Li, C. 2005. Large Mesozoic mammals fed on young dinosaurs. *Nature* 433, 149-152.

Huber, H., Hohn, M.J., Reinhard, R., Fuchs, T., Wimmer, V.C. and Stetter, K.O. 2002. A new phylum of Archaea represented by a nanosized hyperthermophilic symbiont. *Nature* 417: 63-67.

Hudler, G.W. 2001. **Magical Mushrooms, Mischievous Molds.** Princetown University Press: London.

Hull, R. 2002. **Matthew's Plant Virology.** 4th Edition. Academic Press: San Diego.

Imhoff, M.L., Bounova, L., Ricketts, T., Loucks, C., Harriss, R. and Lawrence, W.T. 2004. Global patterns in human consumption of net primary production. *Nature* 429, 870-873.

Intergovernmental Panel on Climate Change (IPCC). 2000. IPCC **Special Report on Emissions Scenarios**. Cambridge University Press: Cambridge.

Intergovernmental Panel on Climate Change (IPCC). 2001a. **Climate Change 2001. The Scientific Basis.** Cambridge University Press: Cambridge.

Intergovernmental Panel on Climate Change (IPCC). 2001b. **Climate Change 2001. Impacts, Adaptation and Vulnerability.** Cambridge University Press: Cambridge.

Intergovernmental Panel on Climate Change (IPCC). 2004. **About IPCC.** www.ipcc.ch

International Association for Hydrogen Energy (IAHE). 2004. **Hydrogen.** www.iahe.org

International Atomic Energy Agency (IAEA). 2005. **Country Profiles.** www.iaea.org

International Council for Science (ICSU). 2004. ICSU's mission. www.icsu.org

International Energy Agency (IEA). 2004a. **Biofuels for Transport: An International Perspective.** IEA: Paris.

International Energy Agency (IEA). 2004b. **Energy Statistics/Country Profiles.** www.iea.org

International Geosphere Biosphere Programme (IGBP). 2004. **Structure.** www.igbp.kva.se

International Institute for Applied Systems Analysis (IIASA). 2004. **The RAINS Model.** www.iiasa.ac.at

International Sub-Commission on Cambrian Stratigraphy. 2003**. The Cambrian.** www.uni-wuerzburg.de

Internet World Stats. 2005. **World Internet Usage Statistics – The Big Picture.** www.internetworldstats.com

IUCN – *see* World Conservation Union

Irigoien, X., Hufsman, J. and Harris, R.P. 2004. global biodiversity patterns of marine phytoplankton and zooplankton. **Nature** 429, 863-867.

Irish Peatland Conservation Council. 2003. **Peatlands Around the World.** www.ipcc.ie

James, C. 2003. Global status of commercialized transgenic crops. *International Service for the Acquisition of Agri-Biotech Applications* (ISAAA) *Briefs* No.30.

James, C. 2004. Global Status of Commercialized Biotech/GM Crops. *ISAAA Briefs* No. 32.

Jenkins, D. (Ed.), 2003. **The Cambridge History of Western Textiles.** Cambridge University Press: Cambridge.

Jenssen, T.K. and Kongshaug, G. 2003. Energy consumption and greenhouse gas emissions in fertilizer production. *Proceedings of the International Fertilizer Society* No. 509, pp.28.

Johannesburg Summit. 2002. **Johannesburg Summit.** www.johannesburgsummit.org

Jones, P.D., New, M., Parker, D.E., Martin, S. and Rigor, I.G. 1999: Surface air temperature and its changes over the past 150 years. *Reviews of Geophysics*, 37, 173-199.

Jones, P.D. and Moberg, A. 2003. Hemispheric and large-scale surface air temperature variations: An extensive revision and an update to 2001. *Journal of Climate* 16, 206-223.

Judd, W.S., Cambell, C.S., Kellogg, E.A., Stevens, P.F. and Donoghue, M.J. 2002. **Plant Systematics: A Phylogenetic Approach.** Sinauer Associates: Sunderland, Massachusetts.

Kaufman, A.J. and Xiao, S. 2003. High CO_2 levels in the proterozoic atmosphere estimated from analyses of individual microfossils. *Nature* 425, 279-282.

Kemp, T.S. 2004. **The Origin and Evolution of the Mammals.** Oxford University Press: Oxford.

Kerp, H. 2001. A History of Palaeozoic Forests. www.uni-muenster.de

Kazlev, M.A. 2002a. The Paleozoic. www.palaeos.com

Kazlev, M.A. 2002b. The Silurian. www.palaeos.com

Keeling, C.D. and Whorf, T.P. 2004. Atmospheric carbon dioxide record from Mauna Loa. In Oak Ridge Laboratory **Trends: A Compendium of Data on Global Change.** Carbon Dioxide Information Analysis Center, Oak Ridge National Laboratory, US Department of Energy: Oak Ridge, Tennessee. Available at http://cdiac.ornl.gov

Kemp, D. 2004. **Exploring Environmental Issues.** Routledge: London.

Kendrick, B. 2001. **The Fifth Kingdom.** 3rd Edition. Focus Publishing: Newburyport, Massachusetts. Also available at www.mycolog.com (dated 2003).

Kerp, H. 2001. **A History of Palaeozoic Forests.**

Kershaw, A.P. 1986. The last two glacial – interglacial cycles from northeastern Queensland: implications for climatic change and Aboriginal burning. *Nature* 322, 47-49.

King, S. 2001. **Making Sense of the Industrial Revolution: English Economy and Society, 1700-1850.** Manchester University Press: Manchester.

Körner, C. and Paulsen, J. 2004. A world-wide study of high altitude treeline temperatures. *Journal of Biogeography* 31, 713-732.

Landis, E.R. and Weaver, J.N. 1993. Global coal occurrence. In Law, B.E. and Rice, D.D. (Eds.), **Hydrocarbons from Coal.** *American Association of Petroleum Geologists Studies in Geology* 38, pp. 1-12.

Leakey, R. and Lewin, R. 1996. **The Sixth Extinction.** Weidenfeld and Nicholson: London.

Leigh, G.J. 2004. **The World's Greatest Fix: A History of Nitrogen in Agriculture.** Oxford University Press: Oxford.

Leonard, J.A., Wayne, R.K., Wheeler, J., Valadez, R., Guillen, S. and Vilà, C. 2002. Ancient DNA evidence for Old World origin of New World dogs. *Science* 298, 1613

Lévêque, C. and Mounolou, J.-C. 2003. **Biodiversity.** John Wiley and Sons: Chichester.

Lewin, R. and Foley, R. 2004. **Principles of Human Evolution.** 2nd Edition. Blackwell: Malden, Massachusetts.

Lillesand, T.M., Keifer, R.W. and Chipman, J.W. 2004. **Remote Sensing and Image Interpretation.** 5th Ed. John Wiley and Sons: New York.

Linares, O.F. 2002. African rice (*Oryza glaberrima)*: history and future potential. *Proceedings of the National Academy of Sciences USA* 99, 1360-1365.

Lomborg, B 2001. **The Skeptical Environmentalist. Measuring the Real State of the World.** Cambridge University Press: Cambridge.

Lorius, C., Jouzel, J., Ritz, C., Merlivat, L., Barkov, N.I., Korotkevich, Y.S. and Kotlyakov, V.M. 1985. A 150,000-year climatic record from Antarctic ice. *Nature* 316, 591-596.

Lovelock, J. 1972. Gaia as seen through the atmosphere. *Atmospheric Environment* 6, 579-580.

Lovelock, J. 1991. **Gaia. The Practical Science of Planetary Medicine.** Gaia Books Limited: Stroud.

Lovelock, J. 1995. **The Ages of Gaia.** 2nd Edition. Oxford University Press: Oxford.

Lovelock, J. 2004. Reflections on Gaia. In Schneider, S.H., Miller, J.R., Crist, E. and Boston, P.J. (Eds.), 2004. **Scientists Debate Gaia.** MIT Press: Cambridge, Massachusetts, pp. 1-6.

Lu, H., Liu, Z., Wu, N., Berne, S., Saito, Y., Liu, B. & Luo, W. 2002. Rice domestication and climatic change: phytolith evidence from East China. *Boreas* 31, 378-385.

Lumley, S. and Armstrong, P. 2004. Some of the nineteenth century origins of the sustainability concept. *Environment, Development and Sustainability* 6, 367-378.

Lynch, W. and Lang, A. 2002. **The Great Northern Kingdom.** Fitzhenry and Whiteside: Markham, Ontario.

Maber, J. 2004. **Introduction to Glycolysis.** www.jonmaber.demon.co.uk

Mack, M.G., Schur, E.A.G., Bref-Harte, M.S., Shaver, G.R. and Chapin III, F.S. 2004. Ecosystem carbon storage in arctic tundra reduced by long-term nutrient fertilization. *Nature* 431, 440-442.

McConnell, J.R. Lamorey, G.W. and Hutterli, M.B. 2002. A 250-year high resolution record of Pb flux and crustal enrichment in central Greenland. *Geophysical Research Letters* 29, (45) 1-4.

McDonald, G.W. and Patterson, M.G. 2004. Ecological Footprints and interdependencies of New Zealand regions. *Ecological Economics* 50, 49-67.

McNeil, J.R. 1992. **The Mountains of the Mediterranean World: An Environmental History.** Cambridge University Press: Cambridge.

McDougall, I., Brown, F.H. and Fleagle, J.G. 2005. Stratigraphic placement and age of modern humans from Kibish, Ethiopia. *Nature* 433, 733-736.

Mackay, A., Battarbee, R., Birks, J. and Oldfield, F. 2003. **Global Change in the Holocene.** Arnold: London.

Mackintosh, B. 2000. **The National Parks: Shaping the System.** www.cr.gov

Makarova, K.S. and Koonin, E.V. 2003. Comparative genomics of Archaea: how much have we learnt in six years and what's next? *Genome Biology* 4 (8), No.115. available at www.genomebiology.com

Malhi, Y., Meir, P. and Brown, S. 2002. Forests, carbon and global climate. *Philosophical Transactions of the Royal Society of London* A360, 1567-1591.

Mannion, A.M. 1995. **Agriculture and Environmental Change.** John Wiley and Sons: Chichester.

Mannion, A.M. 1997a. **Global Environmental Change**. 2nd Edition. Longman: Harlow.

Mannion, A.M. 1997b. Climate and Vegetation. In Thompson, R.D. and Perry, A.H. (Eds.), **Applied Climatology. Pronciples and Practice.** Routledge: London, pp. 123-140.

Mannion, A.M. 1999. Acid precipitation. In Pacione, M. (Ed.), **Applied Geography.** Routledge: London, pp. 36-50.

Mannion, A.M. 2000. Carbon: the link between environment, culture and technology. Department of Geography, University of Reading, *Geographical Paper* No.154, pp.25.

Mannion, A.M. 2002. **Dynamic World. Land-cover and Land-use Change.** Arnold: London.

Mannion, A.M. 2005. Environmental impact of the railways. In Hillstrom, K. and Hillstrom, L.C. (Eds.), **Industrial Revolution in America. Iron and Steel, Railroads, Steam Shipping.** ABC-Clio: Santa Barbara, California, in press.

Margulis, L.1981. **Symbiosis in Cell Evolution.** W.H.Freeman: New York.

Margulis, L. and Dolan, M.F. 2002. **Early Life.** 2nd Edition. Jones and Bartlett: Boston.

Margulis, L. and Sagan, D. 2002. **Acquiring Genomes: A Theory of the Origins of Species.** Basic Books: New York.

Margulis, L. and Schwartz, K.V. 1998. **The Five Kingdoms: An Illustrated Guide to the Phyla of Life on Earth.** W H Freeman: New York.

Martin, W., Rotte, C., Hoffmeister, M., Theissen, U., Gelin-Dietrich, G., Ahr, S. and Henge, K. 2003. Early cell evolution, eukaryotes, anoxia, sulphide, oxygen, fungi first (?) and a tree of genomes revisited. *IUBMB Life* 55, 193-204.

Mather, A.S. 1992. The forest transition. *Area* 24, 367-379.

Mather, A.S., Neddle, C.I. and Fairbairn, J. 1998. The human drivers of global land-cover change; the case of forests. *Hydrological Processes* 12, 1983-1984.

Maxwell, A.L. 2004. Fire regimes in north-eastern Cambodian monsoonal forests, with a 9,300 year sediment charcoal record. *Journal of Biogeography* 31, 225-239.

Mayle, F.E. and Beerling, D.J. 2004. Late Quaternary changes in Amazonian ecosystems and their implications for global carbon cycling. *Palaeogeography Palaeoclimatology Palaeoecology* 214, 11-25.

Metz, M. (Ed.), 2003. ***Bacillus thuringiensis. A Cornerstone of Modern Agriculture.*** Hawthorn Press: Binghampton, New York.

Middleton, N. 2003. **The Global Casino. An Introduction to Environmental Issues.** 3rd edition. Arnold: London.

Millennium Ecosystem Assessment. 2005. **Millennium Ecosystem Assessment Synthesis Report**. www.millenniumassessment.org

Miller, C.B. 2004. **Biological Oceanography.** Blackwell: Oxford.

Miller, G.H., Magee, J.W., Johnson, B.J., Fogel, M., Spooner, N.A., McCulloch, M.T. and Ayliffe, L.K. 1999. Pleistocene extinction of *Genyornis newtoni*: human impact on Australian megafauna. *Science* 283, 205-208.

Miller, P. and Westra, L. (Eds.), **Just Ecological Integrity.** Rowman and Littlefield: Lanham, USA.

Missouri Botanical Garden. 2002. **Deserts of the World.** www.mobot.org

Mistry, J. 2000. **World Savannas.** Prentice Hall: Harlow.

Molina, M.J. and Rowland, F.S. 1974. Stratospheric sink for chlorofluormethanes: chlorine atom-catalysed destruction of ozone. *Nature* 249, 810-812.

Mooney, H., J. Roy and B. Saugier (Eds.), 2001. **Terrestrial Global Productivity: Past, Present and Future**, Academic Press, San Diego.

Morawetz, H. 1995. **Polymers: The Origin and Growth of a Science.** Dover Publications: Mineola, New York.

Morgan, S. 2002. **Alternative Energy Sources.** Heinemann: Oxford.

Morwood, M.J., Soejono, R.P., Roberts, R.G., Sutikna, T., Turney, C.S.M., Westaway, K.E,, Rink, W.J., J.-X. Zhao, van den Bergh, G.D., Due, R.A., Hobbs, D.R., Moore, M.W., Bird, M.I. and Fifield, L.K. 2004. Archaeology and age of a new hominin from Flores in eastern Indonesia. *Nature* 431, 1087- 1091.

Museum of Palaeontology. 2004. **Geology.** www.ucmp.berkeley.edu

Mycological Society of San Francisco. 2004. **Truffles**. www.mssf.org

Naeem,S. and Wright,J.P. 2003. Disentangling biodiversity effects on ecosystem functioning: deriving solutions to a seemingly insurmountable problem. *Ecology Letters* 6, 567-579.

Nanotechnology Now. 2005. **Nanotechnology Basics, News, and General Information.** www.nanotech-now.com

National Aeronautics and Space Administration (NASA). Wilkinson Microwave Anisotropy Probe (WMAP). www.lmbda.gsfc.nasa.gov

National Center for Biotechnology Information (NCBI) **Archaea.** www.ncbi.nlm.nih.gov

Natural Resources Canada. 2004. **The state of Canada's forests 2003-2004.** www.nrcan.gc.ca

Natural Gas Organization. 2004. **What is natural gas?** www.naturalgas.org

Newell, J. 2004. **The Russian Far East.** Daniel and Daniel Publishers Inc.: McKinleyville, California.

Newton, D.E. 2002. **Recent Advances and Issues in Molecular Nanotechnology**. Greenwood Publishing Group: Westport, USA.

Nisbet, E.G. and Sleep, N.H. 2001. The habitat of early life. *Nature* 409, 1083-1091.

Odum, E.P. 1975. **Ecology.** Holt, Rinehart and Winston: New York.

Olah, G.A. and Molnar, A. 1995. **Hydrocarbon Chemistry.** John Wiley and Son: New York.

Oldstone, M.B.A. 2000. **Viruses, Plagues and History.** Oxford University Press: Oxford.

Olson, J.M. and Blankenship, R.E. 2004. Thinking about the evolution of photosynthesis. *Photosynthesis Research* 80, 373-386.

O'Neil, P.M. and Posner, G.H. 2004. A medicinal chemistry perspective on artemisinin and related endoperoxides. *Journal of Medicinal Chemistry* 47, 2945-2960.

Oren, A. 2004. Prokaryote diversity and taxonomy: current status and future challenges. *Philosophical Transactions of the Royal Society* **B**359, 623-638.

Orkustofnun. 2005. **Iceland Energy Resources.** www.os.is

O'Riordan, T. 2000. The sustainability debate. In O'Riordan, T. (Ed.), **Environmental Science for Environmental Management.** Prentice Hall: Harlow, pp. 29-62.

O'Riordan, T. 2004. Environmental science, sustainability and politics. *Transactions of the Institute of British Geographers* NS 29, 234-247.

Otto, D.; Rasse, D.; Kaplan, J.; Warnant, P. and Francois, L. 2002. Biospheric carbon stocks reconstructed at the Last Glacial Maximum: comparison between general circulation models using prescribed and computed sea surface temperatures. *Global and Planetary Change* 33, no. 1, 117-138.

Oxford Hammond Atlas of the World. 1993. Oxford University Press: Oxford.
Palaeos. 2002. **The Trace of Life on Earth.** www.palaeos.com
Parekh, S. (Ed.), 2004. **The GMO Handbook: Genetically modified animals, microbes and Plants in Biotechnology.** Humana: Totowa, New Jersey.
Parmesan, C. and Yohe, G. 2003. A globally coherent fingerprint of climatic change impacts across natural systems. *Nature* 421, 37-42.
Pearce, F. 2005a. A most precious commodity. *New Scientist* 185 (8 January), 6-7.
Pearce, F. 2005b. Act now before it's too late. *New Scientist* 185 (12 February), 8-11.
Pellant, C. 1990. **Rocks, Minerals and Fossils of the World.** Pan Books: London.
Persley, G.J. and MacIntyre, L.R. (Eds.), 2002. **Agricultural Biotechnology: Country Case Studies – A decade of Development.** CABI Publishing: Wallingford.
Pesticide Action Network (PAN) 2004. **Paraquat.** www.pan-uk.org
Petit, J.R., Jouzel, J., Raynaud, D. and 16 others. 1999. Climate and atmospheric history of the past 420,000 years from the Vostok core, Antarctica. *Nature* 399, 429-436.
Petrochemistry Net. 2004. **What is Petrochemistry?** www.petrochemistry.net
Pfeiffer, P. 2004. **Eating Fossil Fuels.** www.copvcia.com
Pfizer. 2003. **Viagra**. www.viagra.com
Piperno. D., Weiss, E., Holst, I. and Nadel, D. 2004. Processing of wild cereal grains in the Upper Palaeolithic revealed by starch grain analysis. *Nature* 430, 670-673.
Point Carbon. 2004. **Company Profile.** www.pointcarbon.com
Prados de la Escoura, L. (Ed.), 2004. **Exceptionalism and Industrialisation: Britain and its European Rivals, 1688-1815.** Cambridge University Press: Cambridge.
Population Reference Bureau. 2004. World Population Data Sheet 2004. www.prb.org
Prell, H.H. 2001. **Plant-fungal Interaction: A Classical and Molecular View.** Springer: New York.
Prusiner, S.B. 2004. **Prion Biology and Diseases.** 2nd Edition. Cold Spring Harbour Laboratory Press: Cold Spring Harbour, New York.
Pyne, S.J. 2001. **Fire: A Brief History.** University of Washington Press: Seattle.
Ramankutty, N. and Foley, J.A. 1999. estimating historical changes in land cover: croplands from 1700 to 1992. *Global Biogeochemical Cycles* 13, 997-1027.
Ramsar. 2004. **The Ramsar Convention on Wetlands.** www.ramsar.org
Raymond, J., Siefert, J.L., Staples, C.R. and Blankenship, R.E. 2004. The natural history of nitrogen fixation. *Molecular Biology and Evolution* 21, 541-554.
Rees, W. 1992. Ecological footprints and appropriated carrying capacity: what urban economics leaves out. *Environment and Urbanization* 4, 120-130.
Reid, T.R. 2005. Caffeine. It's the World's most popular drug. *National Geographic* 207, 2-32.
Renberg, I., Bindler, R. and Brännvell, M.-L. 2001. Using the historical atmospheric lead deposition record as a chronological marker in sediment deposits in Europe. *The Holocene* 11, 511-516.
Rincon, P. 2004. **Broken bones hint at first use of fire**. www.bbc.co.uk
Rivera, M.C. and Lake, J.A. 2004. The ring of life provides evidence for a genome fusion origin of eukaryotes. *Nature* 431, 152-155.
Root, T.L., Price, J.T., Hall, K.R., Schneider, S.H., Rosenzweig, C. and Pounds, J.A. 2003. Fingerprints of global warming on wild animals and plants. *Nature* 421, 57-60.
Rose, M. 1999. **The importance of Monte Verde**. www.archaeology.org
Roth, D. and Williams, G.W. 2003. The Forest Service in 1905. **www.fs.fed.us**
Ruddiman, W.F. 2003. The anthropogenic greenhouse era began thousands of years ago. *Climatic Change* 61, 261-293.
Rukang, W. and Lanpo, J. 1994. China in the period of *Homo habilis* and *Homo erectus.* In S.J. deLaet, A.H. Dani, J.L. Lorenzo and R.B. Nunoo, (Eds.), **History of Humanity, Volume 1. Prehistory and the Beginnings of Civilization.** Routledge: London and UNESCO: Paris, pp.86-88.

Rundel, P. W., Montenegro, G. and Jaksic, F. M., (Eds.), 1998. **Landscape Disturbance and Biodiversity in Mediterranean Type Ecosystems**. Springer, Berlin.
Sallares, R. 1991. **The Ecology of the Ancient Greek World.** Duckworth: London.
San-Blas, G. and Calderone, R.A. 2004. **Pathogenic Fungi: Host Interactions and Strategies for Control.** Caister Academic Press: Wymondham, Norfolk.
Savolainen, P., Zhang, Y., Luo, J., Lundeberg, J. and Leitner, T. 2002. Genetic evidence for an East Asian origin of domestic dogs *Science* 298, 1610-1613.
Schlegel, M. 2003. Phylogeny of eukaryotes recovered with molecular data: highlights and pitfalls. *European Journal of Protistology* 39,113-122.
Schneider, S.H., Miller, J.R., Crist, E. and Boston, P.J. (Eds.), 2004. **Scientists Debate Gaia.** MIT Press: Cambridge, Massachusetts.
Scott, N. 2004. **Reduce, Reuse, Recycle: An Easy Household Guide.** Green Books: Totnes.
Seife, C. 2003. Breakthrough of the year: illuminating the dark universe. *Science* 302, 2038-2039.
Sellars, R.W. 1997. **Preserving Nature in the National Parks.** Yale University Press: New Haven, Connecticut.
Semaw, S., Renne, P., Harris, J.W.K., Feibel, C.S., Bernor, R.L., Fesseha, N. and Mowbray, K. 1997. 2.5-million-year-old stone tools from Gona, Ethiopia. *Nature* 385, 333-336.
Senese, F. 2004. **Fire and Spice.** http://Antoine.frostburg.edu
Sierra Club. 2004. **Sierra Club History.** www.sierraclub.org
Simmons, I.G. 2001. **An Environmental History of Great Britain.** Edinburgh University Press: Edinburgh.
Singh, A. and Ward, O.P. (Eds.), 2004a. **Biodegradation and Bioremediation.** Springer: Berlin.
Singh, A. and Ward, O.P. (Eds.), 2004b. **Applied Bioremediation and Phytoremediation.** Springer: Berlin.
Slack, C. 2002. **Noble Obsession: Charles Goodyear, Thomas Hancock, and the Race to Unlock the Greatest Industrial Secret of the Nineteenth Century.** Texere: New York.
Smith, B.D. 1995. **The Emergence of Agriculture.** Scientific American Library: New York.
Smith, B.D. 2001. Documenting plant domestication: the consilience of biological and archaeological data. *Proceedings of the National Academy of Sciences USA* 92, 1324-1376.
Smith, J.M. 2004. **Seeds of Deception.** Yes! Books: Fairfield, Iowa.
Smith, R.L. and Smith, T.M. 2001. **Ecology and Field Biology.** 6th Edition. Benjamin Cummings: San Francisco.
Smith, R.L. and Smith, T.M. 2001. **Ecology and Field Biology.** 6th Edition. Benjamin Cummings: San Francisco.
Soaps and Detergents Association 2004. **Soap Chemistry.** www.sdahq.org
Sobel, D. 1995. **Longitude: The True Story of a Genius who Solved the Greatest Scientific Problem of His Time.** Walker: New York.
Sorori Rice Cyber Museum. 2003. Sorori seed rice unearthed in peat soil layers. www.sorori.com
Stainforth, D.A plus 15 others. 2005. Uncertainty in predictions of the climate response to rising levels of greenhouse gases. *Nature* 433, 403-406.
Stephenson, S.L. and Stempsen, H. 2000. **Mykomycetes: A Handbook of Slime Molds.** Timber Press: Portland.
Stiling, P. 2002. **Ecology.** 4th Edition. Prentice Hall: Upper Saddle River, New Jersey.
Stockton, D. 1994. The founding of the empire. In Boardman J., Griffin J. and Murray O. (Eds.), **The Oxford History of the Classical World.** Oxford University Press: Oxford, pp. 531-559.
Stork, N.E. 1997. Measuring global biodiversity and its decline. In Reaka-Kudla, M.I, Wilson, D.E. and Wilson, E.O. (Eds.), **Biodiversity II.** Joseph Henry Press: Washington D C, pp.41-68.
Stott, P.A., Stone, D.A. and Allen, M.R. 2004. Human contribution to the European heatwave of 2003. *Nature* 432, 610-614.

Stuessy, T.F., Hörandl, F. and Mayor, V. (Eds.), 2001. **Plant Systematics: A Half Century of Progress (1950 – 2000).** International Association for Plant Taxonomy: Vienna.

Summerhayes, V.S. and Elton, C.S. 1923. Contributions to the ecology of Spitzbergen and Bear Island. *Journal of Ecology* 11, 214-286.

Tajika, E. 2003. Faint young Sun and the carbon cycle, implications for the proterozoic global glaciations. *Earth and Planetary Science Letters*214, 443-453.

Thieme, H. 1997. Lower Palaeolithic hunting spears from Germany. *Nature* 385, 807-811.

Thomas, C.D. plus 18 others. 2005. Extinction risk from climatic change. *Nature* 427, 145-148.

Thomas, L.P. 2002. **Coal Geology.** John Wiley and Sons: New York.

Tice, M.M. and Lowe, D.R. 2004. Photosynthetic microbial mats in the 3,416-Myr-old ocean. *Nature* 431, 549-552.

Tilman, D. and Downing, J.A. 1994. Biodiversity and stability in grasslands. *Nature* 367, 363-365.

Times Concise Atlas of the World. 1995. Times Books: London.

Tishkov, A. 2002. Boreal forests. In Shahgedanova, M. (Ed.), **The Physical Geography of Northern Eurasia.** Oxford University Press: Oxford, pp. 216-233.

Tomlin, C. (Ed.), 2004. **The Pesticide Manual.** 13thEdition. British Crop Protection Council: Alton, UK.

Townsend, M. 2002. Ecological refugees. *The Ecologist* June 26th.

Tree of Life (ToL). 2004. **Archaea.** www.tolweb.org

Treitz, P. (Ed.), 2004. Remote Sensing for Mapping and monitoring Land-cover and Land-use Change. Pergamon: Oxford.

Tu, M, Hurd, C. and Randall, J.M. 2001. **Weed Control Handbook.** The Nature Conservancy: Arlington.

Turner, J. 2004a. Adventures in the spice trade. *Geographical* 76, 45-50.

Turner, J. 2004b. **Spice: The History of an Obsession.** HarperCollins: London.

United Nations. 1992. **Non-legally Binding Authoritative Statement of Principles for a Global Consensus on the Management, Conservation and Sustainable Development of All Types of Forests.** www.un.org

United Nations. 1997. **UN Conference on Environment and Development.** www.un.org

United Nations. 2001. **World Population Prospects, the 2000 Revision.** United Nations: New York.

United Nations. 2002. **World Urbanization Prospects: The 2001 Revision.** United Nations: New York.

United Nations Economic Commission for Europe (UNECE). 2004. The Convention on Long-Range Transboundary Air Pollution (LRTAP). www.unece.org

United Nations Educational, Scientific and Cultural Organization (UNESCO). 2004. **The MAB Programme.** www.unesco.org

United Nations Environment Programme (UNEP). 2002. **Vital Water Graphics. An Overview of the World's Fresh and Marine Waters.** www.unep.org/vitalwater

United Nations Environment Programme (UNEP). 2004a. **Environment for Development.** www.unep.org

United Nations Environment Programme (UNEP). 2004b. **The Montreal Protocol.** www.unep.org

United Nations Environment Programme - World Conservation Monitoring Centre (UNEP-WCMC). 2004. **UNEP World Conservation Monitoring Centre.** www.unep-wcmc.org

United Nations Framework Convention on Climate Change (UNFCCC). 2004. **Kyoto Protocol.** www.unfccc.int.org

United Nations Office on Drugs and Crime (UNODC). 2004. **Topics in the Chemistry of Cocaine.** www.unodc.org

United Nations Population Fund. 2004. **State of the World Population Report.** www.unfpa.org

United States Department of Agriculture (USDA). 2003. Global distribution of wetlands map. www.soils.usda.gov

United States Department of Agriculture (USDA) Forest Service. 2004. **Mission.** www.fs.fed.us

United States Geological Survey (USGS) 2003. www.wr.usgs.gov

Vallelonga P.; Van de Velde K.; Candelone, J.-P.; Morgan, V.I.; Boutron, C.F.; Rosman, K.J.R. 2002. The lead pollution history of Law Dome, Antarctica, from isotopic measurements on ice cores: 1500 AD to 1989 AD. *Earth and Planetary Science Letters* 204, 291-306.

Van Krankendonk, M.J., Webb, G.E. and Kamber, B.S. 2003. Geological and trace element evidence for a marine sedimentary environment of deposition and biogenicity of 3.45 Ga stromatolite carbonates in the Pilbara Craton, and support for e reducing Archaean ocean. *Geobiology* 1, 91-108.

Vennerstrom, J.L. and 18 others. 2004. Identification of an antimalarial synthetic trioxolane drug development candidate. *Nature* 430, 900-904.

Victorian London. 2004. **Lighting.** www.victorianlondon.org

Vila, C., Savolainen, P., Maldonado, J. E., Amorim, I. R., Rice, J. E, Honeycutt, R. L., Crandall, K. A., Lundeberg, J., and Wayne, R. K. 1997. Multiple and ancient origins of the domestic dog. *Science* **276,** 1687-89.

Viljoen A., Kunert K., Kiggundu A. and Escalant J.V. 2004. Biotechnology for sustainable banana and plantain production in Africa: the South African contribution. *South African Journal of Botany* 70, 67-74.

Vitousek, P.M., Ehrlich, P.R., Erhlich, A.H. and Matson, P.A. 1986. Human appropriation of the products of photosynthesis. *BioScience* 36, 368-373.

Vitousek, P.M., Mooney, H.A., Lubchenko, J. and Melillo, J.M. 1997. Human domination of earth's ecosystems. *Science* 277, 494-499.

Vogiatzakis, I.N., Mannion, A.M. and Griffiths, G.H. 2005. Mediterranean ecosystems: problems and tools for conservation. *Progress in Physical Geography*, in press.

Volk, T. 2003. **Gaia's Body. Toward a Physiology of Earth.** MIT Press: Boston.

Wackernagel, M., Onisto, L., Linares, A.C., Falfan, I.S.L., Garcia, J.M., Guerrero, A.I.S. and Guerro, M.G.S. 1997. Ecological footprints of nations: how much nature do they use and how much nature do they have? Commissioned by the Earth Council of Costa Rica for the Rio + 5 Forum. International Council for Local Environmental Initiatives: Toronto.

Wackernagel, M., Monfreda, C., Erb, K.-H., Haberl, H. and Schulz, N.B. 2004. Ecological footprint time series of Austria, the Philippines, and South Korea for 1961-1999: comparing the conventional approach to an 'actual land area' approach. *Land Use Policy* 21, 261-269.

Walters, E.D. 2001. **Aspartame, A Sweet-tasting Dipeptide.** www.chm.bris.ac.uk

Wang, D.Y.-C., Kumar, S. and hedges, S.B. 1999. Divergence time estimates for the early history of animal phyla and the origin of plants, animals and fungi. *Proceedings of the Royal Society Series B* 266, 163-163.

Watson, J.P. and Crick, F.H.C. 1953. Molecular structure of nucleic acids. *Nature* 171, 737-738.

Wayne, R.P. 2003. Chemical evolution of the atmosphere. In Hewitt, C.N. and Jackson, A.V. (Eds.), **Handbook of Atmospheric Science. Principles and Applications.** Blackwell: Oxford, pp. 3-34.

Weinberg, B.A. and Bealer, B.K. 2002. **The World of Caffeine: The Science and Culture of the World's Most Popular Drug.** Routledge: New York.

West S., Charman D.J., Grattan J.P., Cherburkin A.K. 1997. Heavy Metals in Holocene peats from South West England: detecting mining impacts and atmospheric pollution. *Water Air and Soil Pollution* 100, 343-353.

Weyler, R. 2001. **Waves of Compassion. The Founding of Greenpeace. Where are They Now?** www.utne.com

White, T., Asfaw, B., Degusta, D., Gilbet, H., Richards, G.D., Suwas, G. and Howell, F.C. 2003. Pleistocene *Homo sapiens* from Middle Awash, Ethiopia. *Nature* 423, 742 – 747.
Whitmore, T.C. 1999. **Tropical Forests.** Oxford University Press: Oxford.
Wielgolaski, F.E. (Ed.). 1997. **Polar and Alpine Tundra.** Elsevier: Amsterdam.
Wikibooks. 2004. **Respiration.** http://wikibooks.org
Wikipedia. 2004. **Oil Crises.** www.en.wikipedia.com
Williams, C. and Millington, A.C. 2004. The diverse and contested meanings of sustainable development. *The Geographical Journal* 170, 99-104.
Williams, M. 2003. **Deforesting the Earth.** University of Chicago Press: Chicago.
Wilson, E.O. 2002. **The Future of Life.** Knopf: New York.
Wilson, H.M. and Anderson, L.I. 2004. Morphology and taxonomy of Paleozoic millipedes (Diplopoda: Chilognatha: Archipolypoda) from Scotland. *Journal of Paleontology* 78, 169-184.
Wilson, R.C.L., Drury, S.A. and Chapman, J.L. 2000. **The Great Ice Age.** Routledge and the Open University: London.
Wissenschaftlicher Beirat der Bundesregierung Globale Umweltveränerungen (WBGU). 1988. **Die Anrechnung biolischer Quellen und Senken in Kyoto-Protokoll: Fortschritt oder Rückschlang für den globalen Umweltschutz Sondergutachten 1988.** Bremerhaven, Germany, 76pp.
Woese, C.R. and Fox, G.E. 1977. Phylogenetic structure of the prokaryotic domain: the primary kingdoms. *Proceedings of the National Academy of Sciences USA* 74, 5088-5090.
Woese, C.R., Kindler, O. and Wheelis, M.I. 1990. Towards a natural system of organisms: proposal for the domains Archaea, Bacteria, and Eucarya. *Proceedings of the National Academy of Sciences USA* 87, 4576-4579.
World Bank. 2001. **Trade Blocs.** World Bank: Washington D.C.
World Bank. 2004a. **Data and statistics**. www.worldbank.org
World Bank 2004b. **Poverty (Issue Brief).** www.worldbank.org
World Coal Institute. 2002. **Coal types.** www.wci-coal.com
World Commission on Environment and Development (WCED) 1987. **Our Common Future** (also known as the Brundtland Report)**.** Oxford University Press: Oxford.
World Commission on Forests and Sustainable Development (WCFSD). 1999. **Our Forests. Our Future.** Cambridge University Press: Cambridge.
World Conservation Union (IUCN). The World Conservation Centre. 2004. **Overview.** www.iucn.org
World Meteorological Organization. 2004. **WMO Statement on the Status of the Global Climate in 2003.** www.wmo.ch
World Resources Institute (WRI). 2003. PAGE **Agroecosystems: soil resource condition.** www.pubs.wri.org
World Resources Institute (WRI). 2004. **A Brief History of WRI.** www.wri.org
World Trade Organization (WTO). 2004. **Understanding the World Trade Organization.** www.wto.org
Worldwatch Institute. 2002. **Vital Signs 2002.** W.W.Norton: New York.
Worldwatch Institute. 2004. **Mission.** www.worldwatch.org
Worldwide Fund for Nature (WWF). 1998. **Living Planet Report.** WWF: Gland, Switzerland.
Worldwide Fund for Nature (WWF). 2004a. **Living Planet Report.** WWF: Gland, Switzerland.
Worldwide Fund for Nature (WWF). 2004b. **A History of WWF.** www.panda.org
Worldwide Fund for Nature (WWF) 2005. **Going, Going, Gone! Climate Change and Global Glacier Decline.** www.panda.org
Worsley, T.R., Nance, R.D. and Moody, J.B. 1991. Tectonics, carbon, life, and climate for the last three billion years: a unified system. In Schneider S.H., and Boston P.J. (Eds.), **Scientists on Gaia.** MIT Press: Cambridge.

Wright, R.T. and Nebel, B.J. 2002. **Environmental Science. Toward a Sustainable Future.** 8th Edition. Pearson Education: Upper Saddle River, New Jersey.

Wright Jr., H, E. and Thorpe, J. 2003. Climate change and the origin of agriculture in the Near East. In Mackay, A., Battarbee, R., Birks, J. and Oldfield, F. (Eds.), **Global Change in the Holocene.** Arnold: London, pp.49-62.

Yalcin, K. and Wake, C.P. 2001. Anthropogenic signals recorded in an ice core from Eclipse icefield, Yukon Territory, Canada. *Geophysical Research Letters* 28, 4487-4490.

Yoon, H.S., Hackett, J.D., Pinto, G. and Battacharya, D. 2002. The single, ancient origin of chromist plastids. *Journal of Phycology* 38, 40.

Zeder, M.A., Decker-Walters, D., Bradley, D. and Smith, B.D. (Eds.), **Documenting Domestication: New Genetic and Archaeological Paradigms**. University of California Press: Berkeley (in press).

Zhao, Z.J. 1998. The middle Yangtze region in China is one place where rice was domesticated: phytolith evidence from the Diaotonghuan Cave, northern Jiangxi. *Antiquity* 278, 885-897.

Index